Drosophila

The Practical Approach Series

SERIES EDITOR

B. D. HAMES
Department of Biochemistry and Molecular Biology
University of Leeds, Leeds LS2 9JT, UK

★ **indicates new and forthcoming titles**

Affinity Chromatography
★ Affinity Separations
Anaerobic Microbiology
Animal Cell Culture
(2nd edition)
Animal Virus Pathogenesis
Antibodies I and II
Antibody Engineering
★ Antisense Technology
★ Applied Microbial Physiology
Basic Cell Culture
Behavioural Neuroscience
Bioenergetics
Biological Data Analysis
Biomechanics—Materials
Biomechanics—Structures and
Systems
Biosensors
★ Calcium-PI signalling
Carbohydrate Analysis
(2nd edition)
Cell–Cell Interactions
The Cell Cycle
Cell Growth and Apoptosis

Cellular Calcium
Cellular Interactions in
Development
Cellular Neurobiology
★ Chromatin
Clinical Immunology
★ Complement
Crystallization of Nucleic
Acids and Proteins
Cytokines (2nd edition)
The Cytoskeleton
Diagnostic Molecular Pathology
I and II
DNA and Protein Sequence
Analysis
DNA Cloning 1: Core
Techniques (2nd edition)
DNA Cloning 2: Expression
Systems (2nd edition)
★ DNA Cloning 3: Complex
Genomes (2nd edition)
★ DNA Cloning 4: Mammalian
Systems (2nd edition)
★ Drosophila (2nd edition)
Electron Microscopy in Biology

Drosophila

A Practical Approach
Second Edition

Edited by

D. B. ROBERTS
Genetics Unit, Biochemistry Department, South Parks Road,
Oxford OX1 3QU, and Magdalen College Oxford OX1 4AU

⬭IRL PRESS
———at———
OXFORD UNIVERSITY PRESS
Oxford New York Tokyo

Oxford University Press, Great Clarendon Street, Oxford OX2 6DP

Oxford   New York

Athens   Auckland   Bangkok   Bogota   Bombay   Buenos Aires
Calcutta   Cape Town   Dar es Salaam   Delhi   Florence   Hong Kong
Istanbul   Karachi   Kuala Lumpur   Madras   Madrid   Melbourne
Mexico City   Nairobi   Paris   Singapore   Taipei   Tokyo   Toronto   Warsaw

and associated companies in
Berlin   Ibadan

Oxford is a trade mark of Oxford University Press

Published in the United States
by Oxford University Press Inc., New York
First edition published 1986
Second edition published 1998

A catalogue record for this book is available from the British Library

Library of Congress Cataloging in Publication Data

Drosophila: a practical approach/edited by D. B. Roberts.—2nd ed.
— (Practical approach series; 191)
Includes bibliographical references and index.
1. Drosophila.   2. Drosophila—Genetics.   3. Insects—Genetics.
I. Roberts, D. B. (David B.)   II. Series.
QL537.D76D75   1998   595.77'4—dc21   97–44376

ISBN 0 19 963661 3 (Hbk)
ISBN 0 19 963660 5 (Pbk)

Typeset by Footnote Graphics, Warminster, Wilts
Printed in Great Britain by Information Press, Ltd, Eynsham, Oxon.

To Barbro.

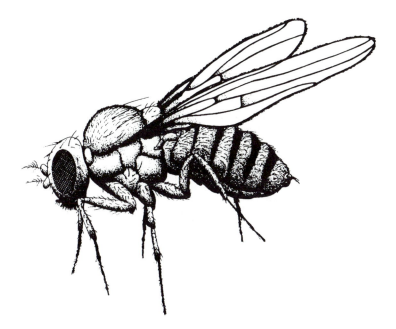

Preface

In the ten years before the publication of the first edition FlyBase lists 14516 papers with *Drosophila* in their title; in the ten years since 19004 papers. This increase does not necessarily show a causal relationship to the first edition, but it does show a continued and growing interest in *Drosophila* as an experimental organism. With more people being attracted to *Drosophila* it seemed that a new edition of this book might be needed, and moreover since 1986 new techniques have become established.

The new edition retains, with modifications, chapters on basic biology and genetics, mutagenesis (with an added chapter on *P* element-mediated mutagenesis), the techniques for looking at developing embryos, and the preparation of nucleic acids.

The new chapters consist of omissions from the earlier edition on advanced genetics, cell culture, and population and ecological genetics; extensions of old techniques, antibody staining; newly developed techniques, enhancer traps; or techniques in a rapidly developing field of *Drosophila* research, behaviour genetics.

Chapters have been omitted because the techniques have been published recently elsewhere; because they have become such standard techniques as to be found in many techniques manuals, or even to have dedicated manuals; or they have been subsumed by new techniques.

The aim of this book has not changed, it is to provide a basic set of techniques necessary to exploit *Drosophila* as a research organism.

Oxford   D. B. R.
November 1997

Contents

Contents

Contents

Contents

7. Immunolabelling of *Drosophila*

Robert A. H. White

8. Population and ecological genetics 241

J. F. Y Brookfield

9. Behaviour, learning, and memory 265

John B. Connolly and Tim Tully

Contents

Contents

10. Cell culture   319

Lucy Cherbas and Peter Cherbas

11. Preparation of nucleic acids   347

T. Jowett

Contents

Contributors

J. F. Y. BROOKFIELD
Department of Genetics, University of Nottingham, Queens Medical Centre, Nottingham 7NG 2UH, UK.

LUCY CHERBAS
Department of Biology, Indiana University, Jordan Hill, Bloomington, IN 47405, USA.

PETER CHERBAS
Department of Biology, Indiana University, Jordan Hill, Bloomington, IN 47405, USA.

JOHN B. CONNOLLY
Beckman Neuroscience Center, Cold Spring Harbor Laboratory, 1 Bungtoan Road, Cold Spring Harbor, NY 11724, USA.

T. A. GRIGLIATTI
Department of Zoology, University of British Columbia, 2354-6270 University Boulevard, Vancouver V6T 1Z4, British Columbia, Canada.

DAVID GUBB
Department of Genetics, University of Cambridge, Downing Street, Cambridge CB2 3EM, UK.

T. JOWETT
Department of Biochemistry and Genetics University of Newcastle, Framlington P1, Newcastle upon Tyne NE2 4HH, UK.

CHRISTIANE NÜSSLEIN-VOLHARD
Max-Planck-Institut für Entwicklungs biologie, Spemannstrasse 35/111, D-72076 Tübingen, Germany

C. J. O'KANE
Department of Genetics, University of Cambridge, Downing Street, Cambridge CB2 3EM, UK.

DAVID B. ROBERTS
Genetics Unit, Biochemistry Department, South Parks Road, Oxford OX1 3QU, UK, and Magdalen College, Oxford OX1 4AU.

GRAEME N. STANDEN
Genetics Unit, Biochemistry Department, South Parks Road, Oxford OX1 3QU, UK.

Contributors

TIM TULLY
Beckman Neuroscience Center, Cold Spring Harbor Laboratory, 1 Bungtoan Road, Cold Spring Harbor, NY 11724, USA.

ROBERT A. H. WHITE
Department of Anatomy, Cambridge University, Downing Street, Cambridge CB2 3DY, UK.

ERIC WIESCHAUS
Department of Molecular Biology, Princeton University, Washington Road, Princeton, NJ 08544, USA.

Abbreviations

ARM	anaesthesia-resistant memory
BDGP	Berkeley *Drosophila* Genome Project
BSS	balanced salt solution
CS	conditional stimulus
CsTFA	caesium tetrafluoroacetate
DAB	diaminobenzidine
DEPC	diethylpyrocarbonate
DMSO	dimethyl sulfoxide
DTS	dominant temperature-sensitive
EDGP	European *Drosophila* Genome Project
EM	electron microscopy
EMS	ethyl methanesulfonate
ENU	*N*-ethyl-*N*-nitrosourea
FITC	fluorescein isothiocyanate
GD	gonadal dysgensis
GFP	green fluorescent protein
HRP	horse-radish peroxidase
LMP	low melting point
LRS	lifetime reproduction success
MLE	maximum likelihood estimates
MR	male recombination
mtDNA	mitocondrial DNA
MTM	middle-term memory
OTU	operational taxonomic unit
PBS	phosphate-buffered saline
PFGE	pulsed-field gel electrophoresis
PI	performance index
PLL	poly-L-lysine
PTFE	polytetrafluoroethylene
PTU	phenylthiourea
QTL	quantitative trait loci
RI	response index
SDS	sodium dodecyl sulfate
STM	short-term memory
STS	sequence-tagged sites
TEM	triethylenemelamine
TRITC	rhodamine isothiocyanate
ts	temperature-sensitive
TSP	temperature-sensitive period
UAS	upstream activating sequence

UPGMA	unweighted pair group method with arithmetic averages
US	unconditional stimulus
X-gal	5-bromo-4-4chloro-3-indolyl-β-D-galactoside

1

The elements of *Drosophila* biology and genetics

DAVID B. ROBERTS and GRAEME N. STANDEN

1. Introduction

If you want to work with *Drosophila* the best start is to spend a couple of weeks in a good *Drosophila* laboratory learning how to handle and keep the flies. For many reasons this may not be possible and so we present this chapter as an alternative which can only be second best. When you have worked with *Drosophila* for many years there is still much to be gained in visiting other 'fly labs' to see how they have made the keeping and handling of flies easier. Indeed this chapter is a mixture of techniques, protocols, and tricks, many of which have been picked up on visits to colleagues and few if any developed by us. It is impossible at this late date to attribute these techniques to any one person but if you learn to handle flies with care and interest we are sure that this will be considered thanks enough.

We do not pretend that what we write here is the optimal way of handling and caring for flies but it does work and throughout this chapter we have adopted the pragmatic approach. It is up to you to improve on it.

2. *Drosophila melanogaster*

2.1 Classification

Drosophila melanogaster is the major experimental species and was first studied experimentally by Castle in 1901 and was used by Morgan for genetic experiments from 1909. It belongs to the genus *Drosophila* which is divided into a number of subgenera named by Sturtevant in 1942: *Hirtodrosophila*, *Pholadoris*, *Dorsilopha*, *Phloridosa*, *Sophophora*, and *Drosophila*, with *Siphlodora* and *Sordophila* being more recent additions. Some of the subgenera have been divided into species groups. *Sophophora* for example has seven species groups *inter alia saltans*, *willistoni*, and *melanogaster* to which latter *D. melanogaster* belongs.

2.2 Distribution

D. melanogaster originated in Central Africa but is now cosmopolitan and is found in all warm countries. In cooler countries it is established by migrants during the summer and can overwinter in warm places such as bakeries. This species does not have a diapause.

2.3 Sibling species

There are a number of close relatives which have been grouped together into the *melanogaster* sibling species. Practically this means that either they can mate with each other to produce sterile male offspring, and in a few cases fertile female offspring (although such matings are not common and occasionally seem impossible under observed experimental conditions), or species A can mate with species B which in turn can mate with species C although species A and species C do not mate.

3. Books

Over the last ten years many books on various aspects of *Drosophila* biology have been published and their titles and contents are readily found on the Internet. We have found the following selection to be generally useful, and there are other more specialized texts.

(a) *Biology of Drosophila*, edited by M. Demerec, first published in 1950 by John Wiley and Sons, New York and more recently republished by Cold Spring Harbor Laboratory Press, 1994, ISBN 0–87969–441–6. This is still the best guide to the bits and pieces of the fly and how they come together.

(b) *Drosophila: a laboratory handbook*, by Michael Ashburner, Cold Spring Harbor Laboratory Press, 1989, ISBN 0–87969–321–5. This *tour de force* gathers together in one volume most of what you might want to know (and more) about *Drosophila* up until 1989, or points you in the right direction.

(c) *The genome of Drosophila melanogaster*, edited by Dan L. Lindsley and Georgianna G. Zimm, Academic Press Inc., 1992, ISBN 0–12–450990–8. This is commonly known as 'The Red Book' and lists the genes and chromosome aberrations known up to the end of 1989. Each gene is described as follows:

- full and abbreviated name
- genetic location
- origin
- discoverer
- references
- phenotype

- alleles
- cytology
- molecular biology

The book also lists:

- chromosome aberrations
- deficiencies
- duplications
- inversions
- rings chromosomes
- translocations
- transpositions

together with details of their cytology, breakpoints, origin, discoverer, reference, etc. There is a short section on special chromosomes including balancers (see Section 4.3.1), compound chromosomes, multiply marked chromosomes, X–Y combinations, and Y derivatives. It is the standard reference work for the nomenclature of *Drosophila* genes and chromosomes which can be confusing to a beginner confronted with, e.g. $T(2;3;4)bw^{v30k18}$ in the literature; an invaluable book now being replaced by databases on the Internet.

(d) *Fly pushing: the theory and practice of Drosophila genetics*, by Ralph J. Greenspan, Cold Spring Harbor Laboratory Press, 1997, ISBN 0–87969–492–0. This book describes, simply, many of the standard genetic techniques used by those working with *Drosophila*.

4. FlyBase and the Encyclopaedia of *Drosophila*

The introduction of electronic mail and ready access to databases has made available information that would have taken ages to find by more conventional methods as recently as ten years ago. In particular FlyBase covers all aspects of genetics and molecular biology and is available on WWW and some of its contents are also available, together with data from the Berkeley *Drosophila* Genome Project as the Encyclopaedia of *Drosophila* on WWW (http://fruitfly.berkeley.edu.) and as a CD-ROM (eofd-sales@morgan.harvard.edu.)

(a) Access to FlyBase can be obtained through any one of the following addresses using Netscape Navigator or other WWW browsers:

- http://flybase.bio.indiana.edu/
- http://www.embl-ebi.ac.uk/flybase/
- http://astorg.u-strasbg.fr:7081/
- http://www.angis.su.oz.au:7081/
- http://shigen.lab.nig.ac.jp:7081/
- http://cbbridges.harvard.edu:7081/

(b) You can also reach the FlyBase gopher server with any gopher client at:
- `flybase.bio.indiana.edu`
- `gopher.embl-ebi.ac.uk 7071`
- `astorg.u-strasbg.fr 7071`
- `shigen.lab.nig.ac.jp 7071`
- `cbbridges.harvard.edu 7071`

(c) FlyBase includes information on:
- genes
- chromosomal aberrations
- cosmids
- P1s
- YACs
- ESTs
- STSs and related items
- genetic maps of *Drosophila*
- addresses of *Drosophila* workers
- stock lists for *D. melanogaster* and other *Drosophila* species
- a bibliography of publications on *Drosophila*
- pictures and movies of *Drosophila*
- information about transposons and plasmid vectors

(d) The information under each of these categories is linked to the information under other categories and to other databases. For example if, under 'Genes', you search for *Lsp1* the screen shows three matches *Lsp1α*; *Lsp1β*, and *Lsp1γ* together with their genetic and cytological map positions. Click on *Lsp1α* and you get the:
- gene symbol
- full name
- FlyBase ID number
- graphic map (with access to linked genes, chromosomal aberrations, and *P* elements, etc. in this region)
- genetic location
- cytological location
- discoverer(s)
- function(s) of product
- DNA/RNA accessions (with access to these databases)
- protein accessions (with access to Pir; SwissProt, etc.)
- phenotypic information
- molecular biology data
- references (with access to Medline)

However there is no substitute for exploring the database yourself. By the time you have finished your search the only available information you do not have about the gene that interests you is recently published information.

5. *Drosophila* stocks

5.1 Wild-type stocks

In almost any experiment you will need a wild-type fly. This is your reference stock with respect to which any heritable variation is considered to be a mutation. You may choose any stock as your wild-type but generally a wild-type stock is one most representative of flies caught in the wild. This need not be a worry as there are a number of wild-type stocks which have been used as reference stocks for over 50 years. Oregon-R is probably the most common but Canton-S, Samarkand, etc. are used by many laboratories. All Oregon-R flies had the same origin but because of spontaneous mutations, bad housekeeping, etc. not all Oregon-R stocks are, today, genetically identical. Indeed the stock in any one laboratory is frequently polymorphic for enzyme or protein variants. If it is necessary for your experiments that your stock be isogenic either obtain your wild-type stocks from one of the laboratories which have maintained stocks by single brother–sister matings over a number of years, or follow *Protocol 2*.

5.2 Mutant stocks

You will either know what mutant stock you need or you will know the phenotype that interests you. In either case you should find what you want on FlyBase by searching for a particular mutation or using a key word, e.g. muscle or ion channel. Once you have decided on the stock you need you can find out where it is held again using the FlyBase search facility. This searches the three *Drosophila* Stock Centres (see below) and the stocks held by participating laboratories. There may be a charge for stocks from some Stock Centres. Different *Drosophila* species can be obtained from the Species Stock Centres.

(a) Mid-America *Drosophila* Stock Center, Department of Biological Sciences, Bowling Green State University, Bowling Green, Ohio 43403–0212, USA.
Telephone: (1)-419-372-2631
FAX: (1)-419-372-2024
e-mail: `dmelano@bgsuopie.bitnet` or `dmelano@opie.bgsu.edu`

(b) Bloomington Stock Center: Kathy Matthews, Department of Biology, Indiana University, Bloomington, IN 47405–6801, USA.
Telephone: 812-855-5782
FAX: 812-855-2577
e-mail: `matthewk@indiana.edu`

(c) K. A. Yoon, National *Drosophila* Species Resource Center, Department of Biological Sciences, Bowling Green State University, Bowling Green, Ohio 43403–0212, USA.
Telephone: (1)-419-372-2742 or 372-2096
FAX: (1)-419-372-2024
e-mail: `kayoon@bsnet.bgsu.edu`

(d) European Stock Centre, Åsa Rasmuson-Lestander, Department of Genetics, University of Umeå, S901 87 Umeå, Sweden.

Telephone: (46)-90-165-275
FAX: (46)-90-167-665
e-mail: `rasmuson@big.umu.se`

The Stock Centres do not provide stocks for teaching purposes which are available from standard Biological Supply Houses. Together these stock centres carry most *Drosophila* stocks, especially the more common ones, and generally it is one or other of these that you will contact. Occasionally some specific combination of mutations may be needed which is not available in a stock centre but is available in some other laboratory. The community of *Drosophila* workers is a friendly one and with very rare exceptions if you ask for a stock you will get it.

5.2.1 Depositing stocks

During the course of your research you may find mutations or chromosomal aberrations which you feel should be deposited in one or other of the stock centres. Before sending your stock off check with the Stock Centre.

5.2.2 Nomenclature of mutations

A full account of the nomenclature of mutations is given in the Red Book but a brief summary is appropriate here.

Ideally mutations are given a short descriptive name, e.g. yellow (yellow body colour), this is abbreviated to a single letter or to a few letters which are italicized (*y*). Dominant mutations have a capital letter *H* (Hairless) and recessive ones a lower case letter (even if opening a sentence); alleles are signified by a superscript which may be a letter or a number, (y^2 or w^a); in a stock, genes on the same chromosome are separated by a space and genes on homologous chromosomes by a slash (*w m f* and *w m f/y*); genes on non-homologous chromosomes are separated by a semi-colon (*w m f; dp; ru*). The wild-type allele is signified either by a superscript + (y^+) or if it is unambiguous a simple + (*y/+*).

6. Chromosome aberrations

Chromosome aberrations are used extensively in *Drosophila* research and many experiments would not be possible without them. These aberrations have been collected over the last 70 years:

- they have arisen spontaneously in natural and laboratory populations
- they have been induced by mutagens
- they have been constructed by crossing-over between appropriate chromosomes

Here we review the more common aberrations and how and why they are used. The detailed nomenclature of these common aberrations is given in the Red Book.

6.1 Polytene chromosome band nomenclature

Each of the four major chromosome arms is divided into 20 numbered regions (X = 1–20; 2L = 21–40; 2R = 41–60; 3L = 61–80, and 3R = 81–100 with the small fourth chromosome being numbered 101–102). Each numbered division is divided into the letters A–F with the first band in each lettered division being prominent. The bands in each letter division are numbered and this number can vary depending on how close the next prominent band beginning a new lettered division might be.

6.2 Inversions

An inversion is formed when a section of a chromosome is excised and re-inserted in the opposite orientation. Inversions can either include the centromere (pericentric) or exclude the centromere (paracentric). The standard nomenclature for inversions is as follows: *In* followed by the chromosome arm, in parentheses, and letters or numbers for the particular inversion, thus *In(2L)ast* is an inversion of the left arm of chromosome 2 with both breakpoints in the same arm while *In(3LR)TM3* has breakpoints in both arms. When known, the breakpoints are published. Thus *In(1)sc*[4] has breakpoints in 1B3–4 and 19F–20C1 giving a new order to the X chromosome. Instead of 1A1–20F4 it becomes 1A1–1B3 | 19F–1B4 | 20C1–20F4.

In a heterozygote the normal and inverted chromosomes can pair forming an inversion loop. A single crossover between the inverted and normal regions generates two normal gametes, one carrying the normal chromosome and one the inverted chromosome, and two abnormal gametes which either die or lead to the formation of a lethal zygote. The explanation for this is covered in any standard Genetics text. The net result is to suppress crossing-over between the normal and inverted regions. This keeps alleles of linked genes in the region together and inversions are used for this purpose.

However if the inverted region is long two strand double crossovers may be relatively frequent and would be viable. Such crossovers could separate linked genes, and so remove the advantage of having an inversion as a crossover suppressor.

Balancer chromosomes have been constructed with many short inversions, inversions within inversions, etc. to overcome this problem. In addition balancer chromosomes are marked with dominant mutations so that they can be followed through crosses. In autosomal balancers these dominant mutations are generally recessive lethals. The most commonly used balancers are described in *Table 1*. More suitable balancer chromosomes may exist for solving your particular problem; consult the Red Book and FlyBase.

Table 1. Balancer chromosomes[a, b]

Chromosome 1

FM1 first multiple 1
$In(1)sc^8 + dl$-$49, y^{31d} sc^8 w^a lz^s B$
Dominant marker B (bar eyes)
Obvious recessive marker w^a (apricot coloured eyes)
Males viable and fertile, females homozygous viable but sterile because of lz^s

FM7 first multiple 7
$In(1)FM7, y^{31d} sc^8 w^a sn^{X2} v^{0f} g^4 B$
Dominant marker B (bar eyes)
Obvious recessive marker w^a (apricot coloured eyes)
Males viable and fertile, females homozygous viable but sterile due to sn^{X2}
Excellent suppressor with normal sequence X chromosome

Chromosome 2

CyO curly of oster
$In(2LR)O, Cy dp^{lvl} pr cn^2$
Dominant marker *Cy* (curled wings)
This chromosome carries recessive lethal mutations
Heterozygous males and females viable and fertile
Good balancer for all of chromosome 2 except for 2R when occasional double crossovers
occur if the first and third chromosomes are also heterozygous for inversions

SM5 second multiple 5
$In(2LR)SM5, al^2 Cy lt^v cn^2 sp^2$
Dominant marker *Cy* (curled wings)
This chromosome carries recessive lethal mutations
Better balancer than *CyO* but heterozygote viability and fertility less good, and according
to Ashburner (2) there may be problems with the expressivity of *Cy*

Chromosome 3

TM1 third multiple 1
$In(3LR)TM1 Me ri sbd^2$
Dominant marker *Me* (moiré eyes) not easy to classify
This chromosome carries recessive lethal mutations
Suppresses crossing-over on chromosome 3
TM1/TM2 viable

TM2 third multiple 2
$In(3LR)Ubx^{130}, Ubx^{130} e^s$
Dominant marker Ubx^{130} (ultrabithorax) weak dominant marker and not easy to classify
This chromosome carries recessive lethals
Suppresses crossing-over in chromosome 3 except for the unbalanced end of the right
arm. Double crossovers have been reported between 61C and 74 in the left arm. We have
lost a mutant which maps to 68E through double crossovers in a stock which was also
heterozygous for inversions in chromosomes 1 and 2.

Table 1. *continued*

TM3 third multiple 3
In(3LR)TM3, y⁺ ri pᵖ sep bx³⁴ᵉ eˢ

The most useful *TM3* chromosome also carries *Ser* (serrate wings) or *Ser* and *Sb* (stubble bristles) as dominant markers

This chromosome carries recessive lethal mutations

Suppresses crossing-over on chromosome 3 except at the left-hand end in the presence of heterozygous inversions of chromosomes 1 and 2. We have lost mutants of the *LSP-1γ* gene, which maps to the tip of 3L, when balanced with *TM3*.

TM6B third multiple 6B
In(3LR)TM6B, Hu e also carries either *D³* or various combinations of *Tb* with *ca, h* and/or *Hnᵖ*

This is probably the best suppressor of crossing-over on chromosome 3

[a] For a detailed review see Ashburner (2).
[b] Before choosing your balancer chromosome consult the Red Book. Try to ensure that the region that interests you is not in an unbalanced tip nor in a long stretch without a breakpoint or centromere. In these long stretches double crossovers can take place.

Flies heterozygous for a normal chromosome and a balancer chromosome have crossing-over suppressed between regions of homology that include the inversion. This leads to enhanced crossing-over in regions where crossing-over can take place. In a stock heterozygous for two or three balancer chromosomes short unbalanced regions, where crossing-over normally occurs but rarely, are now involved in frequent crossover events. It is particularly important, if you are dealing with a multiply balanced stock, that the chromosome region which interests you is well balanced and is not a short region outside an inversion.

6.2.1 Maintenance of recessive lethal mutations

Recessive lethal mutations must be kept over a balancer chromosome otherwise they would soon be lost (*Figure 1*). X-linked lethal mutations present something of a problem which is generally solved by having a balancer chromosome viable in both males and females but carrying a recessive female sterile mutation which allows the X-linked recessive lethal mutation to be kept (*Figure 1*).

6.2.2 Maintaining the integrity of a mutagenized chromosome

After mutagenesis it is necessary to keep the mutagenized chromosome intact for study. This is done using balancer chromosomes as described in Chapter 2. Because there is no crossing-over in male *Drosophila* the balancer chromosome is only strictly needed in the female for this purpose.

6.2.3 Generating duplications and deletions

This is discussed in Chapter 4.

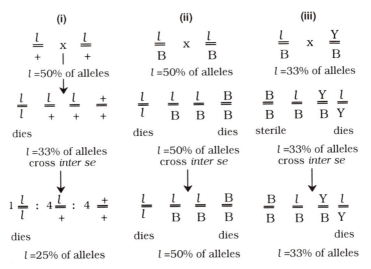

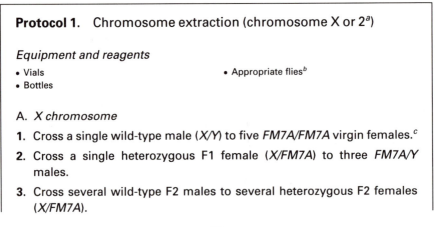

Figure 1. Maintaining lethal mutants in stocks. (i) Shows how a recessive lethal mutation is gradually lost from a stock unless selected. (ii) Shows how the same mutation can be maintained in a stock by being made heterozygous with an appropriate balancer chromosome. (iii) Shows how an X-linked lethal mutation can be maintained using a balancer chromosome. l = lethal mutation; B = balancer chromosome marked with a dominant allele and which, in the case of autosomes (ii), carries a recessive lethal. In the case of the X chromosome (iii) the balancer carries a female sterile mutation.

6.2.4 Chromosome extraction

This is the isolation of a single chromosome from a new stock of flies (*Protocol 1*). It is used to maintain the genetic variability of wild-type populations in the laboratory (see Chapter 8) or to ensure that an experimental stock carries no recessive lethal mutations (or other mutations) on a particular chromosome before carrying out a mutagenesis experiment (see Chapter 2).

Protocol 1. Chromosome extraction (chromosome X or 2[a])

Equipment and reagents
- Vials
- Bottles
- Appropriate flies[b]

A. *X chromosome*

1. Cross a single wild-type male (*X/Y*) to five *FM7A/FM7A* virgin females.[c]

2. Cross a single heterozygous F1 female (*X/FM7A*) to three *FM7A/Y* males.

3. Cross several wild-type F2 males to several heterozygous F2 females (*X/FM7A*).

---

4. Cross several wild-type F3 females to several wild-type F3 males to establish a stock with a single X chromosome.

B. *Second chromosome*

1. Cross a single wild-type male (II^A/II^B) to five *SM5/Sp* virgin females.[c]

2. Cross a single heterozygous F1 ($II^A/SM5$) male to five *SM5/Sp* virgin females.

3. Cross several heterozygous F2 females ($II^A/SM5$) to several heterozygous F2 males ($II^A/SM5$).

4. If the II^A chromosome is homozygous viable maintain a stock, with a single second chromosome, by crossing II^A/II^A males and females.

[a] For third chromosome extraction simply substitute *TM6/Sb* for *SM5/Sp*.

[b] The wild-type stock can be a laboratory stock or flies caught in the wild. An appropriate stock for first chromosome (X) extraction would be *FM7A/FM7A*; a suitable stock for second chromosome extraction would be *SM5/Sp*; a suitable stock for third chromosome extraction would be *TM6/Sb*. Since wild-type flies carry *P* and *I* elements the balancer female stock should also be *PI* to avoid hybrid dysgenesis (see Chapter 7).

[c] The number of crosses set-up depends on the aim of the experiment. If it is to purify a chromosome for mutagenesis five crosses will probably be sufficient, unless the stock is riddled with recessive lethal mutations. If it is to maintain the genetic variability of a wild population then tens of crosses will be set-up depending on how many chromosomes you wish to examine.

6.2.5 Making isogenic lines

This is similar in principle to chromosome extraction, but the aim is to isolate a single variant of the three large chromosomes in a homozygous stock and it is correspondingly more complex. There are several difficulties using female flies with three balancers:

(a) They have low viability (they carry many mutations) (this also affects male flies).

(b) They have low fertility (meiosis is severely disturbed; high non-disjunction and many dominant lethal combinations).

(c) They cannot be maintained as a stock but must be synthesized anew from a single and double balancer cross.

(d) The interchromosomal effects of three heterozygous balancers generates an appreciable frequency of viable double crossover products, which introduces genetic diversity into a programme designed to exclude variation.

If these problems are unacceptable, there is an elegant, but slightly longer procedure to make isogenic lines without using triple balancer females (1), but see ref. 2 for *erratum*.

Protocol 2. Making isogenic lines

Equipment and reagents
- Vials
- Bottles
- Appropriate flies

Method

1. Cross X^A/X^B; II^A/II^B; III^A/III^B females to $In(2LR)bw^{VI}/CyO$; $H/TM3\ Sb$ males *en masse*.

2. At the same time cross $FM7A/FM7A$ females to $SM5/+^a$; $TM6b/+^a$ males *en masse*.

3. Set-up several crosses of three $FM7A/+^a$; $SM5/+^a$; $TM6b/+^a$ F1 females (from step 2) and a single X^A/Y; $II^A/In(2LR)bw^{VI}$; $III^A/TM3$ F1 male (from step 1).

4. If male from step 3 is still alive cross $FM7A/X^A$; $SM5/II^A$; $TM6b/III^A$ F2 females (from step 3) to their fathers X^A/Y; $II^A/In(2LR)bw^{VI}$; $III^A/TM3$ (in step 3).

5. If male from step 3 is dead cross $FM7A/X^A$; $SM5/II^A$; $TM6b/III^A$ F2 females (from step 3) to $FM7A/Y$; $SM5/II^A$; $TM6b/III^A$ F2 males (from step 3).

6. Cross X^A/X^A; II^A/II^A; III^A/III^A females from step 4 to X^A/Y; II^A/II^A; III^A/III^A to establish an isogenic stock.

7. If the male from step 3 was dead (step 5) cross $X^A/FM7A$; II^A/II^A; III^A/III^A females (from step 5) to X^A/Y; II^A/II^A; III^A/III^A males (from step 5).

8. Cross X^A/X^A; II^A/II^A; III^A/III^A females from step 7 to X^A/Y; II^A/II^A; III^A/III^A to establish an isogenic stock.

[a] The genotype of these chromosomes does not matter since it is the marked homologous chromosome that is being followed.

6.2.6 Chromosome substitution

Chromosome substitution combines chromosomes from different stocks to generate new genotypes. A common contemporary example is to combine a transgene (e.g. on chromosome 2) regulated by a GAL4-dependent upstream activating sequence (UAS) (see Chapter 5), with an enhancer trap transgene (e.g. on chromosome 3) expressing GAL4 in a particular set of cells or at a particular stage of development. The standard procedure may involve, as in the case of isogenic lines, a triple balancer fly with all the drawbacks mentioned above. If necessary substitution can be achieved without using a triple balancer female (1, 2).

Protocol 3. Chromosome substitution[a,b]

Equipment and reagents
- Vials
- Bottles
- Appropriate flies

Method

1. Cross X/X; $SM5/+^c$; $TM3\ Sb\ Bd^S/III^c$ females to X^W/Y; II^W/II^W; III^W/III^W males *en masse*.
2. Cross X^Z/X^Z; II^Z/II^Z; III^Z/III^Z females and X/Y; $bw^D/+^c$; $TM3\ Sb/+^c$ males *en masse*.
3. Cross X^W/X; $SM5/II^W$; $TM3\ Sb\ Bd^S/III^W$ F1 females (from step 1) to X^Z/Y; bw^D/II^Z; $TM3\ Sb/III^Z$ F1 males (from step 2) *en masse*.
4. Cross X^d/X^Z; $SM5/II^Z$; $TM3\ Sb/III^W$ F2 females (from step 3) to X^d/Y; $SM5/II^Z$; $TM3\ Sb/III^W$ F2 males (from step 3) *en masse*.
5. Establish the stock X^d/X^d; II^Z/II^Z; III^W/III^W females (from step 4) and X^d/Y; II^Z/II^Z; III^W/III^W males from step 4.

[a] For example, chromosome 2 from stock 'Z' and chromosome 3 from stock 'W'.
[b] Variations on this scheme will allow the combination of, for example, II^W and II^Z.
[c] The genotype of these chromosomes does not matter since the heterozygous marked chromosome is being followed.
[d] The X chromosome can come from any background since no particular X chromosome is required in the final stock.

6.2.7 Making inversions

It is not generally possible to make inversions with specific declared breakpoints and so it is only rarely that you would set out to make inversions. Most of those that you make will be picked up in routine mutagenesis experiments, for example as putative double point mutations. However Roberts (3) has constructed a stock which screens chromosomes 1, 2, and 3 simultaneously for dominant crossover suppressors many of which are inversions.

6.3 Deletions (deficiencies)

A deletion refers to a chromosome from which a segment is missing. Although deletion is the more commonly used term (and will be used here) the standard nomenclature refers to a deficiency. This is signified *Df* followed by the chromosome number and arm and letters or numbers identifying the particular deletion, thus *Df(2L)ast-5* is a deletion of the left arm of chromosome 2, in this case a deletion which uncovers the mutation *ast* and is the fifth of such deletions in the series. Where known, the breakpoints are published. Thus *Df(2L)ast-5* is 21E1–2 – 21F3–22A1 and has a distal (away from the centromere) breakpoint between bands 21E1 and 21E2 and a proximal (close to the centromere) breakpoint between bands 21F3 and 22A1.

Homozygous deletions are generally lethal and heterozygous deletions of over 50 polytene chromosome bands are also lethal. Smaller deletions can be heterozygous viable with the viability depending on the particular deletion. There are however a few regions of the *Drosophila* genome which have never been recovered in heterozygous deletions and these are thought to represent haplo-lethal regions (4).

Many deletions are listed in the Red Book and FlyBase and can be obtained from Stock Centres. It is also worth writing to those working in the region who may have unpublished deletions. Standard techniques for generating deletions are given in Chapters 2 and 4.

6.3.1 Deletion mapping

Classical genetic mapping will provide the genetic position of a gene, with reference to other genes. Its physical position can be determined either by *in situ* hybridization, if it has been cloned, or by deletion mapping (*Protocol 4*). Using this technique it is possible to map genes to one or a few bands depending on what deficiencies are available.

6.3.2 To make mutations of a declared gene

For many reasons you may want to make mutations of a particular gene, one for example that you have mapped by *in situ* hybridization or by deletion mapping. In the most common case and the one discussed here (*Protocol 5* and *Figure 2*) the expected mutation has a lethal phenotype. However simple variants of this protocol can be used to select visible or conditional mutations (see Chapter 2).

Protocol 4.   Deletion mapping[a]

Equipment and reagents

- Bottles and vials of fly medium (see Section 9)
- Equipment for handling flies (see Section 11)
- Appropriate fly stocks (see the Red Book or FlyBase)

Method

1. Use the Red Book or FlyBase to determine the cytological positions of the closest cytologically mapped proximal and distal genes.[b]
2. Make an appropriate allele[c] of the gene heterozygous with a series of deletions which span the region between the proximal and distal limits.
3. Analyse these heterozygotes for the expression of the allele.[d]
4. Consult the Red Book or FlyBase for the breakpoints of deletions which uncover the gene.
5. If these are unknown analyse the polytene chromosomes of the heterozygous deletion and determine which bands are missing.

14

6. Use an overlapping series of deletions for a precise localization of the gene.[e]

[a] This assumes that the gene has been mapped genetically.
[b] This gives the extreme limits of its physical position.
[c] This is usually: (a) recessive visible; (b) lethal mutation; (c) electrophoretic variant of a protein; or (d) restriction fragment length variant.
[d] If (a) the recessive visible mutation is expressed; (b) no heterozygotes survive (lethal allele); (c) one electrophoretic form is displayed (use two different electrophoretic mobility alleles in case the deletion carries an allele with the same mobility as the allele on the chromosome being tested); (d) a single restriction fragment pattern is given (use two alleles with different restriction enzyme fragment patterns in case the deletion carries an allele with the same pattern as the chromosome being tested); then the deletion has uncovered the gene. The allele is included in the deleted portion of the chromosome.
[e] If two overlapping deletions both uncover the gene then the missing bands common to both must carry the gene. Similarly if only one uncovers the gene then bands common to both cannot carry the gene.

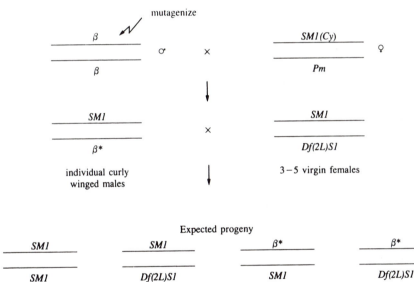

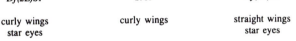

Figure 2. To make lethal mutations in the region of a declared gene. Males of a stock homozygous for β are mutagenized and crossed to females carrying the chromosome 2 balancer *SM1* which is marked with the dominant mutation *Cy* (curly wings) and carries recessive lethal mutations. Single curly winged male progeny are crossed to virgin females heterozygous for the smallest deficiency uncovering β, i.e. *(Df(2L)S1)* and the balancer chromosome *SM1*. If there are no straight winged progeny then the mutagenized chromosome carries a lethal mutant uncovered by the deficiency. (The chromosome should be tested over the deficiency before mutagenizing in case it already carries a lethal mutation in this region. If it does choose a stock that does not.) The induced lethal mutation may be a mutant of β or of a closely linked gene uncovered by the same small deficiency. The mutagenized chromosome is maintained by crossing the curly winged normal eyed flies *inter se.*

15

6.3.3 Mapping by dose response

In general the amount of protein synthesized in *Drosophila* is related to the number of copies of the coding sequence. Deletions, especially the series generated as segmental aneuploids (Chapter 4) can be used to map coding sequences by measuring the amount of enzyme or protein synthesized (50% in a heterozygous deletion).

A problem arises in the case of the X chromosome as it is not normally possible to carry X-linked deficiencies nor X-linked recessive lethal mutations in the male. This can be overcome if the male fly carries a translocation of the X chromosome on an autosome which covers the deficiency of the X chromosome. The account by Judd and colleagues (5) shows how well this technique can be exploited.

Protocol 5. To make mutations of a declared gene[a]

Equipment and reagents
- Bottles and vials of fly medium (see Section 6.1)
- Equipment for handling flies (see Section 8)
- Fly stocks: *Df(2L)S1/CyO*[b], *al dp b pr*[c]

Method

1. Mutagenize males (see Chapter 2) from the multiply marked stock *al dp b pr* and cross to virgin females carrying a second chromosome balancer, e.g. *CyO*.

2. Cross individual curly winged male progeny to five females heterozygous for the smallest deletion uncovering the *Lsp-1β* gene; *Df(2L)S1/CyO*.[d]

3. Examine the progeny for *Cy*[+], (straight) winged progeny; if none then the mutagenized chromosome has a lethal mutation uncovered by *Df(2L)S1*.

4. Cross male and virgin female *al dp b pr/CyO* (phenotypically purple) progeny from the crosses with no straight winged flies *inter se* to maintain the mutant chromosome over a balancer chromosome.

5. Cross each stock carrying a lethal mutation uncovered by *Df(2L)S1* to a series of overlapping deletions (*Df/CyO*) to determine whether the lethal phenotype is associated with the smallest number of bands to which the *Lsp-1β* gene has been mapped (*Protocol 4*).[e]

6. Check that the mutation is a mutation of the *Lsp-1β* gene by either:

   (a) Seeking to rescue the mutant phenotype by transformation with the cloned wild-type DNA.

(b) Comparing the wild-type sequence of the *Lsp-1β* gene with that of the homologous sequence from the mutagenized chromosome.[f]

[a] The example given here is for lethal or visible mutations of the *Lsp-1β* gene.

[b] *Df(2L)S1* is the smallest deletion which uncovers the *Lsp-1β* gene.

[c] A multiply marked second chromosome. A marked chromosome may be invaluable for later experiments. In this case *al* and *dp* are recessive mutations with a readily scored phenotype which flank the *Lsp-1β* gene.

[d] The number of crosses depends on how determined you are. As a very rough guide: after treatment with EMS ~ 50% of each autosome will carry a recessive lethal mutation (Chapter 2, Section 2.1); each autosome carries 40% of the genes; *Drosophila* has ~ 15 000 genes; assume they are equally distributed and have an equal probability of mutating to lethality; then on average a mutation of any one gene will be found once in 12 000 vials. This errs on the pessimistic side but by far less than an order of magnitude.

[e] If there are *Cy*[+] straight winged progeny then the deletion does not uncover the new mutation.

[f] The wild-type *Lsp-1β* sequence should be the sequence from the *al dp b pr* chromosome in order to avoid differences caused by sequence polymorphism in wild-type genes.

6.4 Duplications

Chromosomes with an additional segment, which can be derived from a sister chromatid, a homologous chromosome, or a non-homologous chromosome carry duplications. They are signified by *Dp* followed by the chromosomes involved and letters or numbers signifying the particular duplication, thus *Dp(1;1)zeste* is a tandem repeat of part of chromosome 1 and again the breakpoints are given, 1A–1B4 1E3–20. *Dp(1;3)126* represents a duplication of part of chromosome 1 on chromosome 3.

While duplications have been of considerable importance in evolution their experimental value is so far limited to, for example, mapping genes by studying dose response in the series of segmental aneuploids (heterozygous duplication has 150% enzyme activity or protein).

Many duplications are listed in the Red Book and FlyBase—otherwise write to those working on the region you wish to cover with the duplication. Duplications of a specific gene can be made by variations of the schemes presented in *Figure 3*, in Chapters 2 and 4, and by *P* element-mediated transformation.

6.5 Translocations

Translocations are rearrangements in which non-homologous chromosomes exchange segments. The most simple translocations are reciprocal (e.g. ABCDEF and UVWXYZ giving ABCXYZ and UVWDEF). These are signified by *T* followed by the chromosomes involved and letters or numbers signifying the particular translocation thus *T(2;3)C91* is a translocation involving the second and third chromosomes. When known, the breakpoints on both chromosomes are given, e.g. 28D;69D. Translocations are described in greater detail in Chapter 4, Section 2.

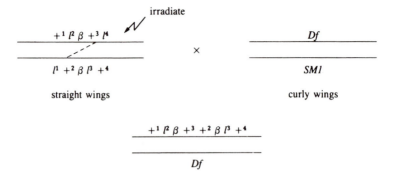

Figure 3. To generate a short duplication. Construct two chromosomes, *a* and *b* by: (i) making a lethal mutation in *a* uncovered by a deficiency to the right of β, and (ii) making a second lethal mutation in this mutagenized chromosome which is uncovered by a deficiency to the left of β. Use the techniques described in *Figure 2*. Repeat to make chromosome *b* which will carry two different lethal mutations. Cross irradiated males or females heterozygous for *a/b* to flies carrying a deficiency which uncovers all four lethal mutations. In this orientation a simple crossover will not survive over the deficiency and so there will only be curly winged progeny. An unequal crossover (dotted line), will generate a duplication which will survive over the deficiency giving straight winged flies and a chromosome carrying a duplication of β. This is a complicated though feasible system and is included to illustrate what can be done in *Drosophila* genetics.

Translocations have been used to cover X-linked deficiencies or lethal mutations in the male fly (see above), and to generate segmental aneuploids (deletions and duplications) see Chapter 4 and ref. 4.

Many translocations are listed in the Red Book and FlyBase—otherwise write to those working on the region you wish to cover with the translocation.

6.6 Compound chromosomes

These are chromosomes in which the homologous arms of a pair of chromosomes are attached to the same centromere. Thus in the case of *C(2L)RM* (compound chromosome 2 left arm reversed metacentric) the two left arms of chromosome 2 are attached to the same centromere. A compound first (X) chromosome [*C(1)*] has both X chromosomes (they have only one arm) attached to the same centromere. These arrangements lead to unusual and useful patterns of inheritance, see *Figure 4*.

Compound chromosomes have been used *inter alia* to demonstrate that crossing-over occurs at the four strand stage and that the gene conversion occurs in *Drosophila* (for review see ref. 6); to select X-linked mutations in the sons of mutagenized fathers (Chapter 2); to construct segmental aneuploids (Chapter 4).

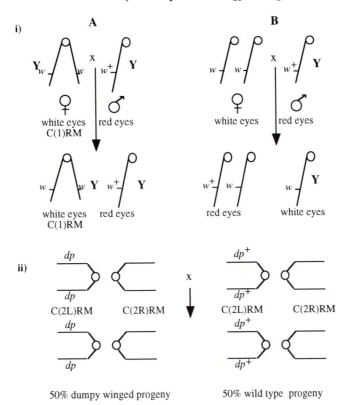

Figure 4. (i) The inheritance pattern of attached-X chromosomes (A) and of normal X chromosomes (B). (A) The daughters receive both X chromosomes from their mothers (matroclinous) and the sons the single X chromosome from their fathers (patroclinous). The presence of the Y chromosome has no effect on female fertility but its absence makes the male sterile. (B) The normal pattern of X chromosome inheritance. If in cross (A) the male parent is mutagenized, recessive visible mutations will be immediately expressed by the son carrying the mutagenized X chromosome. (ii) The inheritance pattern of *C(2L)RM* and *C(2R)RM* chromosomes. The fusion of gametes carrying *C(2L)RM* or *C(2R)RM* with normal gametes gives a zygote with either three chromosome 2 left arms and one right arm or vice versa. Both these conditions are lethal. The fusion of a gamete carrying *C(2L)RM* with a similar gamete gives four chromosome 2 left arms and is lethal. The only viable combination is with the two compound chromosomes forming a zygote. For this reason a cross between a *C(2L)RM/C(2R)RM* stock homozygous for the recessive wing mutant *dp (dumpy)* and a wild-type *C(2L)/C(2R)RM* stock gives 50% *dp* and 50% wild-type progeny, whereas a similar cross involving normal chromosomes would give 100% wild-type progeny because *dp* is recessive.

7. Environmental conditions

7.1 Temperature

Most *Drosophila* experiments are carried out at 25 °C and this is the temperature we assume unless another temperature is indicated. The facilities you need depend on the type of experiment you intend to carry out. If you are only dealing with a few hundred flies of a few stocks then a 25°C incubator is sufficient. (A cooling facility on the incubator is essential in countries, or laboratories, where the ambient temperature is frequently greater than 25°C.) If, on the other hand, you are going to rear flies in large quantities, or keep many stocks, and set-up many crosses then a 25°C room is preferable to a series of incubators.

A 25°C room has the added advantage that where temperature is critical experiments can be carried out in the room. There are many designs of room and many ways of maintaining the temperature. How you achieve your ideal will to a large extent be determined by your budget and it is best to consult with local contractors. If there is sufficient space a small booth within the room is appropriate as a 25°C laboratory. The room should be equipped with shelving. We find metal industrial shelving adequate. This can be readily cleaned and if necessary autoclaved.

7.2 Humidity

In some places low humidity can be a serious problem. Over 70% humidity is necessary. This can be achieved either with wicks dipped in trays of water in the 25°C room or by using an appropriate humidifier.

For details of optimal temperature and humidity see ref. 7.

8. Maintaining stocks

Drosophila stocks can only be kept by transferring to fresh medium at regular intervals. At 25°C this is about every two weeks. To transfer stocks:

(a) Anaesthetize the flies from the old bottle (see Section 11.1).

(b) Examine briefly to check the phenotype.

(c) Select the appropriate parents for the next generation (if necessary).

(d) Use some ten flies of each sex to set-up the next bottle generation.

The number of flies transferred depends on the genotype and more parents are required from stocks with low fertility. If you have any doubts start off with a larger number and decrease with experience. An overcrowded bottle has almost as many problems as one with too few flies.

If your stocks are healthy and do not need checking then you can transfer flies without anaesthetizing them by tapping the flies to the bottom of the old bottle, removing the stoppers from the new and then the old bottles and inverting the new bottle over the old, and then re-stoppering both bottles when sufficient flies have moved into the new bottle. This, although quick, has several disadvantages: escaping flies, transfer of contaminated food, and risk of contamination with already escaped flies. We leave food bottles with stoppers ajar as traps for escaped flies, we also have sticky flypaper. One consequence of this mass transfer, without examining the flies, is that you may lose mutant phenotypes due to the accumulation of modifiers. These can be retrieved by selection.

With few stocks, stock changing is not much work, but as your experiments progress it is common to find the number of your stocks increasing. The time spent keeping stocks can be reduced by growing them at 18°C and slowing their development, so that they need only be transferred every three to four weeks. But take care as some stocks do badly at 18°C and must be kept at 25°C. If you are going to keep many stocks and money is no object then an 18°C room is ideal. Otherwise you can keep stocks in a cooled incubator at 18°C.

8.1 Temperature-sensitive (ts) stocks

Finally you may need to keep temperature-sensitive stocks. Flies carrying ts mutations survive well at one temperature but fail to develop or may express the mutant phenotype at another. In most cases the permissive temperature is 18°C or 25°C and the restrictive temperature 29°C. To use these stocks you will need a 29°C incubator.

9. Food

A detailed analysis of the nutritional requirements of *Drosophila* is dealt with well by Sang (8).

There are many types of fly food used by different laboratories and many of these are considered by Ashburner and Thompson (7). An exhaustive comparison of different media would probably not be rewarding. One stock might fare best on one food, a second stock faring better on another food. Here we will consider seven different media used for different purposes:

(a) A standard all purpose medium used for maintaining flies and for setting-up crosses.

(b) A rich medium used for obtaining large larvae for polytene chromosome work and biochemical studies.

(c) A commercially available instant medium, quick to make up in emergencies.

(d) A cheap potato medium, also quick to make in emergencies.

(e) An apple juice (grape juice) medium for egg laying.

(f) An axenic medium for raising flies under sterile conditions.

(g) A defined synthetic medium for biochemical work.

9.1 Flour–treacle–agar medium

This medium is made according to the recipe in *Protocol 6*.

Protocol 6. Flour–treacle–agar medium

Equipment and reagents

- Vials
- Bottles
- Stoppers
- Saucepan
- Live yeast
- Agar (min 960 gel) (T. P. Drewitt)
- Nipagin solution: 25 g Nipagin M (Tegosept M, *p*-hydroxy benzoic acid methyl ester) in 250 ml ethanol
- Treacle
- Distilled water
- Wholemeal flour

Method

1. Add enough of the first volume of water[a] to the agar to form a smooth suspension.

2. Mix the entire second volume of water[a] with the flour to make a smooth paste.

3. Boil the remainder of the water in the saucepan and add the agar suspension which should dissolve rapidly.

4. Add the treacle by inverting the measuring cylinder over the saucepan and stir as it runs in.

5. Boil again, take off the boil, and finally add the flour paste slowly, stirring vigorously to avoid lumps. Simmer for 10 min and before pouring add the Nipagin.

6. Pour, using a jug, into sterile bottles or vials to a depth of 2 cm and 3 cm, respectively.

7. Tilt the vials at an angle before they set to increase the surface area.

8. Make a yeast suspension with the consistency of a thin paste and pipette a few drops onto the medium in the vials or bottles, or sprinkle a few grains of dry active yeast onto the medium in each vial or bottle.

9. Roll up a piece of cellulose wadding and push into the medium.

[a] No. of bottles	4	20	60	90
No. of vials	30	150	450	675
g of agar	5	25	74	111
1st vol. of distilled water (ml)	200	1000	3000	4500
g of treacle	33	166	500	750
g of flour (wholemeal)	25	125	375	565
2nd vol. of distilled water (ml)	55	275	825	1240
ml Nipagin	4	12	36	54

We find that in sealed bags the vials will last for many weeks at 4 °C. If you are going to store the food for a long time then it is best to add the live yeast the day before use. Reject any vial or bottle with fungal contamination.

9.1.1 Vessels

We use 7.5 × 2.25 cm (3 × 1″) glass or plastic vials. These are cheap and can be considered as disposable. (This may be ecologically unsound but there are ecological hazards in washing, sterilizing, and recycling.) Shop around for the cheapest vials you can get. Traditionally *Drosophila* have been grown in half-pint milk bottles but due to changing habits these are no longer readily available in the UK. Instead we use 230 ml powder bottles which are 11 cm (4.5″) high, with a 6 cm (2.5″) diameter at the base, and a 4.5 cm (1.8″) diameter at the mouth.

9.1.2 Plugs

The vials and bottles must be plugged. Non-absorbent cotton wool plugs are preferable (see below) but polyurethane foam plugs are frequently used. These foam plugs are reusable after autoclaving, require less time to produce, and are more convenient to use than the cotton wool plugs, but they make contamination by mites more likely (see Section 10.3). In some countries cardboard stoppered bottles are common (Chapter 2).

9.1.3 Yeast

While there has been discussion on the type of yeast to use we have always used commercial fresh bakers yeast with no obvious ill effects. We find that dried active yeast is convenient and acceptable for our stocks.

9.1.4 Dispensing medium

For the amount of medium we use (up to 400 bottles and 800 vials a week made by one person working mornings), dispensing the medium by hand using a small jug is efficient when carried out by an experienced operator. Automated and semi-automated devices have been constructed and are discussed by Ashburner and Thompson (7) and should be considered if you require large amounts of medium regularly. For a single 'big' experiment medium can be prepared in advance and stored sealed at 4°C (see above).

9.1.5 Washing up

Both bottles and vials are autoclaved before washing and the used medium, debris, etc. is poured off into a plastic bag for disposal. The vessels are cleaned with a motorized brush, washed in an automatic washing machine, and stored dry in a clean cupboard until used.

9.2 Yeast–glucose–agar medium

This is a rich medium which produces fat third instar larvae ideal for cytological and biochemical work. However not all stocks prosper on this medium. It is simple to make as shown in *Protocol 7*.

Protocol 7.   Yeast–glucose–agar medium

Equipment and reagents

- Vials
- Stoppers
- Bottles
- Agar (min 960 gel) (T. P. Drewitt)
- Distilled water

- Dried yeast
- Glucose
- Nipagin solution: 25 g Nipagin M (Tegosept M, *p*-hydroxy benzoic acid methyl ester) in 250 ml ethanol

Method

1. Mix 10 g of dried yeast and 2 g of agar into a paste in 20 ml of water.
2. Boil 80 ml water, add 10 g glucose, and the yeast–agar paste.
3. Boil and simmer for 10 min.
4. Add Nipagin M, dispense, and seed with live yeast as in *Protocol 6*.

9.3 Instant medium

This medium is available from a number of commercial firms. It is convenient and a good stand-by if you run out of the cheaper medium. It is made according to *Protocol 8*.

Protocol 8.   Instant medium[a,b]

Equipment and reagents

- Bottles
- Vials
- Stoppers

- Distilled water
- Dried yeast granule
- Instant medium

Method

1. Weigh 5 g instant medium into a vial (20 g/bottle).
2. Add 6 ml water.
3. Mix vigorously with glass rod.
4. After 2 min add dried yeast granule and stopper.

[a] For example, Philip Harris Biological Ltd.
[b] We have found that some stocks and at least one species, *D. busckii*, do better on instant medium than on regular medium, other stocks do not fare as well.

9.4 Instant potato medium

Commercial dried mashed potato provides a cheap and reliable instant medium (8) (*Protocol 9*).

Protocol 9. Instant potato medium[a]

Equipment and reagents

- Bottles
- Vials
- Stoppers
- Agar (min 960 gel) (T. P. Drewitt)
- Baker's yeast

- Nipagin M (Tegosept M), *p*-hydroxy benzoic acid methyl ester
- Commercial instant mashed potato[b]
- Distilled water

Method

1. Thoroughly mix 176 g instant mashed potato with 10 g agar and 1.27 g of Nipagin.

2. Mix this quickly with twice its volume of water in the vial or bottle to give the final appropriate depth of medium.

3. Seed with baker's yeast made into a paste with a little water, or sprinkle with granules of dried active yeast.

[a] Some fly strains may not do well on this food; you should test each strain.
[b] We have used instant mashed potato from the local Co-op.

9.5 Defined medium

For certain experiments it is essential to use defined or semi-defined medium (*Protocol 10*). In the former all components are defined; in the simpler semi-defined medium the amino acid mixture is replaced by casein.

Protocol 10. Defined medium[a]

Equipment and reagents

- Ingredients as shown in *Table 2*
- Stock solution as shown in *Table 3*
- Amino acids as shown in *Table 4*

- Boiling tubes
- Cotton wool stoppers

Method

1. Grind the agar, casein, sucrose, cholesterol, and lecithin with a pestle in a mortar (*Table 2*).

2. Make up stock solutions (*Table 3*) of the remainder of the constituents, and add the appropriate amounts to the mortar.[b]

3. Transfer the suspension into boiling tubes and plug with cotton wool before autoclaving at 15 lb for 15 min.[c]

4. Agitate the medium while it sets after autoclaving otherwise the insoluble constituents settle out.

Protocol 11. *Continued*

5. To make a fully defined medium replace the casein by the amino acid solution given in *Table 4*.

[a] See Sang (9) for the nutritional requirements of *Drosophila*. These media were developed by Sparrow and Sang (10), and by Hunt (11).
[b] Because of the variability in the acidity of different batches of casein, the final pH of the suspension is adjusted, when necessary, to pH 6.8 with bicarbonate.
[c] If a large batch of medium is made in the mortar there is a problem of achieving homogeneity when the suspension is dispensed. This can be overcome by making sufficient for one tube at a time and transferring the whole, or by agitating the suspension when dispensing it with a pipette.

Table 2. Constituents of defined medium—casein based

Constituent	Weight (g)	Constituent	Weight (g)
Agar	3.00	Aneurin (thiamine)	0.0002
Casein	5.50	Riboflavins	0.001
Sucrose	0.75	Nicotinic acid	0.0012
Cholesterol	0.03	Ca pantothenate	0.0016
Lecithin[a]	0.40	Pyridoxine	0.00025
Yeast RNA[b]	0.40	Biotin	0.00002
$NaHCO_3$	0.1	Folic acid	0.0003
KH_2PO_4	0.071		
K_2HPO_4	0.373		
$MgSO_4.7H_2O$	0.062		
Water to 100 ml			

[a] Lecithin can be substituted by 0.024 g of chorine chloride.
[b] Suspend 5 g of yeast RNA in 11 ml water; this tends to be very sticky but dissolves when you add 14 ml of $NaHCO_3$ solution at 16.8 g/300 ml. Make up to a final volume of 50 ml with water. Yeast RNA can be substituted by 0.064 g of inosine and 0.057 g of uridine/100 ml.

Table 3. Stock solutions

Salts

KH_2PO_4	0.71 g/litre; use 10 ml/100 ml medium
K_2HPO_4	37.3 g/litre; use 10 ml/100 ml medium
$MgSO_4$	6.2 g/litre; use 10 ml/100 ml medium
$NaHCO_3$	10 g/litre; use 10 ml/100 ml medium

Vitamins

Aneurin (thiamine)	0.05 g/250 ml; use 1 ml/100 ml medium
Riboflavins	0.05 g/150 ml; use 3 ml/100 ml medium
Nicotinic acid	0.05 g/125 ml; use 3 ml/100 ml medium
Ca pantothenate	0.05 g/150 ml; use 5 ml/100 ml medium
Pyridoxine	0.05 g/200 ml; use 1 ml/100 ml medium
Biotin	0.01 g/500 ml; use 1 ml/100 ml medium
Folic acid	0.05 g/167 ml in 20% ethanol; use 1 ml/100 ml medium

Table 4. Amino acids in defined medium—amino acid based

L-Arginine hydrochloride	0.08 g
L-Glutamic acid	0.84 g
L-Histidine	0.10 g
L-Isoleucine	0.30 g
L-Leucine	0.20 g
L-Lysine hydrochloride	0.19 g
L-Methionine	0.08 g
L-Phenylalanine	0.13 g
L-Threonine	0.20 g
L-Tryptophan	0.05 g
L-Valine	0.28 g

9.6 Apple juice medium

This medium promotes egg laying (see Chapter 5).

Protocol 11. Apple juice medium for egg collections[a]

Equipment and reagents

- Petri plates
- Agar min 960 gel (T. P. Drewitt)
- Apple juice
- Sucrose

- Nipagin solution: 25 g Nipagin M (Tegosept M, *p*-hydroxy benzoic acid methyl ester) in 250 ml ethanol
- Distilled water

Method

1. Mix 500 ml distilled water and 17.5 g agar.

2. Autoclave.

3. Mix 250 ml distilled water and 12.5 g sucrose.

4. Microwave to dissolve the sucrose.

5. Add 250 ml apple juice and 20 ml Nipagin solution to the sucrose solution.

6. Mix the sucrose solution with the agar, taking care to avoid the formation of air bubbles.

7. Pour into Petri plates.

8. Store in cold.

9. Warm before use.

[a] The addition of a few drops of food colouring (Supercook Blue, containing Brilliant blue FCF and carmoisine) with the Nipagin allows the eggs to be more readily seen on the agar.

9.7 Axenic medium

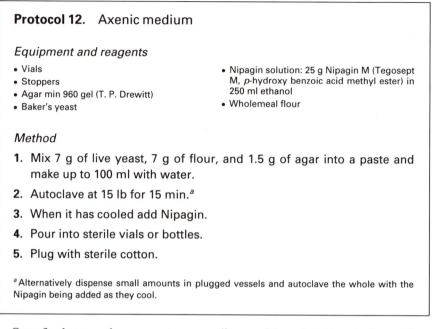

Protocol 12. Axenic medium

Equipment and reagents
- Vials
- Stoppers
- Agar min 960 gel (T. P. Drewitt)
- Baker's yeast
- Nipagin solution: 25 g Nipagin M (Tegosept M, *p*-hydroxy benzoic acid methyl ester) in 250 ml ethanol
- Wholemeal flour

Method

1. Mix 7 g of live yeast, 7 g of flour, and 1.5 g of agar into a paste and make up to 100 ml with water.

2. Autoclave at 15 lb for 15 min.[a]

3. When it has cooled add Nipagin.

4. Pour into sterile vials or bottles.

5. Plug with sterile cotton.

[a] Alternatively dispense small amounts in plugged vessels and autoclave the whole with the Nipagin being added as they cool.

One final general comment on medium, with unhealthy stocks or low density cultures (and especially with propionic acid medium; see below) scoring the medium surface helps the larvae to flourish.

10. Pests

The three major pests of *Drosophila* cultures are fungi, bacteria, and mites (for a more detailed consideration see ref. 2).

10.1 Fungal control

Nipagin M is added routinely to the regular medium, the yeast–glucose–agar medium, and the axenic medium, and it keeps fungal contamination to a minimum. This is especially true if care is taken to prepare the medium under clean conditions and if any stored medium which shows fungal contamination is discarded. Problems can occur especially with slow growing stocks or if stocks are transferred to fresh medium from contaminated medium. To overcome this problem we transfer contaminated stocks to medium containing both Nipagin and propionic acid (1 ml/200 ml medium) added immediately prior to pouring. If this fails to prevent fungal growth it may be necessary to surface sterilize the eggs (see Section 15.3).

10.2 Bacterial control

We have not found bacteria to be a particular problem and on the few occasions when a considerable infection has occurred we have 'cured' it by transferring flies to medium which has been treated according to *Protocol 13*.

Protocol 13. Bacterial control

Reagents
- Dihydrostreptomycin sulfate
- Penicillin G

Method

1. Make a solution containing 20 mg/ml of dihydrostreptomycin sulfate and 6 mg/ml of penicillin G.
2. Add two drops/vial or five drops/bottle after pouring the medium.
3. Yeast the medium the following day.

10.3 Mite control

Mites are the most serious contaminants likely to be found in *Drosophila* cultures. Avoid mites as follows:

(a) Keep all incoming stocks in quarantine.
(b) Make sure they are mite-free before you introduce them to your fly room. This requires a detailed examination.
(c) The stock should be kept for at least two generations before you add it to your other stocks.

For pictures of mites and a detailed consideration of how to deal with the problem see Ashburner and Thompson (7).

Even the precautions of quarantine can break down and you may have a budding mite infection. Good housekeeping will help to reduce the likelihood of an infestation to a minimum.

(a) Autoclave all infected bottles and vials.
(b) If you don't already do it, clean down your bench, microscope, and fly handling equipment at the end of the days work.
(c) Wash down the shelves and benches in the constant temperature rooms and incubators regularly.

If after all this you still have mites you need to take more drastic action.

Two mite pests have been described. The first (Mesostigmatic mites), which fortunately is rare, eats *Drosophila* eggs and can easily kill a culture. The second (Anoetid mites) does not harm the flies in such a direct manner but these mites do multiply rapidly and cling to the adults especially to the genitalia which in effect can render the fly sterile. Ashburner describes the drastic action that was necessary to get rid of an infection of the first class of mites (7), and also describes how to control and eliminate the Anoetid mites. Following this protocol we have successfully got rid of an infection of these mites. Two steps should be taken:

(a) If you are using foam bungs then change to non-absorbent cotton wool which imposes a greater barrier to the spread of mites.
(b) Use Tedion.

Tedion (2,4,5,4'-tetrochloro-diphenyl sulfone) is an acaricide which is manufactured by Duphar B. V. Weesp, and should be available from a local agrochemical supplier—in the UK, Nursery Supplies. It is generally marketed as an 8% emulsion for use in the horticultural industry. Dissolve it in acetone to a final concentration of 5000 p.p.m. and soak filter paper in this, air dry, and put in bottles and vials. This is effective in removing mites but should be used sparingly for fear of developing Tedion-resistant mites. According to the manufacturers in the UK Tedion is 'harmless to the operator, wild life and beneficial insects and it has been cleared for safe use on all edible and ornamental crops'.

Benzyl benzoate has also been used as an anti-mite agent either by treating filter paper with a 50% solution in ethanol and putting this in vials/bottles or by treating the foam plugs with a 20% solution. This has the advantage of puffing up the plugs and rejuvenating them as well as being an anti-mite agent. However benzyl benzoate suffers from the serious disadvantage that at the concentrations used it can kill some stocks of flies! It can also cause severe skin irritation if handled without gloves.

Finally a plea! If you have mites and are asked for stocks warn whoever you are sending the stocks. Rather admit to mite infection than wreak havoc elsewhere.

11. Handling flies

In order to handle flies you will need the following:

- one smooth white glass plate (15 × 8 cm)
- two pairs of fine forceps (we use watchmaker's forceps but these are expensive)
- one fine camel hair artists brush
- some empty but plugged vials
- 2 cm thick rubber pad
- marking pen for writing on glass

- morgue (containing paraffin oil or industrial methylated spirits)
- dissecting microscope with lamp
- anaesthetizer

11.1 Anaesthetizers

The two most commonly used anaesthetics are ether and carbon dioxide.

11.1.1 Ether

Many different etherizers have been developed and two of these are illustrated in *Figure 5*. The first is simple to make but it requires frequent addition of ether. The second is less simple but adding ether once a day is sufficient for most work.

(a) Tip the flies to the bottom of the vial or bottle by tapping on the rubber pad.
(b) Remove the plug and rapidly invert over the mouth of the etherizer.
(c) Tap the flies gently into the etherizer (vigorous tapping may dislodge the medium).
(d) Replug the vial or bottle.
(e) Wait until all flies have stopped moving.
(f) Tip them on to the glass plate for examination.

The time required to anaesthetize rather than kill flies varies with the strain used, and how long they remain unconscious is best left to experience.

11.1.2 Carbon dioxide

If you use carbon dioxide you will need a carbon dioxide cylinder with the correct outlet valve which will allow a gentle flow of gas without freezing. The gas is passed through a water trap to humidify it and then either into the apparatus in which the flies are anaesthetized (*Figure 6a*) or to a porous polyethylene film covering a Perspex box into which the CO_2 passes (*Figure 6b*). A two-way tap will divert the gas to either apparatus.

(a) Tip flies into the anaesthetizer as before. (They soon stop moving since CO_2 has a more rapid effect than ether.)
(b) Transfer to the porous ethylene film with the carbon dioxide flowing through. Flies can remain unconscious in the carbon dioxide atmosphere for a considerable time (up to 20 min) and still recover but this is strain-dependent. Once the carbon dioxide is removed the flies recover rapidly.

Carbon dioxide has a number of advantages:

(a) It is safer and more pleasant for the user.
(b) Flies are anaesthetized more quickly.
(c) They recover consciousness rapidly and will mate whereas ether delays mating.

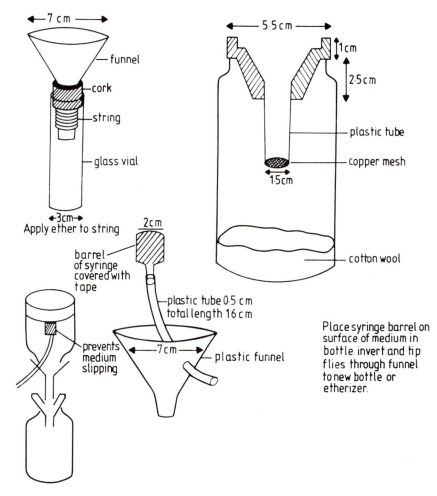

Figure 5. Two etherizers and an apparatus designed to prevent food from falling from the bottom of bottles. Before constructing the etherizers make sure that the components are not affected by ether!

The main disadvantages are:

(a) It is less convenient to use.

(b) It is toxic to some flies, those infected with sigma virus.

Both ether and CO_2 can affect the behaviour of flies (see Chapter 9).

11.2 Transferring flies

Most people at some time or other experience the problem of the medium slipping down to the neck of the bottle when collecting flies. This occurs especially if you make a series of collections from the same bottle. The apparatus

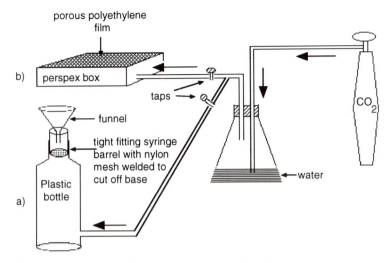

Figure 6. An apparatus using CO_2 for anaesthetizing flies. The taps divert the flow of gas either to the anaesthetizer, or to the Perspex box with the porous polyethylene film, for examining the flies.

shown in *Figure 5* is simple to make and largely overcomes this problem. In some cases where the medium has shrunk adding a little water may remedy the problem.

Once the flies have been anaesthetized they can be tapped on to the glass plate or the CO_2 plate, and examined under the microscope. It is advisable from the beginning to learn to use the paintbrush to handle flies as much as possible since forceps can cause damage and should be used sparingly. Occasionally eggs/ larvae/ pupae/ flies stick to anaesthetizers/ brushes/ plates, etc. and if overlooked can be transferred into the wrong bottle or vial contaminating a stock or a cross. Take care!

11.3 Distinguishing the sexes

The most important features of the fly for most experimental purposes are those which distinguish the sexes. While this distinction is easy to make with practice, care must be taken at the beginning.

(a) The females are generally larger.

(b) They have a pointed abdomen which becomes expanded prior to egg laying and is somewhat concave immediately after.

(c) The tip of the male abdomen is more rounded and is dark dorsally compared with the striped tip of the female abdomen. Both shape and colour may be deceptive in newly eclosed flies and in some body colour mutants.

(d) The genitalia of the two sexes are very different; learn to know these differences by examining the genitalia of flies of known sex.

(e) *D. melanogaster* can be sexed unambiguously by looking for the presence of sex combs, a row of thick dark bristles on the tarsus of the first pair of legs. These are only found in males.

11.4 Collecting virgins

In order to set-up a cross between male and female flies of known genotypes, it is necessary to collect virgin females. This is because after courtship and mating the females store sperm in the ventral receptacle and spermatheca. This stored sperm is sufficient to allow females to lay fertile eggs for many days: under ideal conditions for a week at the rate of 100 eggs per day, and under less favourable conditions for a longer period but with fewer eggs being laid. As male and female flies will eclose in the same bottle or vial then brother–sister mating can occur and the required cross cannot be set-up. This problem is overcome by collecting virgin females.

Under normal conditions flies will not mate for up to 8 h after eclosion. The routine for collecting virgins is to tip off all flies first thing in the morning and to collect and separate into sexes all those flies that emerge in the next 8 h. This is usually done by a second collection at the end of the day. Males can be collected overnight even if they have been promiscuous. They soon recover!

If necessary virgin females can be collected from flies which have eclosed overnight by identifying young newly eclosed flies either by their light body colour (but be careful as some strains are naturally light coloured even when mature), or more accurately by the presence of the dark meconium in the gut, visible through the ventral abdominal wall. Newly eclosed flies are likely to be in a majority if you keep your flies on a 12 hours light, 12 hours dark cycle as flies tend to eclose just before dawn. Keeping bottles at 18°C overnight reduces the rate of eclosion and development such that females eclosing overnight are more likely to be virgin than those kept at 25°C. The bottles can be kept at 25°C during the day.

11.4.1 'Virginator stocks'

For large virgin collections there are 'virginator' stocks in which the sons are killed by virtue of the possession of a conditional recessive X-linked or a Y-linked mutation, see Chapter 2, Section 4.

11.5 Setting-up a cross

Crosses are set-up according to *Protocol 14*.

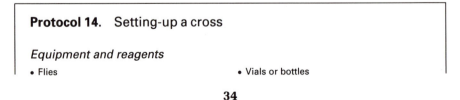

Protocol 14. Setting-up a cross

Equipment and reagents
- Flies
- Vials or bottles

Method

1. Place anything from one to five pairs of flies on the glass inside a vial of medium lying on its side; this prevents the anaesthetized flies from sticking to the medium.

2. When the flies have recovered consciousness stand the vials upright.

3. Alternatively catch unanaesthetized sexed flies from their appropriate vials with a pooter[a] and blow them into the mating vials.

4. Use bottles if more than about five pairs of flies (depending on the cross) are used.

[a] To construct a pooter: connect a mouth piece by flexible tubing to a piece of glass tube 0.7 cm diameter with a taper at the distal end. A constriction at the proximal end close to the flexible tubing holds a piece of cotton wool. Flies can be sucked into the glass tube but are prevented by the cotton wool from being swallowed by the operator.

It is not possible to give reliable estimates of the number of progeny expected which depends *inter alia* on the strain of flies used. Ten single pairs of our Oregon-R flies grown on regular medium and transferred to fresh vials daily over a five week period gave on average 1400 progeny with extremes of 900 and 2000. The most progeny from a 24 hour laying period by a single pair was 230 on day five after eclosion and mating. There have been reports of numbers greater than these. A pair of flies with balanced lethal chromosomes would give much fewer progeny. You will soon learn from experience what to expect with the stocks you use.

12. Simple genetic mapping

12.1 Assigning a gene to a chromosome

You may want to assign a gene to a chromosome for one of several reasons:

(a) A new mutation originating spontaneously in your stocks.

(b) A new mutation from a random mutagenesis experiment (note: mapping is made easier if you include suitable visible markers in your experimental stock before starting a random mutagenesis).

(c) A mutation found in flies caught in the wild.

(d) The site of insertion after *P* element-mediated transformation.

(e) The site of insertion in an enhancer trap experiment (Chapter 5).

One of the most common examples today is to locate the site of an inserted *P* element construct. Most of these carry a marker sequence (usually a variant of w^+) which will, in part, rescue the red eye colour in a homozygous mutant white eyed fly. In effect this acts as a dominant mutation and here we describe how it is assigned to a chromosome (*Figure 7*). Variants of this technique can be used to assign recessive mutations to a chromosome.

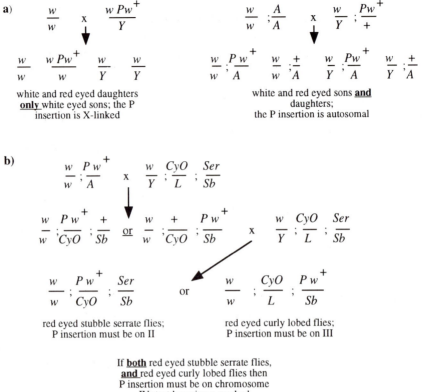

Figure 7. Assigning a *P* element insert marked with w^+ to a chromosome. It is assumed that the *P* element was inserted into a stock homozygous for *w* (white eyes). (a) Cross red eyed males from the insertion stock to white eyed females. If there are only white eyed sons the insertion is X-linked; if there are red and white eyed sons the insertion is autosomal. (b) If the insertion is autosomal cross red eyed females to males of a multiply marked stock: *w/Y; CyO/L; Ser/Sb*. Backcross: red eyed, curly winged, stubble female progeny to *w/Y; CyO/L; Ser/Sb*. If there are red eyed stubble serrate progeny the insertion is on the second chromosome; if there are red eyed curly lobed progeny the insertion is on the third chromosome; if there are flies of both these classes the insertion is on the fourth chromosome. *w* = white eyes and is recessive; A = any autosome; *P* w^+ is a *P* element insertion carrying the wild-type allele of *w*; *CyO* is a second chromosome dominant mutation giving curly wings; *L* is a second chromosome dominant mutation giving lobed eyes; *Ser* is a third chromosome dominant mutation giving nicked margins on the wing; *Sb* is a third chromosome dominant mutation giving stubble bristles.

12.2 Mapping a gene on a chromosome

Once a new mutation has been mapped to a chromosome its position on that chromosome can be located by genetic mapping (i.e. meiotic recombination mapping); an example is given in *Protocol 15*. Multiply marked chromosomes for mapping are available from the stock centres.

Exhaustive genetic mapping is unlikely to be useful—for most purposes today the information you really want is the physical location of your new mutation. A genetic map position will give you an idea as to the gene's cyto-genetic location (refer to the correlated cytogenetic/genetic map, Ashburner, pp. 1117–33 in ref. 12). A more precise physical location can be obtained using deletions from the region. For example a fly with one chromosome carrying the recessive mutation will show the mutant phenotype when the homologous chromosome carries a deletion of the wild-type gene. If the breakpoints of the deletion are known then the physical position of the wild-type gene must lie between these breakpoints.

A few cautionary notes:

(a) If a new mutation occurs in association with a chromosomal rearrange-ment, genetic mapping may be deceptive, because of local suppression of crossing-over (in an inversion heterozygote), and new linkage relation-ships (in the case of a translocation).

(b) The recombination fraction from your data should be treated sceptically when you are using widely-spaced markers because of multiple crossover events. These have the effect of reducing the observed recombination fraction below the genetic map length in the case of intervals with > 10% recombination. An approximation to the true map length from an observed recombination fraction is given by the Kosambi formula (13)—in ref. 2:

$$x = 0.25 \ln[(1 + 2y)/(1 - 2y)]$$

where x is the map length and y is the observed recombination fraction.

(c) The choice of mapping markers may be limited by the phenotype of the mutation being mapped, as it must be possible to distinguish all categories of progeny.

(d) There is no crossing-over in males. There is normally no crossing-over on chromosome 4 in females.

(e) In a stock heterozygous for a rearrangement on another chromosome, crossing-over on the chromosome of interest will be substantially increased by the 'interchromosomal effect' (see ref. 2); this can give mis-leading recombination fractions. However, if you wish to make otherwise rare recombinants between tightly linked loci, this crossover-promoting effect can be useful (exploit it by making heterozygous females carrying balancers on other chromosomes, or by using females carrying one of the *C(1)* chromosomes with a strong interchromosomal effect—see ref. 2).

Protocol 15. Mapping a new mutation on the second chromosome

Equipment and reagents
- Bottles and vials
- *al dp b pr c px sp* flies
- Flies carrying the new mutation

A. *Setting-up the cross*

1. Cross the two stocks in bottles.[a]

2. Culture the bottles at 25 °C.[b]

3. Collect F1 virgin females heterozygous for the chromosome carrying the new mutation and *al dp b pr c px sp*.

4. Set-up new bottles with the F1 virgin females backcrossed to males carrying *al dp b pr c px sp*.[c]

5. Remove the parents after five days.[d]

B. *Dominant mutation*

1. Collect, classify by phenotype, and count all progeny from each bottle daily, either until no more emerge, or until there is some danger that a subsequent generation is about to eclose (this is unlikely to happen until the bottle has been set-up for some 20 days).[e]

2. Analyse results.[f]

C. *Recessive mutation*

1. Collect several males of each class of single recombinant (e.g. ten males which are *al dp b + + + +/al dp b pr c px sp*; ten males which are *al dp b pr c px +/al dp b pr c px sp* etc.) from the bottles set-up in part A, step 3.

2. Set-up new vials, each with one recombinant male and three or four virgin females carrying the new mutation.

3. Look for the new mutant phenotype in the next generation. This should appear in progeny of some males recombinant for the interval carrying the new mutation.[g]

4. A more precise location can then be obtained by repeating the procedure, but only collecting and backcrossing the recombinants from the appropriate interval.

5. Analyse the results.[f]

---

[a] These bottles should not be crowded (five to ten pairs of flies per bottle depending on how fertile the stocks are), as larval stress may alter crossover frequency and distribution.
[b] Lower and higher temperatures increase crossover frequency.

> *c* These cultures should not be crowded (say five to ten pairs of flies per bottle). Competition between the larvae will intensify any differential viability of the various combinations of marker mutations, which could distort the recombination data.
> *d* Crossover frequencies and distribution alter as females age. Use young heterozygous females (up to two-days-old at time of setting-up the bottles).
> *e* Collect all the emergent progeny, rather than just the early ones, as mutant flies often show developmental delays, and this could distort recombination data. If the mutant phenotypes can be distinguished in preserved flies (fixed in 100% ethanol), it may be simpler to kill and fix them all, and then classify a random sample.
> *f* Analysis of recombination data is explained in most introductory genetics textbooks.
> *g* If the new mutation lies between *b* and *pr*, the new mutant phenotype should appear in the F1 of some but not all crosses between *al dp b + + + +/al dp b pr c px sp* and females carrying the new mutation. Some of the males will be *al dp b m + + + +/al dp b pr c px sp* and others will be *al dp b ± + + + +/al dp b pr c px sp* depending on where between *b* and *pr* the new mutation lies. If equidistant then each class will be represented with the same frequency. Rare double recombinants may cause confusion, but the frequencies of the recombinant progeny types will indicate the correct map order.

13. *Drosophila* life cycle

Having set-up the crosses it is appropriate to consider what is going to happen. *Figure 8* summarizes the *Drosophila* life cycle but this summary requires some comment.

13.1 Larval growth

From the time the larva hatches from the egg to the time of puparium formation the larval cells do not divide although the larva undergoes a considerable increase in size. This increase is due almost exclusively to an increase in cell size, which is accompanied by DNA replication with the replicates lying together to form the polytene chromosomes. The adult cells, represented by the cells of the imaginal discs, do divide during this period.

13.2 Metamorphosis

During metamorphosis most of the larval cells undergo autolysis, exceptions being the cells of the Malpighian tubules, some nerve cells, the imaginal rings of the salivary gland and the gut, and the imaginal islands of the gut. The adult fly is constructed from imaginal cells which in the case of those of the imaginal discs have been dividing during larval life but cease to divide at pupariation, and in the case of other imaginal cells such as the abdominal histoblasts have been dividing during the early part of metamorphosis. These cells undergo growth and morphogenetic movement until at last they sculpt the adult fly.

13.3 Development times

The time in hours after the eggs are laid (post-lay) for the different developmental stages at different temperatures for our Oregon-R stock are given in *Table 5*.

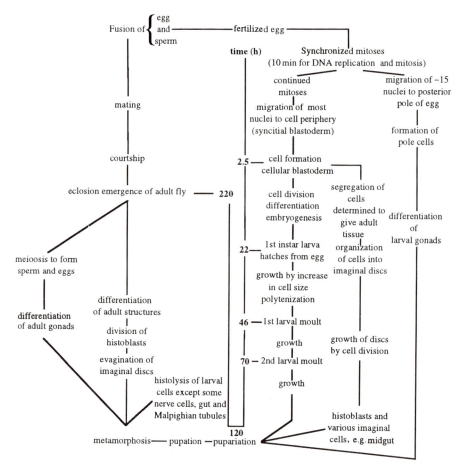

Drosophila development at 25°C

Figure 8. The *Drosophila* life cycle. All times are for 25 °C.

14. Mass culture

Biochemical/molecular studies of *Drosophila* are on the increase and frequently require large amounts of material. The simplest way to obtain lots of eggs, larvae, pupae, or flies is obvious—start with lots of parents. For this milk bottles are inconvenient and so population cages are used. Cages of various degrees of sophistication have been built, browsing through DIS will illustrate many of these. Here we describe two simple cages we have used successfully.

14.1 Small cages

A small cage can be made out of a hard plastic sandwich box, *Protocol 16*.

Table 5. Development times (h) at various temperatures

	18.5°C	25°C	29°C
Hatching	41	21	20
Pupariation	234	118	110
Eclosion	440	221	190

Protocol 16.   Small cage[a]

Equipment
- Hard plastic sandwich box 17.5 cm × 11.5 cm × 6 cm high

Method
1. Use a hot scalpel to cut a hole, with a diameter 5 mm smaller than the external diameter of a disposable Petri dish, in the lid.[b]
2. Cut a second hole in the lid 2 cm in diameter.
3. Plug it with a cotton wool plug.
4. Fasten the lid to the box by taping round the perimeter with sticky tape.
5. Place a Petri dish of yeast–glucose–agar medium seeded with live yeast over the large hole.
6. Hold this in place with rubber bands.
7. Shake flies into the box *via* a funnel through the small hole.
8. Tap the flies to the bottom of the box to change the medium.
9. Kill the flies at the end of the experiment by putting the box into the cold room (4°C) overnight and dispose of them.

[a] This cage is suitable for 500–1000 flies depending on the stock.
[b] This should be done in the fume-hood but, there are more sophisticated methods for cutting holes.

Small cages are usually used for specific and limited experiments and are set-up with flies from bottles. They are not maintained on a regular basis as are the large cages (see below).

14.2 Large cages

Many different types of large cage have been developed. We find the one described in *Protocol 17* to be effective.

Protocol 17. Large cage[a]

Equipment
- PVC sheet and tube[b]
- Black cotton material

Method

1. Bend a 3 mm sheet of PVC 25 cm wide and 120 cm long to form a hollow cube and weld the edges together half-way along one side.
2. Weld a 30 cm square sheet of the same material on the back and a similar sheet with a central hole 15 cm in diameter on the front.
3. Weld a circular tube of Darvic 15 cm in diameter and 4 cm deep to this hole.
4. Slip a 60 cm long black cotton sleeve 15 cm in diameter with an elasticated end over the projection.
5. Tape in place.
6. Close by tying a knot in the material.
7. Fill the cage either by tipping flies through a funnel into a second small hole closed with a cotton wool plug or by allowing washed pupae to eclose in the cage (see below).

[a] A large cage will hold over 10 000 flies.
[b] We use Darvic (marketed by ICI).

Both of these cages are readily cleaned.

14.2.1 Maintenance of large cages

A constant supply of eggs can be obtained from large cages maintained according to *Protocol 18*.

Protocol 18. To maintain the large population cage

Equipment and reagents
- Flies
- Feeding trays

Method

1. Transfer eggs laid overnight[a] or during the day, if not required for other purposes, by lifting the medium out of the polystyrene feeding trays (see below) on to a box[b] with yeast–glucose–agar medium seeded with live yeast.
2. Score the surface of the medium to give the larvae a rough surface for burrowing.
3. Collect pupae a week later (see below) and place them in a shallow box in a population cage where they eclose.

4. Day 1: place pupae into population cage A.

5. Day 7: place pupae into population cage B; start collecting eggs for experimental work from cage A.

6. Day 10: collect eggs from cage A for cage C.

7. Day 14: dispose of flies from cage A.

8. Day 15: start collecting eggs for experimental work from cage B, and place pupae into cage C.

9. Day 21: dispose of flies in cage B, etc.

[a] The original cage can be set-up with flies from bottles.
[b] A 25 cm square 10 cm deep polythene box with a tight-fitting lid, and a hole in the lid plugged with cotton wool allows sufficient air circulation.

15. Eggs

15.1 Egg collecting from small cages

In order to collect eggs from small cages we allow the flies to lay on a fresh Petri dish of yeast–glucose–agar medium seeded with live yeast for as long as is required. Rapid egg laying has been encouraged by making up the yeast paste in a 1:1 mixture of water and wine vinegar, or you can use apple juice plates. For precisely timed experiments this period of laying may be as short as 10 min. For anything much shorter than this it is probably worth putting a spoon of medium into a bottle and watching the flies lay and time from then. In order to see the eggs on the medium it is convenient to add a colourant, we use charcoal or a blue food dye.

The eggs laid in the first hour of the working day are usually discarded because females may retain fertilized eggs which develop in the oviduct, especially if the yeast on the feeding plates has been consumed overnight.

15.2 Egg collecting from large cages

In the large cages we use the same medium as for small cages but in 13×17 cm polystyrene trays (5 mm deep, these are the trays used to package foods in supermarkets) (Glenhurst Ltd.). For egg collection (*Protocol 19*) during the day we find 2% agar seeded with live yeast an adequate medium. In sealed boxes these trays or Petri dishes will keep for up to four weeks at 4 °C and so it is worth making up large numbers at a time.

Protocol 19. Egg collection

Equipment and reagents
- Eggs
- Sieves as described below
- Cetyldimethyl benzylammonium chloride

Protocol 19. *Continued*

A. *Harvesting eggs*

1. Wash eggs and yeast from the surface of the medium using a paint-brush.

2. Filter the suspension through a plastic household sieve which retains the dead flies.

3. Filter this suspension through a second sieve[a] which retains the eggs but not the yeast cells.

4. Wash copiously to remove most of the yeast cells.

B. *Sterilizing eggs*

1. Suspend and agitate the eggs in sterile detergent.[b]

2. Wash in sterile water.

---

[a] Cut the base off a wide-mouthed plastic bottle, and the centre from its screw cap. Cover the mouth of the bottle with nylon mesh which allows yeast to pass through but not *Drosophila* eggs. (We use Swiss Nylon Bolting Cloth, John Staniar and Co.) Hold the mesh in place with the screw cap. This is easy to dismantle and clean and makes an effective sieve.

[b] We have used a 0.1% sterilized solution of cetyldimethyl benzylammonium chloride (now called benzyldimethyl-*n*-hexadecylammonium chloride) (CDBC) (see ref. 8). This removes the remainder of the yeast cells and the eggs are ready for surface sterilization.

15.3 Dechorionating eggs

In many experiments involving eggs it is important to remove the outer membrane or chorion (*Protocol 20*). For most purposes the hypochlorite treatment surface sterilizes the eggs but if necessary they can be treated with CDBC (*Protocol 19*).

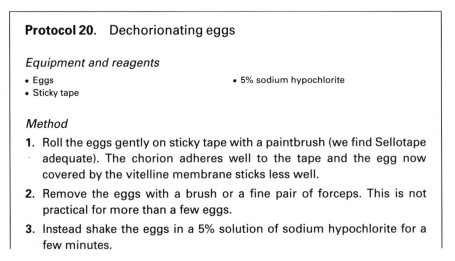

Protocol 20. Dechorionating eggs

Equipment and reagents

- Eggs
- Sticky tape
- 5% sodium hypochlorite

Method

1. Roll the eggs gently on sticky tape with a paintbrush (we find Sellotape adequate). The chorion adheres well to the tape and the egg now covered by the vitelline membrane sticks less well.

2. Remove the eggs with a brush or a fine pair of forceps. This is not practical for more than a few eggs.

3. Instead shake the eggs in a 5% solution of sodium hypochlorite for a few minutes.

4. Examine the eggs periodically.

5. Dilute the hypochlorite solution with sterile water once the chorions have been removed.

6. If necessary wash with CDBC (*Protocol 19*) followed by 80% ethanol.

15.4 Collecting timed eggs

The only effective way of getting large numbers of timed embryos for molecular work is to allow flies which are laying well to lay for a short period and age the embryos from the mid-point of that period. For smaller numbers a more precise age may be obtained by first noting the time an egg was laid and later scrutinizing the egg and ageing it according to morphological features, see Chapter 6.

16. Larvae

16.1 Collecting aged larvae

In order to collect larvae (*Protocol 21*) of a certain age *en masse* the approach depends on how accurate you need to be.

Protocol 21. Collecting larvae

Equipment and reagents
- Eggs
- Larva box (*Protocol 12*)
- Yeasted Petri dish with apple juice medium
- Nylon filter or plastic tea strainer depending on the age of the larvae

Method

1. Allow flies which are laying well to lay eggs for a relatively short period, say 2 h.

2. Transfer the eggs to larvae boxes, as described above.

3. Harvest the larvae at the appropriate time after the mid-laying point.

4. For a more precise collection: allow flies to lay eggs for 1–2 h on a yeasted Petri dish of apple juice medium.

5. Wash the larvae off the surface of the medium with a gentle stream of water after 21 h of development at 25°C (this time depends on the stock and should be determined empirically), when about 5% of the eggs have hatched. Most eggs stick to the medium.

6. Examine the plates and remove any larvae left with a fine forceps.

7. Allow development to continue for a further hour at 25°C.

Protocol 21. *Continued*

8. Wash newly hatched larvae into the nylon filter and rear on a yeasted Petri dish of yeast–glucose–agar medium.

9. Harvest larvae at the appropriate time after the mid-hatching time.

16.2 Harvesting larvae

Protocol 22. Harvesting larvae

Equipment and reagents

- Larvae
- Plastic bucket
- 3 M NaCl
- Sieve (plastic tea strainer)

Method

1. Wash the surface of the larvae boxes with a vigorous stream of water into a plastic bucket. In a healthy culture many small pieces of medium as well as yeast are washed in with the larvae.

2. Flood the box with 3 M NaCl and put on one side.

3. Pour off the supernatant from the bucket. This carries much of the debris. The larvae sink to the bottom of the water in the bucket.

4. Repeat this washing with water until the supernatant is clear. By this time the main contaminants of the larvae are large pieces of agar medium.

5. Half fill the bucket with 3 M NaCl in which larvae and some of the pieces of medium float.

6. Dilute with water until the medium sinks and the larvae remain afloat.

7. Treat the larvae which will have floated to the top of the NaCl solution in the box in the same way.

8. Collect floating larvae by skimming the surface with a sieve.

9. Wash the larvae free of salt by sedimenting in several changes of water.

The procedure in *Protocol 22* takes no more than 30 minutes and has no adverse effect on larvae.

16.3 Staging larvae

Drosophila larvae undergo two moults. At 25 °C the first and second instars each last about 24 hours and the third about 48 hours, depending on the stock used and the conditions of rearing. Larval instars can be recognized by a number of features. One of the most commonly used is the number of teeth on the larval mouth hooks but this requires the sacrifice of larvae. *Figure 9* shows

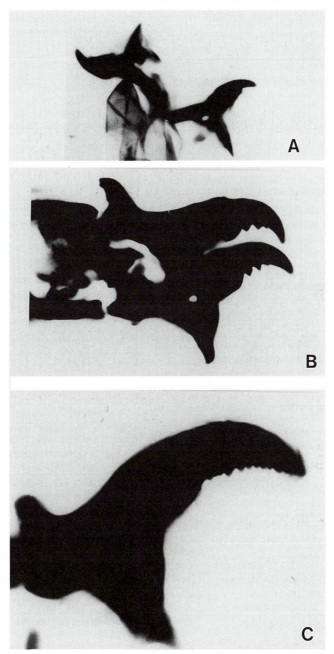

Figure 9. The mouth hooks for identifying the three larval instars. The magnification is the same in all three photographs. (A) First instar, (B) second instar, and (C) third instar. The larvae were squashed between two slides which were then taped together and examined under a microscope.

mouth hooks from first, second, and third instar larvae. The larvae were squashed between two microscope slides and viewed under a high power microscope. The larval instar of live larvae can be recognized by observing the spiracles. First instar larvae lack anterior spiracles; second instars have these spiracles but without the papillae characteristic of the third instar larvae (see ref. 14).

The precise age of late third instar larvae can be determined with reference to the puffing pattern of the salivary gland chromosomes (see ref. 15) but this also requires the sacrifice of the larvae.

16.4 Recognizing chromosomes in larvae

It may be necessary to recognize the presence of a specific chromosome in the larva. There are a number of larval markers which allow this. The list in *Table 6* is not exhaustive.

16.5 Sexing larvae

With practice it is relatively easy to sex third instar larvae (*Protocol 23*). This relies on the male gonad being about three times the size of the female gonad. The gonads are embedded in the opaque fat body in the latter half of the animal.

Table 6. Larval markers

Chromosome	Mutant	Phenotype
1	*w* 1–1.5 (white eyes)	Colourless Malpighian tubules seen through the body wall of second and third instar larvae
	y 1–0.0 (yellow body)	Some alleles have yellow-brown mouthparts distinguishable from the dark brown of wild-type larvae (seen in first–third instar larvae)
2	*Bc* 2–78.2 (black cells)	Numerous pigmented cells under cuticle of head, thorax, and abdomen obvious in larva and pupa; homozygous lethal
	cn 2–575 (cinnabar)	Combined give colourless Malpighian tubules
	bw 2–104.5 (brown)	seen through the body wall of second and third instars
3	*Tb* 3–90 (tubby)	Short fat larvae, pupae, and adults. Adults not always as obvious as larvae or pupae; homozygous viable.
	red 3–53.6 (red Malpighian tubules)	Recognized through body wall. Time of recognition varies but can be made earlier by using different strains of yeast, especially brewers yeast.

Protocol 23. Sexing larvae

1. Wash the larvae clean of medium.

2. Place on a chilled dark glass plate (place the plate on a bed of ice) under the microscope. The cold reduces the problems of larval movement.

3. Bring the tips of the forceps together and place the sides three-quarters the way down the larval cuticle.

4. Allow the forceps to spring apart gently stretching the cuticle. A clear oval area in the fat body is the male gonad. If this is seen then the larva is male, if not then the larva is either female or you have missed it.

5. Try sexing a number of larvae, put them in vials according to sex.

6. Check your success by sexing the adults which emerge.

7. Use transmitted light, until you are confident when you may return to incident light.

---

17. Pupae

17.1 Definition of pupation

The problem of when pupation, as opposed to pupariation, occurs has occupied some insect physiologists for many years. This is not the place to enter the debate. We will take the following approach: pupariation leads to the formation of the prepupa as described below and may be considered as the first part of a moult with the epidermis still attached to the puparium formed from the old larval cuticle. The completion of this moult and the separation of the epidermis from the puparium is pupation proper.

The process of pupariation, pupation, and metamorphosis has been studied by many workers over a long period. The most detailed study in *Drosophila* by Bainbridge and Bownes (16) identifies 51 visible stages during metamorphosis and describes 24 stages for use in experimental analysis. Here we simply note a few convenient criteria for identifying the early stages for the collection of homogeneous pupae.

17.2 Staging pupae

Some 8 h before pupariation begins the third instar larva stops feeding and crawls out of the medium, and is then known as the wandering larval stage. Pupariation begins with the formation of the white prepupa with its everted spiracles and the cessation of movement (time zero). Within the next 20 min–1 h the white prepupa becomes brown. About 4 h after puparium formation an air bubble forms in the abdomen which eventually leads to the organism becoming buoyant. This buoyancy is exploited for collecting timed organisms (see below). Pupation proper begins shortly after this and is completed some 12 h after pupariation begins.

17.3 Harvesting pupae

Most pupae are stuck to the side of the vessel by glue protein secreted by the salivary gland but they are readily removed by wetting either with a wet paint-brush (for a few) or by filling the vessel with water. They may need some coaxing by a jet of water or a large brush. Removal with forceps can damage the pupal case.

17.4 Collecting aged pupae

In order to obtain precisely aged pupae the visible stages of Bainbridge and Bownes can be used if only a few organisms are needed. If many organisms are required then timing is the only feasible approach. You can either use the observation that the white prepupal period lasts for a relatively short time and collect all white prepupae and start timing later stages from then; or you can use the observation that the organisms become buoyant at a particular stage; in which case carry out *Protocol 24*.

Protocol 24. Collecting aged pupae

1. Harvest larvae, prepupae, and pupae from a larva box.
2. Suspend in water.
3. Discard the floating pupae.
4. Dry the remaining organisms.
5. Allow them to continue development.
6. 1 h later immerse them in water.
7. Collect floating pupae; all will be of the same age ± 30 min.

Using this technique large numbers of pupae can be collected and timed for use at later developmental stage.

17.5 Recognizing chromosomes in pupae

If it is necessary to identify pupae carrying particular chromosomes this is probably done best by using chromosomal markers visible in the larvae and segregating these larvae before pupariation although some markers, e.g. *Tubby* can be identified in the pupa.

18. Collecting flies in the wild

There are a number of works describing how flies may be collected from the wild including chapters in Demerec (see Section 3), and Ashburner, Carson, and Thompson (17), and numerous articles in *Drosophila* Information Service (18). These frequently discuss the number of different species caught in different places at different times. Anyone interested in the population biology of these species would do well to consult these sources. Here we outline simple techniques for catching *D. melanogaster* in the wild as a potential source of electrophoretic variants of enzymes, other types of spontaneous mutations, repeated DNA sequences, etc.

There are many ways of collecting flies in the wild some complicated and some simple. We have had success using milk bottles with regular yeasted medium set out on trees or bushes in an orchard. We have also used milk bottles with rotting fruit but these tend to get messy.

18.1 Identification of *Drosophila melanogaster*

Once you have caught the flies how do you know which are *D. melanogaster*? They are readily distinguished from the larger dipterans and from most other Drosophilans. The main problem is to distinguish them from the sibling species which, outside Africa, generally means *Drosophila simulans*. While there may be anatomical differences which can be recognized by experts it is probably not worth bothering with these. The following approach is easy and unambiguous:

(a) If the unknown fly is male then cross it to *D. melanogaster* virgin females, and if the progeny are fertile then the initial fly was *melanogaster*.

(b) If the caught fly is a gravid female you can repeat this with some of her sons, with the remainder of the progeny used to establish a new stock.

(c) If the caught fly is a virgin female then you can mate her with known *melanogaster* males and see whether their progeny are fertile.

Although this approach takes time to get a result, it presents few problems.

19. Preparation of tissues

For biochemical and molecular work it is frequently necessary to isolate specific tissues in bulk. In nearly all cases these have been isolated from wandering third instar larvae and there are many descriptions as to how this might be achieved. Because imaginal discs are the larval precursors of adult tissues much attention has been paid to these. For small scale isolation (up to 100 larvae) of imaginal discs or any other tissue, hand dissection is probably best. For large scale preparations an automated approach is necessary. Eugene and Fristrom (19) describe one such technique for the isolation of imaginal discs and can purify 10^6 discs at a time and anyone aiming for these numbers should consult their protocol.

We describe simpler techniques which have allowed us to purify fat bodies, imaginal discs, Malpighian tubules, and salivary glands in sufficient quantity to raise antisera in rabbits against protein extracts and to examine the proteins from these tissues by electrophoresis.

19.1 Fat body purification

Protocol 25. Fat body purification

Equipment and reagents

- Grinding mill (Van Waters and Rogers)
- 100 and 670 μm nylon mesh (J. Staniar and Co.)
- Phenylthiourea (PTU)-saturated ethanol
- Ringer's solution: 6.5 g NaCl, 0.14 g KCl, 0.2 g $NaHCO_3$, 0.12 g $CaCl_2$, 0.01 g NaH_2PO_4 per litre
- 1% Ficoll (Pharmacia)

Method

1. Grow and harvest larvae as described previously.[a]

2. Suspend the washed larvae in *Drosophila* Ringer's solution.

3. Pour the suspension of larvae (20 g wet weight/100 ml) through the grinding plates of a grinding mill which allows the pressure between the grinding plates to be adjusted.[b]

4. Set these empirically such that 90% of larvae are disrupted.

5. Collect the grindate in a vessel containing a few drops of PTU-saturated ethanol which inhibits tyrosinase activity.

6. Pass the grindate through the nylon mesh which retains intact larvae, large pieces of gut, and cuticle.

7. Pass the filtrate through a 100 μm mesh which retains the intact tissues.

8. Suspend these tissues in 50 ml of a 1% solution of Ficoll in Ringers.

9. Centrifuge at 500 *g* for 5–10 min. The fat body collects on the surface and all other tissues sediment.

10. Skim the fat body off and resuspend in 5 ml of Ringers.

11. Examine under the microscope. This will show that the fat body is only contaminated with trachea.

12. Remove these by degassing the suspension for a few minutes in a vacuum desiccator which removes the air from the trachea and allows them to sink.

13. Recentrifugation and collection gives a pure fat body preparation.

[a] We obtain our best purification of tissues using larvae reared from eggs laid over a 12 h period and harvested when 5–10% of the larvae have pupariated.
[b] This disrupts the larvae by shearing and releases the tissues. This step is critical and shearing leaves the tissues more intact than any cutting motion.

19.2 Imaginal disc purification

Imaginal discs can be purified by the method described originally by Eugene and Fristrom (19) as recently adapted by Natzle and Vesenka (20).

19.3 Purification of other tissues

Protocol 26. Purification of other tissues

Equipment and reagents
- Glassware coated with water repellent (Repellcote BDH)
- Percoll
- Sedimented tissues from *Protocol 25*

Method

1. Centrifuge at 500 *g* the sedimented tissues from the fat body preparation (*Protocol 25*) on a Percoll step gradient[a] in a 30 ml polycarbonate centrifuge tube.

2. Collect the material at each interface with a Pasteur pipette into a small glass Petri dish.

3. Remove the contaminating gut from the Malpighian tubules (which are found mainly in the bottom layer) with a hypodermic needle attached to a vacuum pump.

4. Remove the salivary glands from the 10–20% interface, picking out contaminating tissue with the hypodermic needle.

5. Examine aliquots from the different steps in each experiment.[b]

6. Centrifuge again if necessary or remove and centrifuge on a second step gradient using different values for the steps.

7. Wash the purified tissue in Ringer's solution prior to use.

[a] 5 ml of 40%, 30%, 20%, and 10% Percoll in Ringer's solution.
[b] We have not fathomed why the behaviour of other tissues is not as reliable as discs and fat body but using these basic principles the other tissues are purified empirically.

Acknowledgements

We thank all colleagues erstwhile and present who have helped us enjoy working with *Drosophila* and especially Jean Matthews and Wendy Mulvany.

References

1. Craymer, L. (1984). *Drosophila Inf. Serv.*, **60**, 78.
2. Ashburner, M. (1989). *Drosophila: a laboratory handbook*. Cold Spring Harbor Laboratory Press.

3. Roberts, P. A. (1976). In *The genetics and biology of Drosophila* (ed. M. Ashburner and E. Novitski), Vol. 1a, p. 67. Academic Press Inc., London and New York.

4. Lindsley, D. L., Sandler, L., Baker, B. S., Carpenter, A. T. C., Denell, R. E., Hall, J. C., *et al.* (1972). *Genetics*, **71**, 157.

5. Judd, B. H., Shen, M. W., and Kaufman, T. C. (1972). *Genetics*, **71**, 139.

6. Holm, D. G. (1976). In *The genetics and biology of Drosophila* (ed. M. Ashburner and E. Novitski), Vol. lb, p. 529. Academic Press Inc., London and New York.

7. Ashburner, M. and Thompson, J. N. Jr. (1978). In *The genetics and biology of Drosophila* (ed. M. Ashburner and T. R. F. Wright), Vol. 2a, p. 2. Academic Press Inc., London and New York.

8. Frankham, R. (1973). *Drosophila Inf. Serv.*, **50**, 199.

9. Sang, J. H. (1978). In *The genetics and biology of Drosophila* (ed. M. Ashburner and T. R. F. Wright), Vol. 2a, p. 159. Academic Press Inc., London and New York.

10. Sparrow, J. C. and Sang, J. H. (1975). *Genet. Res. Camb.*, **24**, 215.

11. Hunt, V. (1970). *Drosophila Inf. Serv.*, **45**, 179.

12. Lindsley, D. L. and Zimm, G. G. (1992). *The genome of Drosophila melanogaster.* Academic Press Inc., London and New York.

13. Kosambi, D. D. (1944). *Ann. Eugen. London*, **12**, 172.

14. Bodenstein, D. (1950). In *Biology of Drosophila* (ed. M. Demerec), p. 275. John Wiley and Sons.

15. Ashburner, M. (1967). *Chromosoma (Berl.)*, **21**, 398.

16. Bainbridge, S. P. and Bownes, M. (1981). *J. Embryol. Exp. Morphol.*, **66**, 57.

17. Ashburner, M., Carson, H. L., and Thompson, J. N. Jr. (1986). The genetics and biology of *Drosophila* Vol. 3a–e. Academic Press Inc., London and New York.

18. *Drosophila Information Service* now published by FlyBase.

19. Eugene, O. M. and Fristrom, J. W. (1978). In *The genetics and biology of Drosophila* (ed. M. Ashburner and T. R. F. Wright), Vol. 2a, p. 121. Academic Press Inc., London and New York.

20. Natzle, J. E. and Vesenka, G. D. (1994). In *Drosophila melanogaster: practical uses in cell and molecular biology* (ed. L. S. B. Goldstein and E. A. Fyrberg), p. 110. Academic Press Inc., London and New York.

2

Mutagenesis

T. A. GRIGLIATTI

1. Introduction

Revising this chapter engendered the thought that perhaps the days of large scale screens for mutations had waned. But the genome projects have described many genes (ORFs), or expressed sequences, for which no known function exists, and this provides impetus for the isolation of mutations in specific genes or regions of the genome. Until homologous gene replacement becomes available to make full use of *in vitro* mutagenesis, there will be a need to induce mutations in specific genes in order to analyse or confirm the function of protein domains *in vivo*. In addition, as the basic steps for essential functions in developmental and physiological pathways are being defined, screens for more specific phenotypes such as aberrant chromosome pairing, aberrant migration of specific cell types, aberrant organelle assembly, or aberrant behavioural characteristics are being undertaken. The successful identification of mutants in a particular process has always depended on the specificity of the screen. Therefore, considerable attention needs to be paid in the screening protocol to:

(a) The precision with which the aberrant phenotype targets proteins in the particular process to be dissected.

(b) The efficiency of the screening protocol in detecting mutations with this phenotype.

A large number of mutagenic agents have been used to induce mutations in *Drosophila*. No attempts will be made to cover all of the various chemicals or high energy particles that have been used, but a list of chemicals that are mutagenic in *Drosophila* can be obtained from the Environmental Mutagen Information Center at the Oak Ridge National Laboratory (1). This chapter will focus on those mutagens, excluding transposable elements and viruses, commonly used to induce aberrations which genetically and cytogenetically are characterized as point mutations, and mutagens that induce cytogenetically detectable chromosome aberrations at reasonably high frequencies. For more detailed discussions of the effects of mutagen dose, parental age, gamete stage, and storage on the efficiency of inducing and recovering various types of mutations, and references to specific articles on these subjects see the reviews on radiation genetics in *Drosophila* by Sankaranarayanan and Sobels (2), chemical mutagenesis by Lee (3), and a general review by Ashburner (4).

2. Mutagens and methods of application

2.1 Ethyl methanesulfonate

Ethyl methanesulfonate (EMS) is a monofunctional alkylating agent. It produces point mutations by attacking (ethylation) the *O*-6 position of guanine and the *O*-4 position of thymine, which allows mispairing with thymine and guanine, and results in G:C to A:T and T:A to C:G transitions, respectively. It is probably the most frequently used chemical mutagen in *Drosophila* genetics, especially over the past 25 years. In bacteria (5–7) and bacteriophages (8) EMS has been shown to cause G:C to A:T transitions preferentially, suggesting that the *O*-6 alkylation of guanine is the most frequent alteration, but it can produce other types of mutations. In *Drosophila* it is generally used to induce point mutations. Indeed, some have claimed that EMS induces only point mutations in *Drosophila*. However, chromosomal rearrangements are routinely detected after mutagenesis with EMS. These may result from its ability to attack the *N*-7 position of guanine. This alteration appears to be more cytotoxic and may engender more chromosome breaks than alkylation of the *O*-6. The breeding protocol followed during the screen, after mutagenesis with EMS, can influence the frequency of recovering chromosomal aberrations rather dramatically. Allowing the females to store 'mutagen treated' sperm generally increases the number of cytologically detected aberrations, and so this procedure should be avoided. However, chromosomal rearrangements generally occur at a relatively low frequency in experiments where sperm storage is not an issue. For example, molecular analysis of EMS-induced mutations at the *rosy* locus (9) showed that only one of 36 was a detectable structural aberration (an insertion). In other genes, for example, genes of the *bithorax complex*, the incidence of detectable structural aberrations is much higher (10). The differences in the recovery of aberrations, between experiments in which sperm storage is not an issue, may be due in part to the nature of the gene—that is, the function of the gene product(s) —the regulation of gene expression, and the mutant phenotype for which one selects. Base pair substitutions may fail to produce readily detectable phenotypes in some genes or screening protocols, in which case gross structural alterations may be recovered preferentially. In the latter case, the frequency of recovered mutations at a given dose should be markedly lower than expected, but in most screens controls for the induction of sex-linked lethals are not done so mutation frequency is at best an estimate. In those cases where an antibody for the gene product is available, many of the mutants recovered after EMS treatment produce detectable cross-reacting material. In addition, a reasonably high proportion, approximately 5–10%, of the mutations induced by treatment with EMS are temperature-sensitive. Thus many of the lesions induced by EMS have properties characteristic of missense mutations. Since EMS is a reasonably simple mutagen to handle and is very efficient, it is a good choice for inducing mutations that behave as point mutations and it is an obvious choice for inducing conditional mutations.

EMS is not without its drawbacks as a mutagen. In screens for visible phenotypes, a large proportion of the mutant individuals recovered following treatment with EMS are mosaics (3, 11). Evidently only one strand of the DNA helix is affected by the mutagen and fixation of the mutation is delayed by at least one cell division. The high frequency of mosaicism may further support the axiom that EMS induces point mutations preferentially. However, it has the drawback that many mutations may not be detected in the initial screen. For example, some individuals that are mosaic for a 'lethal' mutation may survive, and individuals mosaic for a visible or behavioural mutation may escape detection. For this reason an extra generation in the mating scheme is frequently included when using EMS. This additional step is time-consuming and, in screens that involve pair matings, limits the number of treated chromosomes that can be examined in a single cycle.

This drawback is offset somewhat by the high frequency of mutations induced by EMS, which varies with the dose administered. While the frequency of mutations/dose may be linear, the amount of mutagen taken up by the flies is somewhat capricious. A concentration of 25 mM EMS routinely produces about 25–40% sex-linked recessive lethals among the progeny of treated males. The X chromosome is approximately half the size of the major autosomes, thus 50–80% of each of the major autosomes should carry a recessive lethal mutation following treatment with EMS. This high frequency of induced mutations suggests that many chromosomes (mutant lines) will possess more than one mutation. In many cases these linked, second site mutations must be removed, since they might confound further analyses. Hence it is useful to induce new mutations on genetically marked chromosomes. If marked stocks are used, choose the markers carefully and limit the number of markers to be carried by the mutagen-exposed chromosome so that the stocks remain viable and fertile. On occasion, we have decreased the concentration of mutagen to 15 mM to reduce the number of double hits per chromosome. However, reducing the concentration of the mutagen has its drawbacks as well; the proportion of recessive lethals induced in a screen—relative to mosaics—increases with increasing dose (12). To establish a true estimate of the mutation frequency, a subset of the mutagen-exposed flies must be used for a sex-linked recessive mutation assay, or similar assay. This is clearly time-consuming and detracts from time spent recovering putative mutants in 'your gene or process' and most elect not to do it.

2.1.1 Method of administering EMS

A denaturing solution (*Protocol 2*) is used to denature EMS and should be made up first, and in reasonably large amounts (500 ml), so that it is available in case of a spill. It should be made up fresh each day.

Deliver the mutagen to adult males by allowing them to feed on a 1% sucrose solution containing EMS at the appropriate concentration (13) (*Protocol 1*). EMS is available from a number of different chemical supply sources. It remains a potent mutagen even after storage for several years at room temperature. Gloves should be worn at all times when handling EMS and all of the operations, including feeding the flies, should be carried out in a fume-hood.

Protocol 1. EMS–sucrose solution

Equipment and reagents
- 500 ml flask
- 1 ml syringe
- 10 ml syringe
- Ethyl methanesulfonate
- Sucrose

Method

1. Make up 100 ml of the 1% sucrose solution in a flask.
2. Use a disposable 1 ml syringe with a long needle to remove 0.24 ml (for 25 mM) of EMS from the bottle.
3. Place the tip of the needle into the sucrose solution and gently dispense the EMS into the solution.[a]
4. Remove the 1 ml syringe and immediately fill it with the denaturing solution and place it in an empty container in the fume-hood, well away from the immediate work space.
5. Slowly take up the EMS droplets along with 4–6 ml of the sucrose solution using a disposable 10 ml syringe with a long, moderate gauge needle (e.g. 22 gauge).[b]
6. Carefully expel the liquid from the syringe keeping the tip of the needle below the level of the sucrose solution. Apply moderate force to the syringe, but be careful not to expel the remaining air in the syringe as the EMS–sucrose solution will splash.
7. Repeat this process five or six times or until droplets are no longer visible (the droplets generally disappear after two or three repeats).

[a] The EMS will form droplets that sink to the bottom of the sucrose solution since its density is 1.167. These 'droplets' can be dispersed easily as EMS is miscible with water.
[b] The same syringe can be used to dispense the EMS–sucrose solution to the bottles containing adult flies (see *Protocol 2*).

In order to administer EMS–sucrose to flies many researchers starve or desiccate the flies prior to treatment. This pre-conditioning entices them to ingest the EMS–sucrose solution which is unpalatable to them; to do this place the flies into empty vials or bottles at room temperature for 8–12 h prior to treatment. Administer the EMS–sucrose as described in *Protocol 2*.

Protocol 2. Administering EMS–sucrose solution to the flies

Equipment and reagents

- Empty half-pint milk bottles with caps
- Whatman No. 1 filter paper circles (or equivalent)
- Disposable 10 ml syringe
- Adult male flies about three-days-old
- Denaturing solution: add 0.5 ml thioglycolic acid[a] to 4 g of NaOH in 100 ml water
- EMS–sucrose solution (*Protocol 1*)

Method

1. **Day one.** Place two pieces of Whatman No. 1 filter paper, or the equivalent, on the bottom of each clean, empty half-pint milk bottle.[b]

2. Add approx. 100 of the males to be treated to each bottle and cap it with the normal cardboard cap or foam plug.

3. Place these bottles in the fume-hood while you make up the denaturing and the EMS–sucrose solutions, by which time the agitated flies will have calmed down and most will have climbed to the top of the bottles.

4. Make up the denaturing and the EMS-sucrose solutions.

5. Fill a 10 ml syringe with the EMS–sucrose solution to dispense EMS–sucrose solution to each bottle of flies.[c]

6. Insert the needle through the thin area beneath the thumb tab, or down the side of the foam plug, and slowly inject 1.1 ml of the solution directly onto the centre of the filter paper.[d] If the bottles are not disturbed during this operation, most of the flies will be at the top of the bottles, not on the filter paper.

7. After adding the EMS–sucrose solution to the treatment bottles, add at least an equal volume of the denaturing solution to the remaining EMS–sucrose solution.

8. Fill the 10 ml dispensing syringe with the denaturing solution.

9. Leave the EMS–sucrose solution and the 10 ml syringe in the fume-hood for 24 h. After 24 h the bottles can be thoroughly washed, and the EMS–sucrose solution and syringe can be discarded following the normal protocols.

10. **Day two.** Make up fresh denaturing solution just prior to removing the flies from the treatment bottles.

Protocol 2. *Continued*

11. Remove the flies from the treatment bottles by removing the caps or plugs and inverting the bottles over bottles with fresh medium and tapping gently.[e]

12. Put the flies aside and allow them to feed and recover for about 24 h prior to mating.

13. Pour fresh denaturing solution into the treatment bottles containing the EMS–sucrose solution and allow them to stand for approx. 24 h.

14. Discard the denatured solution after 24 h and wash the bottles.

[a] Thioglycolic acid (mercaptoacetic acid) is readily oxidized and should be stored at –20°C, and since it decomposes to give hydrogen sulfide it should be used in a fume-hood.
[b] 5.5 cm circles fit easily but securely on the bottom of half-pint bottles, that is, they do not fall away from the bottom of the bottles when the bottle is inverted to remove the flies after the mutagen treatment.
[c] This can be the same syringe used to mix the EMS–sucrose solution.
[d] This is just enough liquid to saturate the filter paper and leave a small puddle in the recess between the bottom of the filter paper and the glass. This small excess of fluid ensures that the filter paper will remain damp throughout the period of treatment, but it will not be so wet that the flies to stick to the paper.
[e] If the correct amount of EMS–sucrose solution was added on the previous day, the filter papers will still be moist but no liquid will run down the side of the bottle when it is inverted.

Many workers allow the flies to feed for 24 h. This is reasonably convenient, but is not absolutely necessary. Most of the mutagen uptake probably occurs during the first few hours of feeding. Once in solution, EMS is hydrolysed rather quickly. Its half-life in solution at 37°C, as measured by the Ames test, is about 2–3 h (14).

2.2 *N*-ethyl-*N*-nitrosourea (ENU)

ENU has recently gained popularity as a chemical mutagen for *Drosophila*. ENU is a monofunctional alkylating agent, like EMS. Its chief target is the O-6 position of guanine. It has a low Swain–Scott value ($s = 0.26$ at 25°C) and therefore is purported to attack areas of low nucleophilicity such as the O-6 of guanine. EMS with a higher s value (0.67 at 20°C) should have an enhanced affinity for sites with higher nucleophilic strength. Indeed, in direct comparison ENU produced fewer translocations (a test for chromosome breaks) than EMS (15). In a direct comparison of mutagenic activity, ENU was about 2.5-fold more potent than EMS, $35 \pm 3\%$ versus $14.1 \pm 0.08\%$ sex-linked recessive lethals at a concentration of 5 mM respectively (16). This should not influence the choice of one of these alkylating agents over the other, since both are very potent. The suggestion that ENU might produce fewer chromosome breaks was substantiated by genetic and molecular studies on ENU mutations induced in the alcohol dehydrogenase region. A concentration of

5 mM ENU produced 35% sex-linked recessive lethals. Eight of ten *Adh* null mutations behaved as point mutations in complementation tests and showed no abnormalities in Southern blot analyses. Two of the ten mutations behaved as deletions (17).

ENU is available from Sigma in isopac vials. It can be fed to adult flies in a sucrose solution following a protocol similar to that described for EMS. All alkylating agents undergo hydrolytic decomposition once in solution. EMS is relatively stable in solution, but nitrosoureas, such as ENU, can decay within minutes to hours. This process is influenced by the pH of the solution. At pH 6.0 ENU has a half-life of about 31 h at 20°C (18). Therefore make up a 0.01 M sodium acetate buffer pH 4.5 and inject this solution into the Sigma isopac vial. The concentrated solution is then diluted with a 1% sucrose solution to a final concentration of 5 mM. This solution can then be delivered to three-day-old flies following the protocol given for EMS. Once again, ENU is an extremely potent mutagen, all experiments should be carried out in a fume-hood and gloves should be worn at all times.

2.3 Triethylenemelamine (TEM)

TEM has been used to induce small deletions and point mutations. Generally, it has been dispensed by injection into the abdomen of adult males. However, we routinely administer TEM by feeding adults in a manner similar to that described in *Protocol 2*. In four separate screens the frequency of X-linked recessive lethal mutations varied from 5–15%. This compares favourably with experiments in which 0.2 mM TEM was administered by injection; in these experiments the X-linked recessive lethal frequencies were 10–12%. In an experiment to induce deletions in the chromosomal region between bands 39A and 40A we recovered four deletions among nine lethal mutations induced by TEM. Since we screened 2035 treated second chromosomes, the frequency of deletions induced, in that region of the genome, was 0.19%. The frequency with which deletions are recovered following treatment with ionizing radiation ranges between 0.005% and 0.15%. Hence TEM compares favourably with X-rays for inducing deletions. The size of the visible deletions we recovered varied from four to 30 bands. Some of the mutations that behave as point mutations, in fact, may be very small deletions. TEM is available from Polysciences.

2.3.1 Administering TEM by feeding

The TEM solution is made up fresh, since at room temperature TEM is polymerized to an apparently inactive substance, by dissolving 15 mg of TEM in 500 ml of 1% sucrose and shaking vigorously for several minutes in a flask tightly stoppered with a ground glass stopper. Adjust to a final concentration of 0.15 mM with 1% sucrose solution.

Administer TEM to adult flies as described for EMS in *Protocol 2*.

2.3.2 Administering TEM by injection

Since we have had repeated success by feeding TEM to adult flies, we no longer administer TEM by injection. The injection technique is more tedious since it involves handling each fly to be treated. In addition, injection involves more risk to the researcher. None the less, since this is the more conventional approach to administering TEM, it is presented in *Protocol 3*.

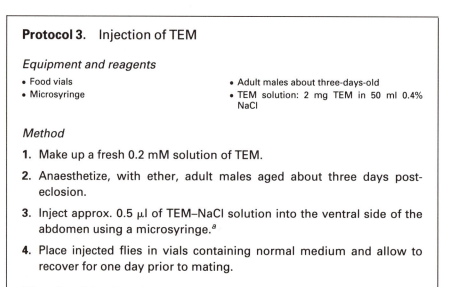

Protocol 3. Injection of TEM

Equipment and reagents

- Food vials
- Microsyringe
- Adult males about three-days-old
- TEM solution: 2 mg TEM in 50 ml 0.4% NaCl

Method

1. Make up a fresh 0.2 mM solution of TEM.

2. Anaesthetize, with ether, adult males aged about three days post-eclosion.

3. Inject approx. 0.5 μl of TEM–NaCl solution into the ventral side of the abdomen using a microsyringe.[a]

4. Place injected flies in vials containing normal medium and allow to recover for one day prior to mating.

[a] The volume injected need not be controlled precisely since the fly often expels variable amounts of haemolymph upon recovery.

2.4 Formaldehyde

The use of formaldehyde as a mutagen was extensively reviewed by Auerbach *et al.* in 1977 (19). Formaldehyde produces a full spectrum of alterations including point mutations and chromosomal aberrations. A number of investigators have used formaldehyde to induce small deletions and duplications since it is purported to induce a higher frequency of deletions relative to translocations and inversions than X-rays (20). This point is somewhat contentious. A large number of deletions have been recovered in the region delimited by bands 34D and 35E which includes the structural gene for alcohol dehydrogenase (*Adh*). These include 20 deletions induced by treatment with X-rays, and 12 by treatment with formaldehyde. In a review of these data, Ashburner *et al.* (21) pointed out that a large proportion of the mutants recovered were deletions, 67% (12/18) and 71% (20/28) after treatment with formaldehyde and X-rays, respectively. Furthermore, the frequency of deletions that removed the *Adh* gene was higher in those screens which used X-rays (0.0218%) than those which used formaldehyde (0.00184%). This suggests that there is little or no advantage in using formaldehyde to induce deletions. However, it should be pointed out that these screens for deletions around the *Adh* gene used different selection procedures and that the frequency with which deletions were recovered in the *Adh* region is very high compared with other experiments (see below). Perhaps of greater consideration is the fact that effective mutagenesis with formaldehyde requires treatment during the larval period. The efficiency of mutagenesis varies with:

- developmental stage
- strain
- duration of exposure
- concentration of formaldehyde
- formulation and condition of the medium (19)

Unlike X-rays, the mutations induced by treatment with formaldehyde often appear as mosaics. Hence, while it appears that formaldehyde is a useful mutagen for inducing a variety of mutations and perhaps particularly deletions, it is somewhat tedious to use. It is not clear that it offers many advantages over X-rays, but it might be considered when X-rays fail to produce the desired results.

2.4.1 Formaldehyde-containing medium

Formaldehyde is administered by feeding. The protocol is described by Auerbach and Moser (22). Collect late first instar larvae and transfer them to the formaldehyde-containing food (*Protocol 4*). The food appears to lose its mutagenic activity as the larvae feed, and so the larvae can be allowed to complete development on the medium containing formaldehyde. Alternatively, the larvae can be removed after 24 h and placed on normal medium. Only males are susceptible to mutagenesis by feeding with formaldehyde. Collect the male flies and mate to the appropriate virgin females.

Protocol 4. Formaldehyde medium

Reagents
- Cornmeal
- Molasses

- Killed brewer's yeast[a]
- Agar

Method

1. Mix 13 g cornmeal, 15 g molasses, 10 g brewer's yeast, and 1.5 g agar in 100 ml water.

2. Melt the agar by heating, and allow the medium to cool to about 65°C.

Protocol 4. *Continued*

3. Add sufficient formaldehyde to bring the final concentration to 0.2% (based on the volume of water in the initial mixture).

4. Allow the food to cool and set.

[a] Avoid adding live yeast to the medium since it provides an alternative food source for the larvae and also may compete for the formaldehyde in the medium.

2.5 Ionizing radiation

Radiation has been used extensively as a mutagen. It induces a full spectrum of mutations. Since EMS appears to induce point mutations preferentially, radiation might best be used to recover chromosomal aberrations. In addition to their traditional cytogenetic applications, translocation and inversion breakpoints are useful for physically mapping a cloned gene on the genome. Many workers will require deletions for dosage studies, saturation mapping, to delimit molecular walks, and so forth. A number of the chemical mutagens described above will induce chromosomal aberrations, especially deletions. The question arises: which mutagen, radiation, TEM, or formaldehyde, should be used to generate deletions? The frequency of chromosomal aberrations induced by radiation appears to vary from one experiment to another. Comparisons between experiments may be somewhat misleading because of the variable efficiency of the selection protocols used to recover aberrations in specific regions. However, comparisons between experiments done in a single laboratory, over a number of years, on the same region of the genome do provide a reasonably accurate estimate of the frequency of radiation-induced chromosomal aberrations. The frequency of confirmed aberrations induced in the region of *Antennapedia* range from 2.4×10^{-3} to 9×10^{-1} (23, 24). A total of 25 chromosomal aberrations were identified in six different experiments and of these seven (28%) were deletions. The proportion of deletions identified, among those experiments in which several mutations were recovered, varied from a low of 1/19 (5.3%) to a high of 2/6 (33%). This is generally consistent with many other experiments in other regions of the genome.

The protocol for irradiating flies varies with the type of equipment used to generate the high energy particles. Exposures of 3000–4000 rad produce a high frequency of mutations with minimum sterility (5–10%). To induce chromosomal rearrangements we typically use 4000 rad to irradiate three- to five-day-old male flies.

3. Protocols for identifying different types of mutations

A number of different protocols for the isolation of both sex-linked and autosomal mutations are given in the following section along with representative flow diagrams. While all possibilities are not covered, examples are given for the recovery of lethal and visible mutations, conditional mutations, duplications, and deletions. In practice the types of mutations that can be recovered are limited only by the researcher's ingenuity in designing an efficient screening protocol.

3.1 Sex-linked recessive lethal mutations

A typical protocol for isolating sex-linked recessive lethal mutations is described in *Protocol 5* and shown in *Figure 1*.

Protocol 5. Isolation of sex-linked recessive lethal mutations[a]

Equipment and reagents
- Milk bottles with food
- One labelled food vial for each mutagenized chromosome tested
- Several hundred mutagenized male flies
- Several hundred virgin *Binscy/Binscy* females

Method

1. Place 30 pairs of mutagen treated males and virgin *Binscy/Binscy* females (the *Binscy* chromosome survives and is fertile in both males and females) in milk bottles with food and allow them to deposit eggs for three days.

2. Transfer the parents to another set of bottles for egg deposition (three days) and then discard.

3. Number a set of vials corresponding to the number of chromosomes to be screened for X-linked recessive lethal mutations.[b]

4. Place a single F1 female in each vial along with several of their brothers. (Note these females need not be virgins.) It is important that only one female is placed in each vial, since the descendants from each F1 female represent progeny from a single treated X chromosome.

5. Allow the F1 females to lay eggs for about five days.

6. Discard the parents and allow the offspring to develop.

7. When the F2 progeny have eclosed examine each vial for the absence of non-*Binscy* (wild-type in this example) males.

8. Absence of non-*Binscy* (wild-type) males suggests that an X-linked recessive lethal mutation was induced in the gamete produced by the mutagen treated male grandparent.

9. Retain those vials with no, or a low number of, non-*Binscy* males and

---

Protocol 5. *Continued*

establish stocks from the heterozygous (mutagen treated)/*Binscy* females.[c]

10. Place the mutations into balanced stocks according to *Figure 1B*.

[a] This protocol is often incorporated into a genetic screen for mutations to determine the efficiency of mutagenesis. In such cases about 1000 mutagen treated chromosomes (single Fl females) are examined.

[b] The mutagenized X chromosome can be identified in subsequent generations or crosses if the mutagenized male carried a recessive visible X-linked mutation. Avoid mutations such as *yellow* or *white* since they have behavioural abnormalities. Markers such as *cross-veinless*, *vermilion*, *scalloped*, or *forked* are easily scored.

[c] These can be distinguished from their *Binscy*/*Binscy* sisters on the basis of the Bar eye phenotype, *B*/+ versus *B*/*B*, respectively.

At times it may be important to recover mutagen-exposed chromosomes from different stages of spermatogenesis, for example, to determine the stage during which mutagen treatment is most efficacious, or to optimize recovery of a specific class of mutations (e.g. chromosomal aberrations). This is easily done by successively mating the mutagen treated males to several harems of virgin females. Typically:

(a) Mutagen treated males are mated with virgin females for one or two days and the embryos collected.

(b) The males are separated from the females and mated with another group of virgin females for a second two day period.

(c) The process is then repeated.

The number of broods and the duration of each brood period (the sperm sampling period) can be modified to suit the requirements of the experiment. Once mated the females can be transferred to fresh bottles to increase the number of progeny derived from each brood and reduce crowding of the offspring.

3.2 Autosomal recessive lethal mutations

Autosomal recessive lethal mutations are selected by a similar method to that used in isolating sex-linked recessive lethal mutations. The only difference is that each mutagen treated autosome must be isolated, and then made homozygous. This requires at least one extra generation. The method for selecting second chromosome recessive lethals is described in *Protocol 6* and shown in *Figure 2*. For third chromosome recessive lethals one need only substitute a strain with marked third chromosomes and a strain with a multiply inverted third chromosome, such as *TM3* [third multiple 3 = *In(3LR)TM3, y⁺ ri pᵖ sep l(3)89Aa bx³⁴ᵉ e*] or *TM6B* [*In(3LR)TM6B*].

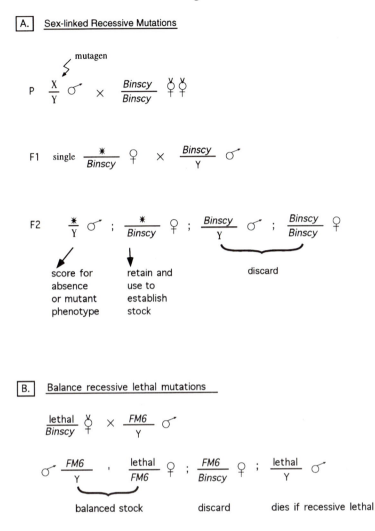

Figure 1. (A) Protocol followed to recover sex-linked recessive lethal mutations. This protocol can be used to estimate mutation frequency in any screeen for mutations. It can also be used to isolate sex-linked recessive lethal or visible mutations. *Binscy = In(1)sc^{S1L}sc^{8R} + dl-49, y sc^{S1} sc^{8} v B* males and homozygous females are viable and fertile. * refers to mutagen treated chromosome which can be marked with various recessive visible mutations. (B) This protocol is often used to balance sex-linked recessive lethal mutations or visible mutations that are weak or sterile as homozygotes. *FM6 = In(1)FM6, y^{31d} sc^{8} dm B*, it is viable and fertile in males, viable but sterile in females when homozygous. In this example 'lethal' refers to a recessive, sex-linked lethal mutation.

Autosomal Recessive Mutations

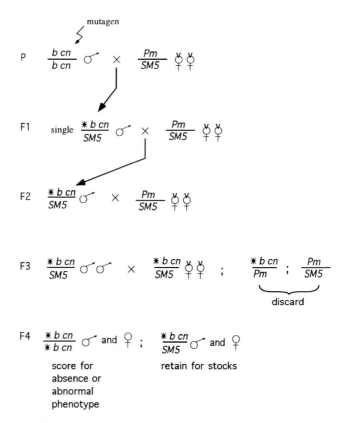

Figure 2. Protocol for isolating second chromosome recessive lethal mutations. The F2 cross is identical to the F1 cross and is done when the mutagen produces a high frequency of mosaic individuals, alterations that require one or more cell divisions to become established (for example EMS especially at low concentrations). The chromosome to be treated is often marked with recessive mutations. *b* = black body; *cn* = cinnabar eyes; *Pm* = *bw^{v1}* = plum eye colour; *SM5* = *In(2LR)SM5, al² Cy lt^v cn² sp²*; * refers to mutagen treated chromosome (marked with *b cn* in this example).

Protocol 6. Isolation of autosomal (2nd chromosome) recessive lethal mutations

Equipment and reagents

- Milk bottles with food
- One labelled food vial for each mutagenized chromosome tested

- Several hundred mutagenized male flies
- Several hundred virgin *Pm* or *Gla/SM5* females

Method

1. Place 30 pairs of mutagen treated males and virgin females hetero-zygous for an inversion and a homologue bearing a dominant mark-er[a] in milk bottles with food and allow them to deposit eggs for three to four days.

2. Transfer the parents to another set of fresh bottles for an additional three to four days.[b]

3. After the final egg deposition, remove the parents and discard.

4. Collect the F1 males that are heterozygous for the mutagen treated chromosome and the inversion.

5. Label a set of vials corresponding to the number of chromosomes to be tested.

6. Mate single F1 males to about five virgin females that are hetero-zygous for the appropriate inversion and dominant marker (e.g. *Pm* or *Gla/SM5* for the second chromosome and *Gl/TM3* for the third chromosome).[c]

7. Collect males and virgin females that are heterozygous for the muta-gen treated chromosome and the inversion (* *b cn/SM5 Figure 2*) from the progeny of the F1 generation and discard the other geno-types. (If the F1 generation is repeated these are the progeny derived from the F2 generation.)[d]

8. Mate these in fresh vials, numbered appropriately (same 'line' as parental), and allow them to lay eggs for about five days, and discard the parents.

9. Examine the progeny from this cross for very low numbers or com-plete absence of homozygotes (* *b cn/* b cn, Figure 2*).

10. Use the surviving heterozygotes from those lines (vials) that lack homozygotes to establish balanced stocks by mating siblings.

[a] For mutations on chromosome 2 the inversion *SM5* and the marker *Pm* or *Gla* are useful. (For mutations on chromosome 3 the marker *Gl* or *Cxd*, for those who have trouble seeing the *Gl* phenotype, and the inversion *TM3* are useful.)
[b] Alternatively, the male parents can be separated from the females (which are discarded) and mated to fresh virgin females in new bottles. The latter procedure is often done to sample gametes that were in earlier stages of spermatogenesis during the exposure to the mutagen.
[c] As each F1 male represents a single treated autosome it is important that only one male is placed in each vial, and it must be maintained and followed as such in all subsequent generations.
[d] The F1 cross is often repeated to eliminate mosaicism (see ref. 3 for a complete discussion). If this step is eliminated, some mutants may go undetected. However, eliminating the F2 genera-tion does save time, allowing more lines to be established in the F1 generation. (We usually eliminate this step.)

A variation of this general protocol for isolating autosomal lethal mutations is shown in *Figure 3*. This variation uses a dominant temperature-sensitive

Autosomal Recessive Mutations using a DTS Selection Scheme

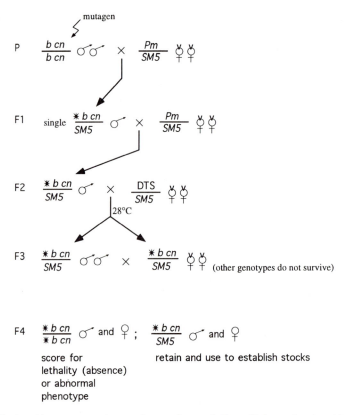

Figure 3. Protocol to recover autosomal recessive mutations. Using a dominant temperature-sensitive mutation (DTS) in the F2 cross eliminates the need to collect virgin females among the progeny of this cross. For the designation of other markers and symbols see *Figure 2*.

(DTS) lethal mutation to eliminate the need for collecting virgins when establishing the F3 cross. The protocol is the same except that the F2 cross is made and the F3 generation is raised at 28 °C. All flies heterozygous for the DTS fail to survive. Only those individuals heterozygous for the mutagen treated chromosome and the inversion survive. These are simply inbred to establish the final generation.

3.3 Non-lethal mutations

Protocols 5 and *6* describe the isolation of lethal mutations but they can be used to isolate visible mutations. For visible mutations each vial is examined for the appropriate mutant phenotype. Stocks of visible mutations can be established from the homozygotes (if fertile) or from the heterozygotes.

For recessive mutations with less obvious phenotypes, such as male or female sterile mutants, maternal effect lethals, behavioural mutants, mutagen-sensitive mutants, and so forth, the homozygotes from each vial (line) must be tested for the appropriate phenotype. This might require another generation in the screening protocol, for example for mutagen-sensitive or maternal effect lethal mutations. Stocks can be established from lines that give a positive result for a particular mutant phenotype, using either the remaining homozygotes or the heterozygotes.

3.4 Conditional mutations

A variety of different types of conditional mutations can be recovered (25). Temperature-sensitive mutations are perhaps the most frequently isolated form of conditional mutation in *Drosophila*. They allow survival of mutations that otherwise would be difficult or perhaps impossible to recover, for example dominant lethal mutations. Temperature-sensitive mutations also allow one to delimit the time during development when a particular gene product is required. Other forms of conditional mutants have been isolated (for review see ref. 25). This section will focus on temperature-sensitive (ts) mutations. However, any set of restrictive versus permissive conditions may be used in place of temperature.

3.4.1 Sex-linked recessive conditional lethals

A simple protocol for isolating sex-linked recessive conditional lethals is described in *Protocol 7* and shown in *Figure 4*.

Protocol 7. Isolation of sex-linked recessive conditional lethal mutations

Equipment and reagents
- Milk bottles with food
- One labelled food vial for each mutagenized chromosome tested
- Several hundred mutagenized male flies
- Several hundred virgin attached-X females

Method

1. Place approx. 30 pairs of males, exposed to the appropriate mutagen, and attached-X virgin females into bottles and allow them to lay eggs for three to five days.

2. Discard the parents, or transfer them into fresh bottles to amplify the number of progeny produced. The F1 males that result are patroclinous (that is, they receive their X chromosome from their paternal parent), and represent a single mutagen-exposed X chromosome. Each is treated as a single line.

Protocol 7. *Continued*

3. Label a set of vials to accommodate the total number of X chromosomes to be examined.

4. Mate each male singly to about five virgin attached-X females, i.e. one male and five females per vial.

5. Allow the females to lay eggs for about three to five days at the permissive temperature.[a]

6. Transfer these parents to a replicate set of fresh vials, place them at the restrictive temperature (or under the restrictive condition), and allow them to lay eggs for another three to five days.

7. Discard the parents.

8. Examine the F2 progeny from those cultures that develop at the restrictive temperature for the absence of males (lethal mutation), or visible phenotype, or test for a particular phenotype such as a behavioural abnormality, sterility, genomic imprinting effects, etc.

9. Establish lines bearing the putative mutation using the males that survive, or lack the abnormal phenotype, in the replicate cultures that develop at the permissive temperature.

10. Maintain the mutation in males by mating them to attached-X/Y females.

11. Homozygous stocks can be established as shown in *Figure 4B*.

[a] For heat-sensitive lethal mutations the permissive temperature is usually 18°C or 22°C, and the restrictive temperature is usually 28°C or 29°C (28°C is often used for ts sterility mutants since 29°C can cause partial sterility in some multiply marked strains). For cold-sensitive mutations the low temperature would be the restrictive condition. From bacteriophage studies, self-assembly processes are particularly susceptible to cold-sensitive mutations.

3.4.2 Autosomal recessive temperature-sensitive mutations

The procedure for isolating ts autosomal recessive mutations is shown in *Figure 5*. Note that the procedure is the same as that described for non-conditional mutations up to the F3 generation. At this point replicate cultures of each line are established, first at the permissive and subsequently at the restrictive temperature. For ts recessive lethal mutations homozygotes should appear among the progeny of the F3 cultures maintained at the permissive temperature but should be absent from the replicate cultures that developed at the restrictive temperature. For visible mutations, abnormal phenotypes should appear only among the homozygotes that eclose in cultures maintained at the restrictive condition. For non-visible phenotypes (such as sterility or maternal effects) the homozygotes from replicate cultures kept at the restrictive temperature must be tested for the aberrant phenotype, and those at the permissive temperature must be tested for absence of that phenotype.

Conditional Sex-linked Recessive Mutations

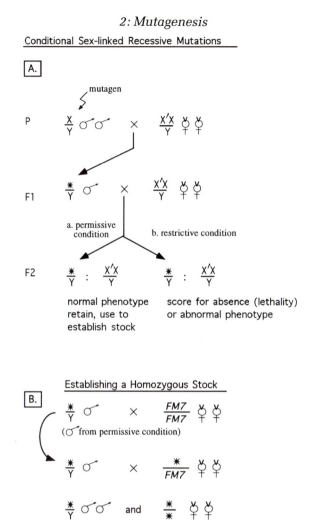

Figure 4. (A) Protocol used to recover conditional sex-linked recessive mutations. (B) Protocol used to establish a homozygous line of the conditional mutation. X^X refers to an attached-X strain, that is, two X chromosomes that share a single centromere. *C(1)DX* is often used; it is stable and marked with yellow body (*y*) and forked bristles (*f*). *FM7* = *In(1)FM7, y³¹ᵈ sc⁸ wᵃ snˣ² vᵒᶠ g⁴ B* ;* refers to mutagen treated chromosome.

3.4.3 Autosomal dominant conditional mutations

Autosomal dominant temperature-sensitive mutations can be recovered by following a protocol similar to the one shown in *Figure 6*. It is self-explanatory.

3.4.4 Y-linked conditional mutations

A simple protocol for the recovery of Y-linked mutations is shown in *Figure 7*. It has been used successfully to recover ts alleles of the Y-linked fertility factors (26).

Conditional Autosomal Recessive Mutations

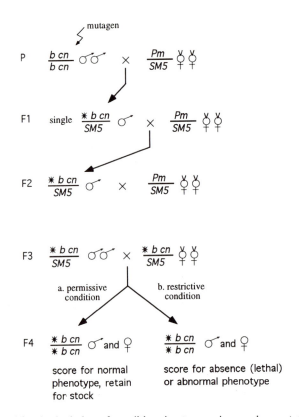

Figure 5. Protocol for the isolation of conditional autosomal recessive mutations. For the designation of markers and symbols see *Figure 2*.

4. Systems to generate virgin females

Screens for mutations often require a large number of virgin females at certain steps in the protocol. Collecting and sorting 50 000 virgin females, within a reasonably narrow period of time, requires a substantial effort. Furthermore, this task generally coincides with the time-consuming task of single fly matings (establishing single lines). As a result, the decision is taken to screen fewer treated chromosomes, which reduces the likelihood of recovering the desired mutants. The effort directed at recovering mutants can be maximized by automating the collection of virgins. Examples of genetic schemes that result in lethality of all males are described in *Protocol 8* and shown in *Figures 8* and *9*. Both schemes have been used successfully in our laboratory, and a variety of other schemes are available. If one requires autosomes with certain

2: Mutagenesis

Conditional Autosomal Dominant Mutations

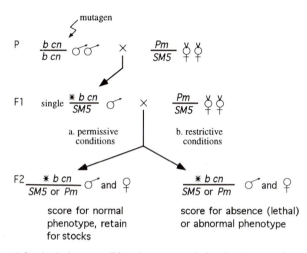

Figure 6. Protocol for isolating conditional autosomal dominant mutations. Many conditional mutations are unconditional recessive lethal or sterile, in such cases they are simply retained as balanced lethal (or sterile) heterozygotes. $Pm = bw^{V1} =$ plum eye colour; $SM5 = In(2LR)SM5, al^2 Cy lt^v cn^2 sp^2$; * refers to mutagen treated chromosome (marked with *b cn* in this example).

Conditional Y-linked Mutations

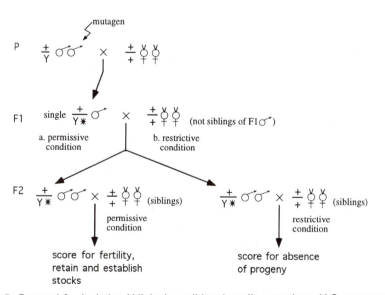

Figure 7. Protocol for isolating Y-linked conditional sterile mutations. Y S represents a mutagen treated Y chromosome.

Eliminating Males with a Temperature-sensitive Lethal Mutation

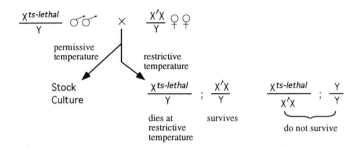

Figure 8. A sex-linked recessive ts lethal used to eliminate male progeny. The strain is maintained at 22 °C where both males and females survive. Males fail to survive when development occurs at 29 °C. The trisomy X females ($X^{ts}/X^{\wedge}X$) generally fail to survive at any temperature. Occasionally they do survive especially at lower temperatures (< 1%); they have abnormal morphology, usually crumpled wings with the inner margin incised and malformed hind legs, and are sterile. Any autosomal markers or inversion may be added to the system as required. $X^{ts\text{-}lethal}$ refers to a sex-linked temperature-sensitive lethal mutation.

genetic markers, or inversions, these can be put into the strains used to produce females. Those systems that utilize temperature-sensitive lethal mutations to eliminate one sex are perhaps the easiest to use (*Figure 8*).

Protocol 8.   Selection of virgin females

Equipment and reagents

- Milk bottles with food
- Several hundred virgin attached-X females
- Several hundred male flies carrying an X-linked ts lethal mutation[a]

Method

1. Maintain and expand the stocks at the permissive temperature.

2. Place approx. 30 pairs of males carrying an X-linked ts lethal mutation and attached-X females in bottles and allow them to lay eggs for three to five days.

3. Transfer the parents to fresh culture bottles.

4. Shift the bottles with developing cultures to the restrictive temperature prior to the temperature-sensitive period (TSP) and allow them to develop at that temperature until all the adults have eclosed (or at least well beyond the TSP).

5. Collect the virgin females as they eclose.

[a] It is convenient to use sex-linked ts lethals that have at least one TSP late during development. We have used the ts lethals *ras*[E6] which has a TSP for lethality during the late third instar and early pupal period, and *shi*[ts1] which has several TSPs for lethality including one in the early pupal period. It is important to note that occasionally the attached-X chromosome in the female will break down, regardless of the temperature at which the strain is kept. In this case fertile males can be produced. These will survive at both the restrictive and the permissive temperature since they lack the ts lethal mutation. Fortunately these males carry recessive visible markers and are easily detected (for example yellow body and forked bristles). Any culture bottle in which such males appear is simply discarded from the stocks maintained at the permissive temperature. At the restrictive temperature, each culture bottle need only be checked for absence of males.

The second protocol (*Figure 9*) requires two strains, and virgin females must be collected from the attached-X/Y *bb⁻* strain to establish the parental cross to produce F1 progeny lacking males. None the less, one need collect only 1000 virgin females for the parental generation to produce 15 000 to 20 000 virgin females in the next generation.

5. Protocol for isolating deletions of specific regions

There are a number of protocols for making deletions for a specific region. One is discussed here, but see Chapter 3.

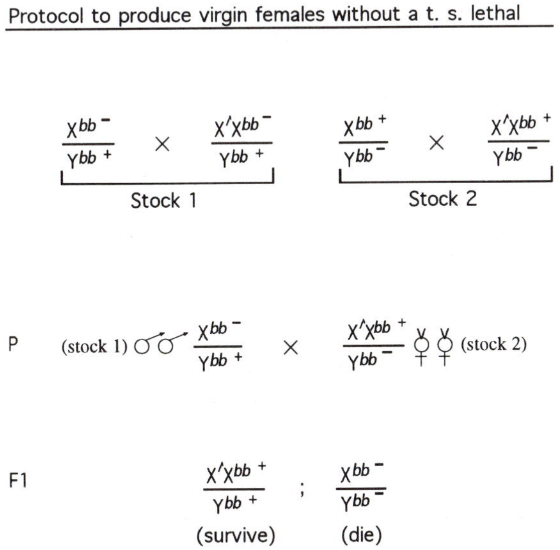

Figure 9. Protocol to eliminate male progeny without using temperature-sensitive mutations. *bb⁺* and *bb⁻* refer to presence and absence of the nucleolus organizer region respectively. The X^X strain *C(1)DX [C(1)DX, In(1)dl-49 — In(1)sc⁸ ◊ yf, y̅ sc⁸ f]* lacks the nucleolus organizer region on the X and thus is lethal as X^X/0.

5.1 Pseudodominance of recessive allele, or reversion of a neomorph or antimorph

Stable deletions for a specific region of the genome are generally produced by inducing chromosome breakage and selecting for pseudodominance of a recessive allele (i.e. expression of a recessive allele in the hemizygous condition), or reversion of a dominant neomorphic allele (deletion of dominant mutation). Examples of both protocols are shown in *Figures 10* and *11*, respectively. In both types of screens deletions are detected visually among the F1 progeny. Hence, a large number of mutagen treated chromosomes can be screened.

While the actual protocol for isolating deletions by either pseudodominance of a recessive mutation or reversion of a dominant mutation is simple, complications can occur once the putative deletion is detected. If the putative

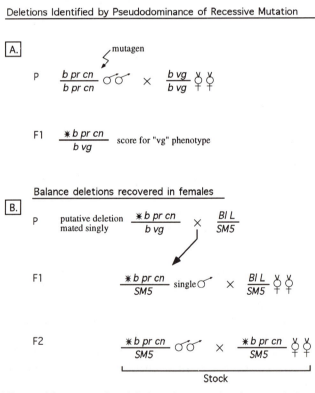

Figure 10. (A) Protocol for recovering deletions by screening for pseudodominance of a recessive mutation. (B) Protocol to balance deletions recovered in F1 females. *b* = black body; *pr* = purple eyes; *cn* = cinnabar eyes; *vg* = vestigial wings; *Bl* = Bristle; *L* = Lobe eyes; *SM5* = In(2LR)SM5 al², Cy, lt^v, cn², sp² ; * represents a mutagen treated chromosome.

78

deletion is recovered in a female, recombination may make it difficult to follow the deletion-bearing chromosome in subsequent generations. This problem can be overcome (*Protocol 9*) by establishing several lines (single chromosomes) for each putative deletion and testing each for the presence of the deletion (*Figure 10B*).

Protocol 9. Recovering induced deletions from female flies

Equipment and reagents

- Vials with food
- Females with putative deletion

- Males heterozygous for a balancer chromosome and a chromosome with a dominant marker

Method

1. Mate females with the putative deletion to males heterozygous for a balancer chromosome (*SM5* in this example) and a chromosome carrying a dominant mutation.

2. Set-up 20–25 crosses of single males bearing the original mutagen-exposed chromosome (absence of the paternal dominant marker) and the balancer chromosome, and five virgin females heterozygous for a chromosome marked with a dominant mutation and the balancer.

3. From the progeny establish lines of the original mutagen-exposed chromosome and the balancer.

4. Test each established line for the presence of the putative deletion by outcrossing to strains homozygous for the recessive marker (*vg* in this example).

5. Retain strains that show pseudodominance of the recessive marker, and examine cytologically. Discard the others.

The absence of recombination in males obviates the need for this extensive set of crosses. Hence, if the putative deletion is recovered in a male, that individual is simply mated with several virgin females heterozygous for an appropriate balancer (multiply inverted chromosome) and a balanced lethal stock is established from the offspring that are heterozygous for the chromosome carrying the putative deletion and the balancer (for example see *Figure 11B*). These are subsequently re-tested, and then examined cytologically.

Figure 11 shows an example of a selection protocol for identifying deletions using reversion of a dominant marker. It works best with dominant markers that are either antimorphs (disturbs the function of a multimeric complex, in modern laboratory jargon these are called dominant negatives) or neomorphs

Deletions Produced by Reverting a Dominant Mutation

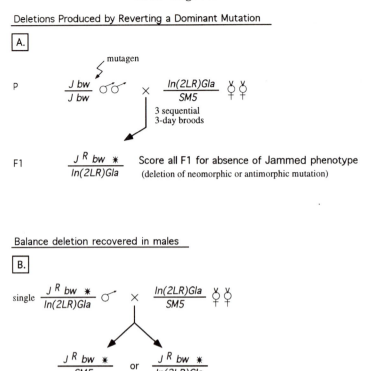

Figure 11. (A) Protocol for recovering deletions by screening for reversion of a dominant mutation. (B) Protocol to balance deletions recovered in F1 males. *J* = Jammed; *J^R* = Jammed revertant (wild-type phenotype); *bw* = brown eyes; *In(2LR)Gla* has glazed eye phenotype; *SM5* = *In(2LR)SM5, al² Cy lt^v cn² sp²*; * refers to mutagen treated chromosome (marked with *J* and *bw* in this example).

(new phenotype). The protocol is a simple F1 screen for loss of a dominant phenotype, therefore many chromosomes can be examined.

Dosage-sensitive loci can be used to select deletions or duplications directly by scoring for the dosage-sensitive phenotype. This might mean that either a duplication or a deletion already exists for the region and one need only isolate the opposite aberration. Alternatively, at least a subset of the loci that share a similar phenotype are dosage-sensitive. For example, deletions for loci the encode ribosomal proteins may be identified by scoring for a *Minute* phenotype (short thin bristles) since several of the *Minute* loci are dosage-sensitive, or one might select for deletions in regions abutting *Su(var)* loci by scoring for suppressors of position effect variegation.

When the appropriate visible markers are not available deletions can be

Deletions Identified by Psuedodominance of a lethal mutation

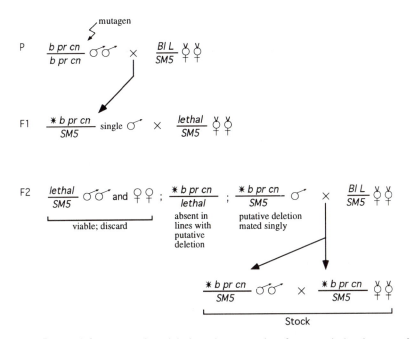

Figure 12. Protocol for recovering deletions by screening for pseudodominance of a recessive lethal mutation. For the designation of markers and symbols see *Figure 10*.

detected by failure to complement another deletion or a recessive lethal mutation (pseudodominance of a recessive lethal allele). The screen for pseudodominance of a lethal mutation (*Figure 12*) is self-explanatory, with the emphasis that single males are used in the F1 crosses and treated as independent lines in subsequent generations.

It is obvious that failure to complement a pre-existing deletion, while laborious, can be used repeatedly to collect a series of overlapping deficiencies and thus 'walk' down a chromosome to a region of interest.

6. Duplications

A tandem duplication for a specific region can be made by inducing unequal exchange and screening for chromosome aberrations that phenotypically suppress the expression of a dominant mutation. This approach requires that a dominant mutation, whose phenotype is markedly reduced in the presence of two wild-type alleles (hypomorphic or antimorphic alleles can often be suppressed), exists in the region in which one wishes to recover the duplication. This is described in *Protocol 10* and shown in *Figure 13*.

Protocol to Isolate Tandem Duplications

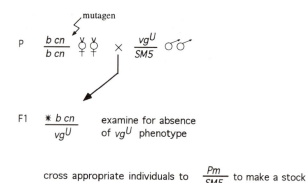

Figure 13. Protocol for recovering tandem duplications by screening for suppression of a dominant mutation. b = black body; vg^U = vestigial ultra; $SM5$ = $In(2LR)SM5$ al^2, Cy, lt^v, cn^2, sp^2; * represents a mutagen treated chromosome.

Protocol 10. Generation of duplications

Equipment and reagents

- Milk bottles with food
- Several hundred males heterozygous for the dominant mutation (e.g. vg^U) and a multiply inverted chromosome marked with a distinct dominant mutation

- Several hundred virgin females wild-type for the dominant mutation used in the screen but marked with other markers that allow the chromosome to be followed in subsequent generations

Method

1. Irradiate the virgin females with a low dose of X-rays (1000–2000 rads).

2. Mate *en masse* with the males carrying the dominant mutation.

3. Examine the progeny for absence or reduced expression of the dominant phenotype.

4. Cross these progeny to a stock carrying a balancer chromosome.

5. Establish a stock heterozygous for the duplicated chromosome and a multiply inverted balancer chromosome.

6. Re-test for suppression of the dominant mutation by crossing this stock to the stock carrying the dominant mutation.

7. Examine cytologically for the presence of the appropriate duplicated segment.

References

1. Environmental Mutagenesis Information Center, Information Division, Oak Ridge National Laboratory, Post Office Box Y, Oak Ridge, TN 37830, USA. In Europe contact the Environmental Mutagenesis Information Center—United Kingdom, Genetic Toxicology Group, Department of Genetics, University College Swansea, Singleton Park, Swansea SA2 8PP, UK.
2. Sankaranarayanan, K. and Sobels, F. (1976). In *The genetics and biology of Drosophila* (ed. M. Ashbumer and E. Novitski), p. 1090. Academic Press, London and New York.
3. Lee, W. R. (1976). In *The genetics and biology of Drosophila* (ed. M. Ashbumer and E. Novitski), p. 1299. Academic Press, London and New York.
4. Asburner, M. (1989). *Drosophila: a laboratory handbook.* Cold Spring Harbor Laboratory Press.
5. Swartz, N. M. (1963). *Genetics*, **48**, 1357.
6. Cloulandre, C. and Miller, J. H. (1977). *J. Mol. Biol.*, **117**, 577.
7. Foster, P. L. and Davis, E. F. (1983). *Proc. Natl. Acad. Sci. USA*, **80**, 2695.
8. Dreig, D. R. (1963). *Genetics*, **48**, 561.
9. Coté, B., Bender, W., Curtis, D., and Chovnick, A. (1986). *Genetics*, **112**, 769.
10. Karch, F., Weiffenback, B., Peifer, M., Bender, W., Duncan, I., Celniker, S., *et al.* (1985). *Cell*, **43**, 81.
11. Jenkins, J. B. (1967). *Genetics*, **57**, 783.
12. Epler, J. L. (1966). *Genetics*, **54**, 31.
13. Lewis, E. B. and Bacher, F. (1968). *Drosophila Inf. Serv.*, **43**, 193.
14. Jensen, E. M., La Polia, R. J., Kirby, P. E., and Haworth, S. R. (1977). *J. Natl. Cancer Inst.*, **59**, 941.
15. Vogel, E. and Natarajan, A. T. (1979). *Mutat. Res.*, **62**, 51.
16. Lee, W. R., Beranek, D. T., Byrne, B. J., and Tucker, A. B. (1990). *Mutat. Res.*, **231**, 31.
17. Batzer, M. A., Tedeschi, B., Fossett, N. G., Tucker, A., Kiroy, G., Arbour, P., *et al.* (1988). *Mutat. Res.*, **199**, 255.
18. Druckney, H., Preussmann, R., Ivancovik, S., and Schmah, D. (1967). *Z. Krebforsch.*, **69**, 103.
19. Auerbach, C., Moutschen-Dahmer, M., and Moutschen, J. (1977). *Mutat. Res.*, **39**, 317.
20. Slizynska, H. (1957). *Proc. R. Soc. Edinburgh, Sect. B*, **66**, 288.
21. Ashbumer, M., Aaron, C. S., and Tsubota, S. (1982). *Genetics*, **102**, 421.
22. Auerbach, C. and Moser, H. (1953). *Z. Indukt. Abstamm. Vererb.*, **85**, 479.
23. Duncan, I. W. and Kaufman, T. C. (1975). *Genetics*, **80**, 733.
24. Hazeirigg, T. and Kaufman, T. C. (1983). *Genetics*, **105**, 581.
25. Suzuki, D. T., Kaufman, T., Falk, D., and UBC *Drosophila* Research Group. (1976). In *The genetics and biology of Drosophila* (ed. M. Ashbumer and E. Novitski), Vol. 1a, p. 207. Academic Press, London and New York.
26. Ayies, G. B., Sanders, T. G., Kiefer, B. I., and Suzuki, D. T. (1973). *Dev. Biol.*, **32**, 239.

3

Transposons—gene tagging and mutagenesis

T. A. GRIGLIATTI

1. Introduction

The discovery of mobile genetic elements and their association with 'spontaneous' mutations in nature has led to novel ways of cloning genes, of germline transformation, and has added methods for creating certain types of mutations. To date, the *P* element is the most exploited transposable element in *Drosophila*, but there are many others, several of which have been characterized sufficiently to allow similar applications with perhaps only a little more effort. For example, the *hobo* element has been used for both transformation and gene tagging, the *IS* elements have been used for mutagenesis and gene tagging, and *mariner* has been used for transformation. These other mobile elements may have properties which, in the future, make them better for certain applications, such as target site specificity, efficiency of movement, or homologous target integration (gene replacement). But, this chapter focuses on some of the common applications of the *P* transposable element technology for creating useful mutations. Specifically it focuses on:

(a) Tagging uncharacterized genes (for the purposes of cloning) by screening for a particular phenotype.

(b) Tagging a gene whose map position is well defined, but which is not yet cloned.

(c) Creating useful new mutations in a gene in which a *P* element insert already exists.

(d) Creating deletions of a defined region of the genome.

The Genome Project (Chapter 1) in *Drosophila* is in the process of defining and providing single *P* element inserts throughout the genome. The ultimate goal is to establish several thousand single *P* insert stocks. Each of these *P* element lines contains a single insert at a unique location within the genome. In all cases the *P* transposon construct incorporates a cloning vector and selectable markers to facilitate cloning the adjacent genomic sequence.

Furthermore, many inserts have a few hundred base pairs of genomic DNA, adjacent to the *P* element insert, sequenced and entered into the database (see Section 4.3.1). This resource is growing rapidly and should be the first source of information for those wishing to clone a defined gene.

2. Historical background—hybrid dysgenesis

The *P* transposable element was discovered as a consequence of its association with male recombination (MR) and later with a panoply of traits termed 'hybrid dysgenesis'. Hybrid dysgenesis describes a collection of traits, generally presumed to be deleterious, that occurs in the hybrids formed from matings between females of certain laboratory strains and males from natural populations of *Drosophila melanogaster*. These characteristics include:

(a) High rates of sterility, or partial sterility, among both males and females.

(b) Recombination in males (normally male *Drosophila melanogaster* do not undergo meiotic recombination).

(c) A high 'spontaneous' mutation rate (often 1×10^{-3} or 1×10^{-4} instead of 1×10^{-5} or 1×10^{-6}).

(d) High frequencies of chromosome aberrations and non-disjunction.

This suite of genetic anomalies occurs in the germline of the hybrids and is accompanied by morphologically visible abnormalities such as poorly formed or deteriorated ovaries and testes among the mature F1 hybrid individuals. Hiraizumi (1) made the first observations of hybrid dysgenesis when he discovered that certain strains isolated from Texas (T-007) induced male recombination when heterozygous with marked laboratory chromosomes (1, 2). This 'Texas' chromosome was called MR for male recombination. MR strains, and chromosomes derived from these strains, were also found in natural populations taken from different geographical locations in the world, for example OK1 from Oklahoma and MR-h12 from Haifa, Israel. The phenomenon was characteristic of crosses between strains carrying chromosomes derived from wild populations and those derived from laboratory stocks (those kept in the laboratory for many decades).

It was soon discovered that these MR chromosomes cause not only male recombination but also cause a high frequency of chromosome aberrations and mutations among the F1 (hybrid) individuals (3–7). Mutations were associated with a number of different loci which were easily assayed, for example *yellow*, *singed*, and *raspberry* on the X chromosome. In addition to the high 'spontaneous' mutation frequency (10^{-4} to 10^{-3}), these new mutations were unstable.They reverted to wild-type at very high frequencies compared to their chemical- or radiation-induced alleles. These high mutation rates and the instability of the new mutant alleles, closely resembled the phenotypes caused by insertion sequences and *mu* elements of bacteria, and the connection to transposable elements was made by Green and others (5–7). The term hybrid dysgenesis was coined by Sved (8) and the phenomenon, termed the *PM* system of hybrid dysgenesis, was popularized by Kidwell, Engels, and others (9, 10, 18). The mobilization of *P* transposable elements within the germline of the hybrid offspring (F1 individuals) was the underlying cause of this suite of abnormal traits (11).

2.1 *P-M* hybrid dysgenesis

The complete *P* transposable element is 2907 bp long (11). It has a 31 bp inverted repeat sequence at each terminus (12). There are two broad categories of *P* elements, complete or intact *P* elements which are called *P* factors, and defective or deleted *P* elements. Complete *P* elements contain a single gene of four exons (0, 1, 2, and 3). It produces a single 2.7 kb mRNA in the germline which is translated into a 87 kDa 'transposase' protein (13, 14) essential for mobility. The primary transcript is properly processed in the germline, but in somatic tissue the third intron is not removed. This differential processing allows movement in the germline but not in somatic tissue. Defective *P* elements usually contain small to large deletions which disrupt the transposase gene. Thus the defective elements cannot transpose autonomously nor can they mobilize other *P* elements. However, defective *P* elements, which retain both *P* element ends (about 400 bp), can be mobilized if transposase is provided by a single complete *P* element elsewhere in the genome.

In general terms, hybrid dysgenesis occurs when *P* strain males (*P* for paternal contributor) are mated to *M* strain females (*M* for maternal contributor), but not vice versa and not in *P* × *P* or *M* × *M* crosses (*Figure 1*). Those familiar with bacteria and bacteriophage genetics will immediately see the parallels between the *P-M* hybrid dysgenesis and the phenomenon known as 'zygotic induction' in bacteria. In bacteria, lambda phage are released when a lysogenic donor strain (male-like) is crossed to non-lysogenic recipient strain (female-like), but not vice versa and not when two lysogenic strains are mated. The lysogenic strains carry a repressor for lambda phage replication, which is absent in non-lysogenic strains. When the DNA from the donor strain (male) is transferred to the non-lysogenic recipient (female), which lacks phage repressor, the lambda phage enters the replicative cycle. An analogous phenomenon occurs in *P* and *M* strains of *Drosophila melanogaster*. *P* strains carry one or more active *P* transposable elements (capable of making transposase) and have built up a 'repressor' which prevents movement. *M* strains lack either lack *P* elements completely or contain only defective elements which do not make the 'repressor' protein (the latter are called *M'* strains). When a *P* bearing sperm fuses with an *M* egg the *P* element encoded transposase is produced, since no repressor is present in the egg. This mobilizes the *P* elements. No movement occurs in hybrid offspring formed by matings

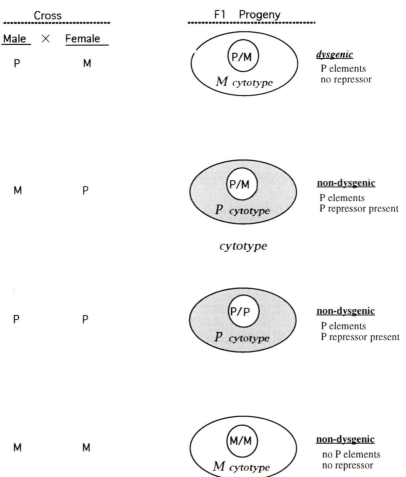

Figure 1. Crosses between *P* and *M* strains and the genotype and cytoplasm type of their offspring. *P* is a *P* strain which contains at least one, and usually several, complete *P* factors. *M* is an *M* strain, which completely lacks any type of *P* transposable element.

between *M* males and *P* females or from *P* × *P* crosses because the *P* element containing female parents contribute the *P* 'repressor' to the egg cytoplasm, which effectively impedes *P* element movement.

    P strains usually carry about 25–50 *P* elements and are a mixture of complete and defective *P* elements, a few of the latter make the repressor protein. True *M* strains have no detectable *P* elements. There are a variety of intermediates that occur such as *M'* and *Q* strains. We will not concern ourselves with these. For reviews of *P-M* hybrid dysgenesis and *P* element function see refs 15–20.

3. Assays for hybrid dysgenesis

There are two simple assays to determine whether a strain contains functional *P* elements (complete *P* factors). One uses morphological criteria and the other is a genetic test.

3.1 Gonadal dysgenesis—a morphological assay

The gonads of dysgenic flies are often abnormal. Gonadal development is disrupted early and either one or both ovaries or testes may fail to develop fully. The frequency and extent of the abnormal phenotype depends on the temperature at which the flies are raised ($> 25\,°C \sim 20\%$ of the gonads show an obvious morphological defect; $< 24\,°C$ striking morphological abnormalities are rare). The abnormal phenotype is very obvious in females and less obvious in males, so it is simpler to just examine ovaries. Age the adults to be examined (F1 individuals from a *P* male $\times$ *M* female cross) for at least three days. Remove and examine the ovaries from about 30 females. Either one or both ovaries will be incompletely developed (*Figure 2*).

Gonadal dysgenesis can be used to test for functioning *P* elements. Cross males of the suspected *P* strain to known *M* strain females. Raise the progeny at $> 25\,°C$ and examine the F1 females for abnormal ovaries (the number of females that need to be examined depends in part on the temperature at which the flies are raised). Offspring of *M* strain males $\times$ *M* strain females (Oregon-R or Canton-S) serve as non-dysgenic controls. Offspring of a known *P* strain (such as Harwich) crossed to *M* strain females serve as dysgenic controls.

% gonadal dysgenesis (GD) =

$$\frac{\text{number of flies with nullo-ovaries} + \tfrac{1}{2}\,\text{number with one ovary}}{\text{the total number of flies examined}} \times 100$$

3.2 *snᵂ*—a genetic assay

Singed mutations (*sn⁻*) have singed looking bristles and are easily scored. An allele called *singed-weak* (*snᵂ*) contains a *P* insert that is highly unstable and, when mobilized, gives rise to either *singedᵉˣᵗʳᵉᵐᵉ* (*snᵉ*) or wild-type (*sn⁺*) at very high frequencies ($\approx$ 40–50%). The weak allele has mildly yet obviously bent bristles, whereas the *snᵉ* allele has very short twisted bristles. The difference is very obvious. This is a simple test and one need only score a small number of flies. The appropriate crosses are shown in *Figure 3*.

4. Using *P* elements to produce novel mutations

P elements create mutations by either inserting into genes or by imprecise excision from genes. Consequently, they can be used either for cloning a gene

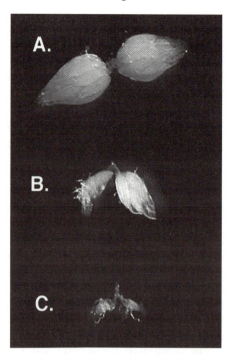

Figure 2. Ovaries dissected from normal and dysgenic adult females. In all cases the ovaries were dissected from females three days post-eclosion. The magnification is identical in all three sets (the images were captured by a video camera attached to a dissecting microscope, so these are typical views through a dissecting scope). (A) Ovaries from normal (Canton-S) females; note the presence of a large number of developing eggs associated with each ovary. (B) Ovaries from a partially dysgenic female; in this case, note that the ovary on the left produces virtually no eggs while the ovary on the right produces some eggs. (C) Ovaries from a completely dysgenic female; note the complete absence of eggs associated with either ovary. The ovaries in (B) and (C) were taken from F1 females derived from a cross between a Canton-S strain female and a π2 strain male, which is a classic *P*-type strain.

(*P* element tagging) or for making new mutations in a *P* tagged gene (imprecise excision). Many derivatives of *P* strains have been made to aid these two aims. Two modifications have been particularly useful:

(a) Separating the source of transposase from the mobile elements.

(b) Engineering defective transposons that carry a cloning vector and selectable markers.

Anchoring the source of transposase at a particular site in the genome, using a *P* element missing one or both termini but encoding functional transposase, and separating it from mobile *P* elements was a major advance. This allows the investigator to either add transposase to stocks and mobilize a

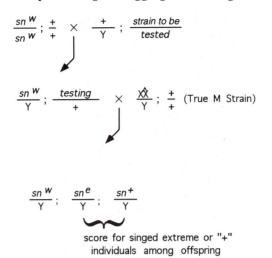

$$\frac{sn^{W}}{sn^{W}} ; \frac{+}{+} \times \frac{+}{Y} ; \frac{strain\ to\ be}{tested}$$

$$\frac{sn^{W}}{Y} ; \frac{testing}{+} \times \frac{XX}{Y} ; \frac{+}{+} \text{ (True M Strain)}$$

$$\frac{sn^{W}}{Y} ; \frac{sn^{e}}{Y} ; \frac{sn^{+}}{Y}$$

score for singed extreme or "+"
individuals among offspring

Figure 3. *Singed-weak (sn^{W})* test for the presence of *P* elements. *sn^{W}/sn^{W}* has slightly singed or shriveled bristles. *singed extreme (sn^{e})* homozygotes have severely shriveled bristles. *sn^{+}* homozygotes have normal bristle phenotype; long slender tapered bristles. X^X is an attached X strain, with the two X chromosomes sharing a single centromere, e.g. *C(1)DX* which is stable and marked with yellow body (*y*) and forked bristles (*f*). In this case the strain must be an *M* strain. The words *'strain to be tested'* or *'testing'* refer to any strain to be tested for mobile *P* elements or the chromosomes derived from such a strain.

single element or a set of elements, or remove the source of transposase from an individual and retain a particularly useful new mutant strain. This anchored source of transposase can be added to, or removed from strains by a simple cross.

Placing cloning vectors and a visible or selectable marker within a defective *P* element to create a marked and recoverable transposon was a second major advance. The visible or selectable marker allows the investigator to follow or select for *P* element jumps. The presence of a bacterial cloning vector allows the genomic DNA adjacent to the insert, to be cloned directly using a restriction endonuclease that cuts at only one position within the transposon. The DNA fragment, which must contain some flanking genomic sequence (between transposon and closest genomic restriction site), can be circularized and transformed into the appropriate bacterial host (21).

The third major advance has been the incorporation of the *P* element system into the *Drosophila* Genome Project. A number of laboratories have inserted a highly engineered single *P* element into large numbers of unique sites in the genome. The site of insertion in each of these single *P* element lines has been mapped by *in situ* hybridization. The genomic fragment adjacent to the *P* insert has been cloned and sequenced. These single *P* element strains are readily available through the *Drosophila* Stock Centers and their corresponding genomic sequences are found in the *Drosophila* Genome Project Database. This is a tremendous resource for *Drosophila* workers.

Several categories of *P* strains are particularly useful either for cloning genes or inducing mutations in genes.

4.1 *P[ry⁺(Δ2–3)]*

Several strains have been created which contain a single *P* element that is incapable of movement but produces transposase. Defective *P* elements are mobilized, in the F1, by simply crossing the transposase encoding strain to a strain which carries one or more defective *P* elements which cannot produce the transposase. Likewise, transposition is halted by removing the transposase source.

The most commonly used source of transposase is a modified *P* element called 'Δ2–3' in which the third intron (between exons 2 and 3) has been removed. This allows the 87 kDa transposase protein to be made in both somatic and meiotic tissues. In addition, in the *pP[ry⁺(Δ2–3)]* one of the *P* element ends has been deleted, so that it can no longer move (20). This element is located at position 99B (tip of the right arm of chromosome three). It is generally used on a chromosome comprising a *rosy* null mutation (such as *ry⁵⁰⁶*) and the dominant marker *Stubble (Sb⁻)* bristles. In this configuration the Δ2–3 gene can be followed by selecting for stubble bristles, and confirmed by the presence of a *ry⁺* phenotype in a *ry⁻* genetic background. Other single *P* transposase sources have been created with very similar features (22, 23), but the Δ2–3 is the most popular. It is both convenient to use and is a potent source of transposase. It is such a potent source of transposase that the F1 progeny carrying this construct and one or more *P* elements are usually raised and mated at low temperatures (less than 20 °C) to reduce the dysgenic effects.

4.2 Screening for genes by selecting for specific phenotypes—tagging genes that are not characterized

Screens are still carried out for genes that have not been identified and described. This is generally done by screening for a 'class' of mutations, that is mutations with a common phenotype. For example, one might screen for flies with a Minute phenotype to look for mutations in ribosomal proteins or one might screen for flies that are mutagen-sensitive to identify genes that participate in DNA repair processes.

The simplest screen uses a single *P* element insert that carries a visible or

---

Figure 4. (A) Protocol used to recover dominant mutations caused by *P* element insertion into a gene. (B and C) Protocol used to recover recessive mutations caused by a *P* element insertion into a gene on the second chromosome; (C) recessive mutations on the third chromosome. P[*ry⁺*] refers to a defective *P* element. In this case the *P* transposon carries a functional xanthine dehydrogenase gene, the *rosy* locus = *ry⁺*. This element can be followed by either looking for rosy⁺ phenotype (normal eye pigment) in a *ry⁻/ry⁻* genetic background or by selecting for purine resistance in such a background. Δ2–3, Sb refers to P[*ry⁺*(Δ2–3)] which a stable source of transposase in both the germline and somatic tissue (Section 4.1). *TM6, Tb = In(3LR)TM6B* with makers *Hu e* and *Tb*, and sometimes others. *ry⁵⁰⁶* = a specific rosy null allele which produces no functional protein product. *Pm = bwᵛ¹* = plum eye colour; *SM5 = In(2LR)SM5, al² Cy Itᵛ cn² sp²*.

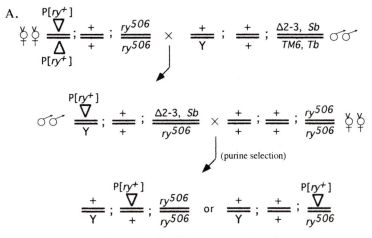

A. P[ry^+]

$\female\female \dfrac{\triangledown}{\triangle} ; \dfrac{+}{+} ; \dfrac{ry^{506}}{ry^{506}}$ × $\dfrac{+}{Y} ; \dfrac{+}{+} ; \dfrac{\Delta 2\text{-}3,\ Sb}{TM6,\ Tb}$ $\male\male$

P[ry^+]

$\male\male \dfrac{\triangledown}{Y} ; \dfrac{+}{+} ; \dfrac{\Delta 2\text{-}3,\ Sb}{ry^{506}}$ × $\dfrac{+}{+} ; \dfrac{+}{+} ; \dfrac{ry^{506}}{ry^{506}}$ $\female\female$

(purine selection)

P[ry^+]   P[ry^+]

$\dfrac{+}{Y} ; \dfrac{\triangledown}{+} ; \dfrac{ry^{506}}{ry^{506}}$ or $\dfrac{+}{Y} ; \dfrac{+}{+} ; \dfrac{\triangledown}{ry^{506}}$

score for ry^+ males for the appropriate dominant phenotype

B.

P[ry^+]

single $\male$
(F2 from A) $\dfrac{+}{Y} ; \dfrac{\triangledown}{+} ; \dfrac{ry^{506}}{ry^{506}}$ × $\dfrac{+}{+} ; \dfrac{Pm}{SM5} ; \dfrac{ry^{506}}{ry^{506}}$ $\female\female$

P[ry^+]   P[ry^+]

$\dfrac{+}{Y} ; \dfrac{\triangledown}{SM5} ; \dfrac{ry^{506}}{ry^{506}}$ × $\dfrac{+}{+} ; \dfrac{\triangledown}{SM5} ; \dfrac{ry^{506}}{ry^{506}}$ siblings

score each line for recessive visible phenotype

C.

P[ry^+]

single $\male$
(F2 from A) $\dfrac{+}{+} ; \dfrac{\triangledown}{ry^{506}}$ × $\dfrac{+}{+} ; \dfrac{Ly\ ry^{506}}{TM3\ Sb,\ ry^{RK}}$ $\female\female$

P[ry^+]   P[ry^+]

$\dfrac{+}{+} ; \dfrac{\triangledown}{TM3\ ry^{RK}}$ × $\dfrac{+}{+} ; \dfrac{\triangledown}{TM3\ Sb,\ ry^{RK}}$ siblings

score each line for recessive visible phenotype

selectable marker and a cloning vector with several discrete restriction endonuclease sites in the polylinker which allows the unique genomic DNA, adjacent to the *P* insert, to be cloned. This engineered transposon is defective, but can be mobilized if transposase is supplied from another source. The *P* transposon is mobilized by creating an F1 hybrid between this strain and a strain which expresses Δ2–3 from a fixed position. A typical example is shown in *Figure 4*. We have found the ry^+ marked *P* element extremely useful. Not only does it provide an eye colour marker, which can be followed in the ry^- background, it is a selectable marker. When the appropriate concentrations of purine are added to the media ry^- flies die, whereas ry^+ flies survive (see below for an example). In screens for novel visible (e.g. Minute) or behavioural phenotypes (e.g. flightless), one can select for jumps from one chromosome to another simply by screening for viability on media containing purine and then screening the survivors for the appropriate visible or behavioural phenotype.

Protocol 1.   Screening for *P* inserts on the autosomes

1. Cross females that are homozygous or heterozygous for the X-linked *pP[ry⁺]* element (labelled *P[ry⁺]* in *Figure 4*) to males that carry *Δ2–3, Sb⁻ry⁺*(both strains homozygous for a *rosy* null mutation at the normal locus).[a]

2. Select F1 males that carry the *P[ry⁺]* element and *Δ2–3, Sb⁻*.

3. Mate these males to *M* strain females (lack *P* elements) that are homozygous for a *rosy* null mutation.

4. Score F2 males that are *rosy⁺* and lack *Δ2–3* (normal bristles).

5. Alternatively, grow the F2 individuals on media supplemented with about 0.08% purine (see below).

6. Select those survivors that are *rosy⁺* and have wild-type bristles.

7. Either:

   (a) Score the F2 for a visible or behavioural dominant mutant phenotype (*Figure 4A*) or cross single males to virgin females that are heterozygous for a dominant marker and Balancer chromosome (*Figures 4B* and *C* for second and third chromosome mutations, respectively).

   (b) Cross *P* insert/Balancer siblings from each line and score for a particular recessive visible phenotype (or test for recessive phenotype, for example a behavioural, mutagen-sensitivity, or sterility phenotype).

8. Collect the appropriate flies (*P[ry⁺]* homozygotes for recessive viable and fertile, or *P[ry⁺]* element/Balancer siblings for recessive sterile or lethal), to establish a *P* insert line(s).

[a] It is possible to follow the *Δ2–3* chromosome at all times since it is marked with the dominant mutation *Sb⁻*.

4.3 Tagging a gene for which mutants are described and genetically and/or cytogenetically mapped

Perhaps the most frequent situation that a new researcher working on *Drosophila* faces is cloning a gene for which one or more mutant alleles exist already. When the P1 phage contig maps have been fully constructed this will be a conceptually simple matter of accurately mapping the gene and then ordering in and dissecting the appropriate P1 phage or phages, especially if among the mutant alleles there is a chromosome aberration (small deletion, inversion, or translocation breakpoint). Until then, it will be necessary to tag the gene with a *P* element, clone the flanking DNA, and use this as a probe to screen cDNA and/or genomic libraries for the complete gene and its RNA product(s).

4.3.1 Complementation analysis with single *P* inserts derived from the *Drosophila* Genome Project

The *Drosophila* Genome Project provides a series of single *P* element insertions which inactivate vital genetic loci. Estimates of the number of essential genes in *Drosophila* range from 3000–6000 (24). The goal of the single *P* element insert project is to collect and define single inserts in at least 1200 different vital loci, or about 30% of the total. This means that there will be at least one *P* insert, and often several, in each lettered unit of the polytene chromosome. Each line should contain a single *P* element insertion which knocks out the function of a vital locus (recessive lethal mutations). The genomic sequences adjacent to these inserts, STSs (sequence-tagged sites), are being defined and linked to both the molecular and chromosome (polytene) map.

Although the *P* element constructs used to make these single element recessive lethal mutations differ somewhat, each insert was made with a *P* element carrying a marker and a cloning construct for plasmid rescue. Each insert will:

(a) Be examined to demonstrate that it is unique (only one *P* insert).

(b) Be mapped by *in situ* hybridization to polytene chromosomes.

(c) Have the flanking genomic DNA cloned (plasmid rescue).

(d) Have 400 bp of genomic DNA immediately adjacent to the *P* insert (STS) sequenced.

(e) Have this STS mapped onto the P1 contig library and the polytene chromosome.

As elements are characterized the *P* insert strain will be deposited into, and made available through, the Bloomington Stock Center. The *P* inserts were derived from several sources including the Jan, Kiss, Laughton, Rubin, Scott, and Spradling laboratories (alphabetical order; for references to the form of the original *P* element see 25–30). Current information about the *P* inserts that have been characterized can be found in FlyBase (*Drosophila* Genome Database) which is regularly updated. The database lists:

- *in situ* hybridization position   26B 1–2
- name of the insertion allele   1(2)02439

- *P* element used $PZ[ry^+ lacZ]$
- laboratory that provided the mutation   A. Spradling
- status   verified or single

'Verified' indicates that there is > 95% probability that the *P* insert is causing the recessive lethality, while 'single' indicates that the *in situ* mapping has been carried out and that there is only a single insert, but that the possibility remains that the insert and recessive lethality are not allelic. The above example was taken directly from FlyBase (Chapter 1).

Prior to embarking on a screen, the researcher is encouraged to map their mutant of interest as precisely as possible, then to query FlyBase for *P* inserts at the same, or in nearby, positions. The appropriate *P* insert line(s) can be obtained from the Bloomington Stock Center. These are tested for allelism to 'your mutant of interest' by a simple complementation test (where appropriate). If your mutation is near, but not allelic to, the insert, you may use the closest insert as the starting point for a local *P* element transposition into 'your gene of interest'.

4.3.2 Mutations by 'local hops'

P elements move locally at a frequency that is higher than transposition to another chromosome. The distance over which local 'hops' occur is about 100 kb to perhaps 200 kb. If you know the position of 'your mutant gene', it is a relatively simple matter to choose a *P* insert stock, from the single lethal *P* element insert library, to act as the start point for a 'local hop' screen for insertions into 'your gene of interest'. An example of such a strategy is given in *Figure 5*. You can screen for lethal or visible alleles of your gene of interest, depending on the phenotype and strength (hypomorph, amorph) of the various loss of function alleles available.

Protocol 2. Screening for recessive visible mutations (*Figure 5A*)

1. Cross balanced *P[ry⁺]* insert stock females to *Δ2–3 Sb⁻* males.
2. Collect *P[ry⁺]* insert/Balancer; *Δ2–3 Sb⁻* F1 males and cross them to your recessive mutant/Balancer *ry⁻* females (from your mutant stock).
3. Screen F2 for *ry⁺* (non-stubble) flies that express 'your mutant phenotype of interest'.
4. Cross males (to avoid rare intragenic crossovers) to females from the recessive mutant/Balancer, *ry⁻* stock; select *P[ry⁺]* insert/Balancer male and virgin female flies (phenotypically, they will be ry⁺) and mate *inter se* to establish a balanced stock (*P[ry⁺]* -*mutant*/Balancer; ry^{506}/ry^{506}).
5. Determine if *P[ry⁺]* -*mutant*/*P[ry⁺]* -*mutant* survives; if so, keep the *P[ry⁺]*-mutation as a true breeding stock.

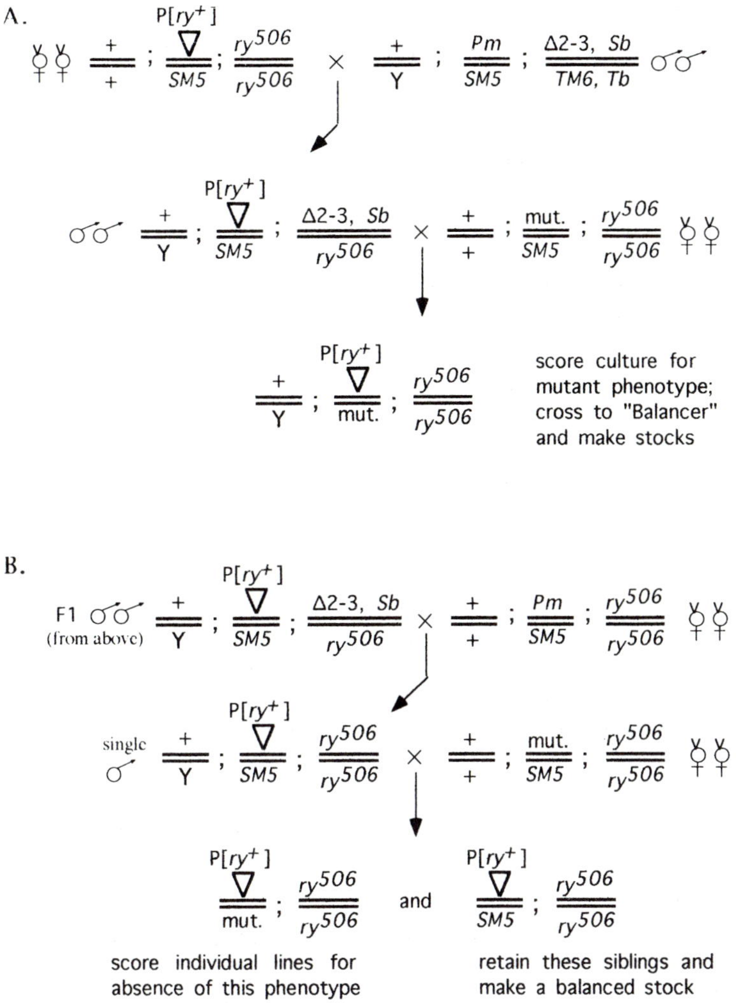

A.

P[ry⁺]

B.

P[ry⁺]

Figure 5. (A) Protocol for the isolation of recessive visible mutations by local hops of a P transposon. (B) Protocol for the isolation of recessive lethal or sterile mutations by local hops of a P transposon. For the designation of markers and symbols see *Figure 4*; the abbreviation 'mut.' refers to the recessive mutation into which you are attempting to insert the P transposon.

Protocol 3. Screening for recessive lethal mutations (*Figure 5B*)

1. Cross females from the balanced $P[ry^+]$ insert ry^- stock (obtained from stock center) to $\Delta2$–3 Sb^- males (as in *Figure 5A*).

2. Collect $P[ry^+]$ insert/Balancer; $\Delta2$–3 Sb^- F1 males and cross them to dominant marker/Balancer; ry^{506}/ry^{506} virgin females.

Protocol 3. *Continued*

3. Collect F2 males that are $P[ry^+]$/Balancer; ry^{506}/ry^{506} and mate single males to three to five virgin females that are: 'your recessive mutant'/Balancer; ry^{506}/ry^{506} (from your mutant stock).[a]

4. Score each F3 line independently for absence (or vast reduction in numbers) of the $P[ry^+]$/mutant class of individuals. Retain their Balanced siblings and establish a stock by *inter se* crosses.

[a] NB: each of these males represents a single *P* 'local hop' exposed chromosome (potentially new mutation in your gene), and therefore each male must be treated and screened as an independent line.

4.4 Creating new alleles in a *P* tagged gene (including reversions)

Frequently genes cloned by *P* tagging have few mutant alleles induced by other means, such as chemical mutagens or gamma rays. Additional mutations are selected by re-mobilizing the *P* insert and screening for altered phenotypes among the offspring. The types of mutations recovered include phenotypic reversion to wild-type, less severe, and more extreme mutant phenotypes.

Screening for complete revertants (most or all of the *P* element removed) is often required in those cases where a rescue fragment is not available. Reversion often provides the key piece of evidence that the mutant phenotype of interest correlates with presence of the *P* element insertion, as opposed to double mutations with the *P* insert genetically near to, but distinct from, the actual mutation of interest. Such two-hit mutations can be up to a megabase apart, if the mutant phenotype is mapped by recombination and the *P* insert is mapped by *in situ* hybridization.

P elements frequently insert into an intron, a 5' UTR, or regulatory region of a gene (often near the TATA box) to produce hypomorphic mutations. Altered phenotypes, especially more extreme phenotypes often result from re-mobilization of the *P* element and insertion into the coding region, or from a partial or complete deletion of the *P* insert which extends into the gene of interest, creating a null mutation (amorph).

A scheme to screen for new mutations by re-mobilization, either revertants or more extreme alleles, is shown in *Protocol 4* and *Figure 6*.

Protocol 4. Inducing reversions or more extreme alleles using *P* inserts (*Figure 6*)

1. Cross $P[ry^+]$ -*mutant*/Balancer; ry^-/ry^- flies (either males or virgin females) to dominant Marker/Balancer; $\Delta 2$–3, Sb^-/Balancer flies (virgin females or males).

2. Collect F1 males that are phenotypically ry^+ and Sb.[a]

3. Mate these $P[ry^+]$-*mutant, Δ2–3, Sb⁻* males to virgin females from your *mutant; ry⁻/ry⁻* stock.

4. Score the F2 for $P[ry^+]$-*mutant*[Revertant]/*mutant; ry⁻/ry⁻* individuals (reversion of the lethal or visible phenotype) if screening for a revertant.

5. Or, score F2 for $P[ry^+]$-*mutant*[extreme]/*mutant; ry⁻/ry⁻* individuals if screening for a stronger visible phenotype.

6. Collect revertant or extreme males, or virgin females from four or five and mate them to a (second or third chromosome) Balancer strain that is *ry⁻/ry⁻*.

7. Recover the $P[ry^+]$-*mutant*[Revertant]/Balancer; *ry⁻/ry⁻* individuals (these will be ry^+ if the appropriate strains were used in the previous cross) and establish stocks by *inter se* matings.[b]

8. Re-test stocks for the novel phenotype.

9. If scoring for the lethal allele, establish independent lines in the F2 generation. Cross single $P[ry^+]$-*mutant**/Balancer; *ry⁻/ry⁻* males to three to five *mutant*/Balancer; *ry⁻/ry⁻* virgin females.

10. Score individual lines for absence, or vastly reduced numbers, of homozygotes ($P[ry^+]$-*mutant**/$P[ry^+]$-*mutant**) among the F3 individuals. In those lines were no homozygotes appear, the $P[ry^+]$-*mutant**/Balancer; *ry⁻/ry⁻* heterozygotes[c] cross *inter se* to establish a balanced lethal stock.

---

[a] These need not be balanced since there is no recombination in males. None the less, we typically use the balanced males since large numbers of males are not required to produce the next generation.

[b] From these crosses establish the viability and phenotype of $P[ry^+]$-*mutant*[extreme] homozygotes.

[c] Identified as ry^+, or by the presence of a selectable marker such as *neo*[R], assuming the rosy gene or selectable marker is not mutated during movement.

4.5 Inducing deletions with defined end-points

It is often useful to have deletions for the region in which your gene of interest is located. These can be used to test whether existing alleles of your gene are hypomorphs or amorphs, for dosage studies, or as a tester strain to screen for new alleles or chemically-induced alleles of the gene in which you are interested, or to define mutations in genes flanking your gene of interest. Deletions with defined end-points can be made by deleting the material between two *P* inserts that map very close together. To generate deletions, with high efficiency, the two *P* inserts must be located on the same homologue.

Choose two *P* elements that are located very close together and flank your gene. Inserts that map near to one another but are either both distal or both proximal to your gene will produce deletions close to, but not including, your gene.

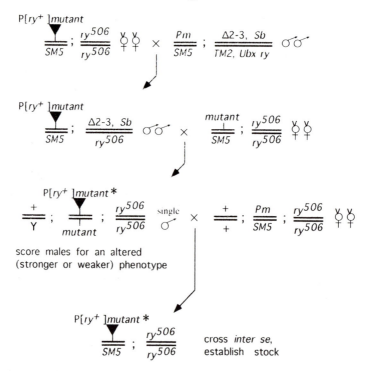

Figure 6. Protocol for inducing new alleles, including new mutant alleles and reversions, in a gene tagged with a *P* element. See *Figure 4* for genetic symbols. *TM2, Ubx ry =* *In(3LR)Ubx^130, emc^2 Ubx^130 ry*. P[*ry^+*]*mutant* S refers to a mutation derived from mobilizing the *P* transposon; it could be a new mutant allele, or a partial, or complete reversion.

Map the *P* inserts relative to your gene of interest by standard recombination mapping since the *P* inserts are usually marked with a visible marker (often *rosy^+* or *white^+*).

Once appropriate *P* inserts have been identified, then these two *P* inserts must be placed *cis* to one another, that is on the same homologue. A simple method for doing this is shown in *Figure 7*. At least one of the *P* inserts should carry a selectable marker; it may be improved if both carry a different selectable marker. Many of the P inserts use *neo^Resistance* as a selectable marker; those inserts that are marked with *ry^+* can be selected for using purine selection (31).

Protocol 5. Inducing small deletions with defined end-points (*Figure 7*)

1. Cross the two *P* insert/Balancer strains together.

2. Select F1 virgin females that carry the two *P* inserts in *trans*, as shown,

and cross these to males that are heterozygous for a dominant marker/Balancer and homozygous ry^-/ry^-.

3. Grow the F2 individuals on media to which neomycin has been added (you must test each new batch from the supplier to determine the effective concentration to use since there are batch to batch variations in effectiveness) and select for ry^+ neo^R/Balancer; ry^-/ry^- virgin females.

4. Mate these virgin females to dominant marker/Balancer; $\Delta2$–3, Sb^-/TM2, ry Ubx^{130} red e males.

5. Select males or virgin females that are P insert-1, P insert-2/Balancer; $\Delta2$–3, Sb^-/ry^- and mate these to individuals that are homozygous for your recessive visible mutation[a] and ry^-/ry^-.

6. Select for the rare offspring that are phenotypically recessive visible mutation and ry^- (they should also be neomycin-sensitive). These should have a deletion which includes your gene (hemizygous).

7. Mate these flies to an appropriate Balancer strain and select deletion over Balancer siblings.[b] Mate *inter se* and establish balanced stock.

[a] A similar approach can be used for a recessive lethal except that the P insert-1 + P insert-2/SM5; $\Delta2$–3, Sb^-/ry^- (in the second to last cross) must be mated to Dominant marker/Balancer; ry^-/ry^- flies. In the subsequent generation 'rosy'/Balancer; ry^-/ry^- male flies are selected (each male represents a putative deletion) and mated singly to *your mutant*/Balancer; ry^-/ry^- virgin female flies. Each line is examined for absence of mutant/mutant flies, and their Balanced siblings (deletion/Balancer, again a recessive marker helps) are mated *inter se* to establish stocks. These are re-tested by crosses to *mutant*/Balancer strains.

[b] It helps if either the original P insert-1 and 2 homologue or the mutant bearing homologue have a recessive mutation that can be selected for or against over the Balancer. The mutations *cinnabar* or *speck* work as markers with *SM1, SM2, SM5*, and *SM6* balancers.

5. Selection for P element movement to non-homologous chromosome

The preceding sections describe schemes for producing P element insertions or mobilizations based on simple visual screening. However, the frequency with which P jumps from one chromosome to another is relatively low. This necessitates sorting through large numbers of flies when performing general screens for new mutants. It is far more efficient to select against those individuals in which P element has not moved. This eliminates most of the offspring. Those that survive selection have acquired the transposon, presumably at a new location on a non-homologous chromosome. These individuals, which are new insertions, are then screened for the appropriate mutant phenotype. 'Selection' drastically reduces the number of flies that must be handled.

There are a number of transposons that contain genes that allow selection. Transposons with the *neomycin*[Resistance] gene, often driven by the heat shock

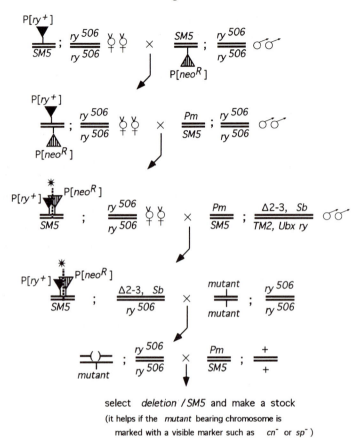

Figure 7. Protocol for isolating a deletion between two *P* element inserts that map near to each another. See *Figure 6* for genetic symbols. P[*neo^R*] refers to a *P* transposon marked with the selectable marker neomycin resistance. Often these are also marked with the visible marker *white^+* = *w^+* which produces red eyes in a *w/w^-* background. In this example the symbol S represents the genomic region between the two *P* inserts that is to be deleted. The symbol -( )- represents the deletion bearing chromosome.

promoter (hs*neo^R*), are commonly used for selection. A variety of other transposons are equally convenient. These include transposons that incorporate *Adh^+* (alcohol dehydrogenase) and *rosy^+* (xanthine dehydrogenase) both of which can be selected for or against. The protocols outlined here use a transposon with both neomycin resistance (*neo^R*) and xanthine dehydrogenase activity (*ry^+*). We have often used the *ry^+* and purine selection. This provides both a chemical selection system and a visual check for movement (the few flies that might escape chemical selection can be eliminated by simple visual inspection).

Regardless of which selection system is used, the amount of chemical to be used for selection must be determined experimentally, since the sensitivity in part depends on:

- the efficiency of the promoter that drives the expression of the resistance gene
- the position of the insert (genomic position effects)
- the size of the containers and amount of media used in each container

Figure 8 shows a typical mating scheme for determining the purine concentration to be used for selection based on *rosy⁺* expression. The cross is self-explanatory. 25 pairs of flies are allowed to deposit eggs on normal media, in half-pint milk bottles, for about two days. The parents are removed, and transferred to fresh bottles for a replicate two-day egg-lay period. Shortly after the adults are removed, 1.5 ml of 1.0% sucrose solution, containing a specific concentration of purine, is pipetted onto the surface of the media (the solution seeps into the medium fairly quickly). The replicate receives the same amount of 1.0% sucrose solution without purine. A series of concentrations from 0.0–0.125% is generally satisfactory. At each concentration, the number of ry^+ F1 individuals in treated containers is scored relative to their untreated siblings. The number of males and females in the 'control' replicate

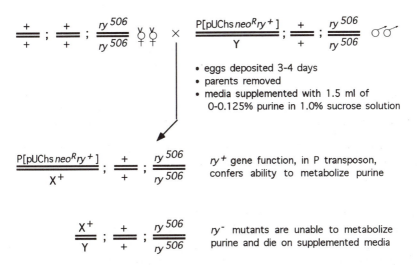

Figure 8. Protocol for determining the purine concentration to be used for selection based on *rosy⁺* expression. P[pUChs*neo*ᴿ *ry*⁺] refers to a *P* transposon carrying a pUC8 or pUC9 or derivative cloning vector, a neomycin resistance gene driven by a *Drosophila* heat shock promoter, and a wild-type xanthine dehydrogenase gene, and which cannot produce transposase on its own but can obviously move in response to transposase provided exogenously. For other genetic symbols refer to *Figure 4*.

103

Table 1. Purine selection data

Purine concentration (%)	Number of ry^+ females	Approximate % survival	Number of ry^- males	Approximate % survival
0.0	906	100	601	100
0.01	1204	100	664	100
0.02	1136	95	610	92
0.03	1316	100	170	26
0.04	1246	100	1	0.2
0.05	1178	98	3	0.5
0.06	1151	96	0	0
0.07	957	80	1	0.2
0.08	789	66	0	0
0.09	614	52	0	0
0.10	531	44	0	0
0.11	513	43	0	0

serves as an estimate of the relative number of individuals expected in the 'treated' container. Of course, the actual number of eggs can be counted to give a more accurate estimate of the total number of individuals expected, but this is usually unnecessary. It is tempting to compare the numbers of ry^+ to ry^- individuals, in this example the relative numbers of females to males, since this should be 1:1. However, in practice the insert often causes reduced viability. Reasonably accurate data can be obtained by examining about 1000 F1 individuals at each concentration tested (about two to three bottles). An example of one such series is shown in *Table 1*. Clearly the viability declines as the concentration of the selective agent (purine in this example) increases. One must choose a concentration that does not allow many flies to 'escape' selection, but is not so high that it would also select against those individuals in which the P insert had jumped from the X chromosome to an autosome (difficult to determine with great accuracy due to genomic position effects). In the example shown 0.08% purine was determined to be adequate. A similar approach can be used with an insert on the autosomes. For example, start with a strain that is heterozygous for the $P[pUChsneo^R ry^+]$ insert and a Balancer chromosome with an easily scored visible phenotype in a ry^- background (ry^- alleles at the normal ry locus) and then score the relative numbers of 'Balancer' to non-balancer, ry^+ flies among the F1 individuals.

Prior to embarking on a screen, it is useful to determine the rate at which 'jumps' between non-homologous chromosomes occur. This in part depends on the P element insert and the amount of activity of the $\Delta2$–3 (transposase construct) which in turn depends on the temperature at which the P insert, $\Delta2$–3 (transposase) individuals are raised (low temperature is often needed to reduce the negative impacts of mobility in the somatic tissue).

Figure 9 shows an example of a scheme to test the frequency at which a particular P insert is either excised from a site or 'jumps' to a non-homologous

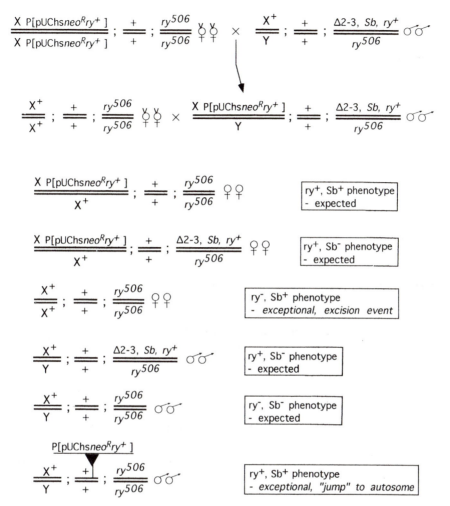

Figure 9. Protocol for testing the frequency with which excision and transposition events occur using any marked selectable transposon. For genetic symbols refer to *Figure 8*.

chromosome. The *P* insert is mobilized in the F1 individuals and the F2 are raised without selection. The F2 individuals are scored for exceptional phenotypes, in this case, ry^- Sb^+ females and ry^+ Sb^+ males. The frequency of excision events can be estimated by as follows:

(a)
$$\frac{\text{number of exceptional females}}{\text{no. exceptional females} + \text{no. females in which exceptions could be detected}} \times 100$$

or, since approximately half of the excision events would occur in flies that are $\Delta 2$–3, Sb^- ry^+ and thus would not be detected:

(b)
$$\frac{2 \times \text{the number of exceptional females}}{\text{total number of females}}$$

These two calculations should give approximately the same results. In trials using 500 flies they did, e.g. 15.9% for calculation 'a' and 16.1% for calculation 'b'.

The number of 'jumps' to a non-homologous chromosome can be calculated as:

$$\frac{\text{the number of exceptional males}}{\text{no. exceptional males} + \text{no. males in which exceptions could be detected}}$$

Again approximately half of the jumps would occur in $\Delta 2$–3, $Sb^- ry^+$ individuals and would not be detected as they are already ry^+. In the example given, the frequency of 'jumps' to a non-homologous chromosome was approximately 3.5%. One would expect that 'jumps' to non-homologous chromosomes are also obtained in females, but they cannot be differentiated from those females that inherited the P insert as a standard segregation even from their father (X P[pUChs$neo^R ry^+$] chromosome). The experiments in this case were done at 22 °C, which accounts for the high frequency of both excision events and 'jumps' to non-homologous chromosomes. At 16 °C these frequencies would be reduced significantly.

References

1. Hiraizumi, H. Y. (1971). *Proc. Natl. Acad. Sci. USA*, **68**, 268.
2. Hiraizumi, Y., Slatko, B., Langley, C., and Nill, A. (1973). *Genetics*, **73**, 439.
3. Slatko, B. E. and Hiraizumi, Y. (1973). *Genetics*, **75**, 643.
4. Slatko, B. E. (1978). *Genetics*, **90**, 105.
5. Golubovsky, M. D., Ivanov, Y. N., and Green, M. M. (1977). *Proc. Natl. Acad. Sci. USA*, **74**, 2973.
6. Green, M. M. (1977). *Proc. Natl. Acad. Sci. USA*, **74**, 3490.
7. Green, M. M. and Sheppard, S. H. Y. (1979). *Genetics*, **92**, 823.
8. Sved, J. A. (1976). *Aust. J. Biol. Sci.*, **29**, 375.
9. Kidwell, M. G. and Kidwell, J. F. (1975). *Genetics*, **84**, 333.
10. Kidwell, M. G. (1979). *Genet. Res.*, **33**, 205.
11. Rubin, G. M., Kidwell, M. G., and Bingham, P. M. (1982). *Cell*, **29**, 987.
12. O'Hare, K. and Rubin, G. M. (1983). *Cell*, **34**, 25.
13. Karess, R. E. and Rubin, G. M. (1984). *Cell*, **38**, 135.
14. Rio, D. C., Laski, F. A., and Rubin, G. M. (1986). *Cell*, **44**, 21.
15. Bregliano, J. C. and Kidwell, M. G. (1983). In *Mobile genetic elements* (ed. J. A. Shapiro), pp. 363–410. Academic Press, New York.
16. Engels, W. R. (1983). *Annu. Rev. Genet.*, **17**, 315.
17. Engels, W. R. (1989). In *Mobile DNA* (ed. D. Berg and M. Howe), pp. 437–84. American Society of Microbiology, Washington DC.
18. Kidwell, M. G. (1983). In *The genetics and biology of Drosophila* (ed. M.

Ashburner, H. L. Carson and J. N. Thompson), Vol. 3c, pp. 125–54. Academic Press, London and New York.

19. Kidwell, M. G. (1986). In *Drosophila: a practical approach* (ed. D. B. Roberts), pp. 59–81. IRL Press, Oxford.
20. Roberston, H. M., Preston, C. R., Phillis, R. W., Johnson-Shlitz, D., Benz, W., and Engels, W. R. (1988). *Genetics*, **118**, 461.
21. Hamilton, B. A., Palazzolo, M. J., Chang, J. H., BijayRaghavan, K., Mayeda, C. A., Whitney, M. A., *et al.* (1991). *Proc. Natl. Acad. Sci. USA*, **88**, 2731.
22. Cooley, L., Berg, C., and Spradling, A. (1988). *Trends Genet.*, **4**, 254.
23. Cooley, L., Kelley, R., and Spradling, A. C. (1988). *Science*, **239**, 1121.
24. Miklos, G. L. G. and Rubin, G. M. (1996). *Cell*, **86**, 521.
25. Karpen, G. H. and Spradling, A. C. (1992). *Genetics*, **132**, 737.
26. Gaul, U., Mardon, G., and Rubin, G. M. (1992). *Cell*, **68**, 1007.
27. Torok, T., Tick, G., Alvarado, M., and Kiss, I. (1993). *Genetics*, **135**, 71.
28. Bier, E., Vassin, H., Shepherd, S., Lee, K., McCall, K. A., Barbel, S., *et al.* (1989). *Genes Dev.*, **3**, 1273.
29. Chang, Z., Price, B. D., Bockheim, S., Boedigheimer, M. J., Smith, R., and Laughton, A. (1993). *Dev. Biol.*, **160**, 315.
30. Mlodzic, M. and Hiromi, Y. (1991). Methods Neurosci **9**, 397–414.
31. Chovnick, A., Ballantyne, G. H., and Holm, D. G. (1971). *Genetics*, **69**, 179.

4

Chromosome mechanics; the genetic manipulation of aneuploid stocks

DAVID GUBB

1. Introduction

The use of chromosomal aberrations for cytogenetic mapping has been critical in the development of *Drosophila* as a model genetic organism. In 1932 Muller (1) designed a series of genetic tests, using chromosomal deletions and duplications, which define five classes of mutation; broadly corresponding to lack of function (amorphic and hypomorphic) and gain of function (hypermorphic, neomorphic, and antimorphic) alleles. Although Muller's terminology is ignored frequently, his basic insight, that duplications and deletions could be used to characterize the nature of mutant alleles, remains true. In recent years, a large number of genes have been mapped between the breakpoints of chromosomal deletions or associated with the breakpoints of translocations, transpositions, or inversions. As a result there is an increasingly fine-scale correlation between the genetic map and cytological positions in polytene larval salivary glands (2). New mutations can be mapped by recombination and then rapidly assigned to a cytogenetic region using the available chromosome aberrations. Deletions are particularly useful for this purpose as failure to complement a mutant phenotype localizes the mutation within the cytological boundaries of the deletion. Translocation and inversion breakpoints can give a precise localization, if one end-point is mutant for the target locus, but breakpoints more frequently fall between transcription units. In this situation, it is often possible to construct a synthetic deletion from the aberration breakpoint. Similarly, the cytogenetic map position of cloned sequences can be identified by *in situ* hybridization to wild-type polytene chromosomes and refined by probing chromosomal aberrations. An up to date catalogue of available aberrations is maintained by FlyBase (Chapter 1).

An additional use of deletions and duplications is to identify epistatic interactions and redundant genetic functions. This last application is increasingly significant as the full extent to which genetic and developmental pathways are conserved between eukaryotic organisms becomes clear.

This chapter will outline the use of translocations and inversions to construct synthetic deletions and duplications. Screens that select for novel aberration breakpoints will be described. The use of aneuploid stocks to identify modifiers of dominant mutations will be discussed.

2. The generation of aneuploids (either deletions or duplications) from translocations by segregation

A simple reciprocal translocation, such as *T(2;3)TE35B-28*, consists of two complementary elements. The breakpoints of *T(2;3)TE35B-28* are at 35B1.2, on the left arm of chromosome 2, and 90C3-6, on the right arm of chromosome 3. The translocated elements of *T(2;3)TE35B-28* correspond to the distal region of chromosome arm 2L attached to the third chromosome centromere and the distal region of chromosome arm 3R attached to the second chromosome centromere. The old convention for describing these products is from the origin of the proximal (*P*) and distal (*D*) elements, relative to the chromosomal centromere, as in $T(2;3)TE35B-28 = T(2;3)TE35B-28^{2D3P} + T(2;3)TE35B-28^{3D2P}$. The FlyBase convention is to refer to these elements as translocation segregants (*Ts*), identified by the origin of their telomeres, either left telomere (Lt) or right telomere (Rt) for the *X* chromosome and autosomes, or long arm telomere (Lt) and short arm telomere (St) in case of the *Y*. Thus, *T(2;3)TE35B-28* consists of *Ts(2Lt;3Lt)TE35B-28* and *Ts(2Rt;3Rt)TE35B-28*; while *T(Y;2)R15* corresponds to *Ts(YSt;2Lt)R15* and *Ts(YLt;2Rt)R15, Figure 1*.

At meiosis the translocation elements will segregate independently from each other, so that only half the gametes from a *T(2;3)TE35B-28* individual carry complete second and third chromosomes, either *+;+* or *T(2;3)TE35B-28*. The remaining gametes carry either a cytologically normal second chromosome plus *Ts(2Lt;3Lt)TE35B-28*, or a normal third chromosome plus *Ts(2Rt;3Rt)TE35B-28*. In an outcross to any normal stock, half the embryos will die as a result of severe aneuploidy of the second and third chromosomes. When outcrossed to a similar translocation, *T(2;3)G40* (with 35F5-7;91E5-6 breakpoints), the outcome is different. The reciprocal products, *Ts(2Lt;3Lt)TE35B-28* + *Ts(2Rt;3Rt)G40* and *Ts(2Lt;3Lt)G40* + *Ts(2Rt;3Rt)TE35B-28*, may survive as flies that are aneuploid for the regions between the second and third chromosomal breakpoints. In this example, *Ts(2Lt;3Lt)G40* + *Ts(2Rt;3Rt)TE35B-28* carries a viable second chromosome deletion, (35B1.2;35F5-7) and third chromosome duplication (90C3-6;91E5-6); the reciprocal aneuploid, *Ts(2Lt;3Lt)TE35B-28* + *Ts(2Rt;3Rt)G40*, carries *Dp* 35F5-7;35B1.2 and the large deletion *Df* 90C3-6;91E5-6, which is lethal. In practice, this is a common problem. It is rarely possible to recover reciprocal duplications and deletions of a given region because only a few pairs of translocation stocks have the necessary similar breakpoints in both autosomes.

T(2;3)TE35B-28/+

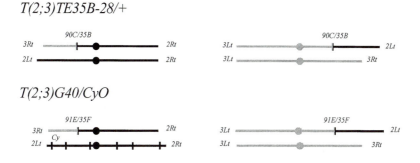

T(2;3)G40/CyO

Gametes from cross of T(2;3)TE35B-28/CyO X T(2;3)G40/CyO

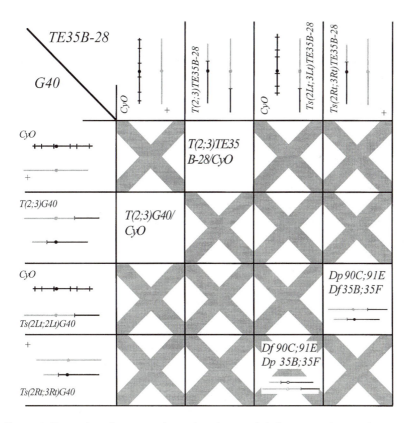

Figure 1. Examples of autosomal translocations and their segregation products.

2.1 The *T(Y;A)* translocation set of Lindsley and Sandler *et al.*

In 1972 Lindsley and Sandler *et al.* (3) published perhaps the single most important contribution to *Drosophila* genetics since the construction of the first balancer chromosomes. The basic technique is to put a few male flies together with a few virgin females in a tube and wait for 11 days. This should leave you sufficient time to read, and understand, the original paper. Like all good ideas, the underlying concept is strikingly simple. In *Drosophila*, the Y chromosome represents a large target of heterochromatin containing no vital genetic functions. It follows that if one of the breakpoints of a translocation is in the Y chromosome, synthetic duplications and deletions of autosomal chromosome segments can be constructed independently of the position of the Y chromosome breakpoint.

In order to construct such 'segmental aneuploids', it is necessary that the *Y;Autosomal* translocation, *T(Y;A)*, stocks should carry the translocated Y in both males and females. This is achieved by using a compound, 'Siamese twin', *X.X* chromosome. Females carrying such compounds, with both chromosomes attached to a single X centromere, also carry a Y chromosome, with the X and Y centromeres disjoining during meiosis. There is a further practical problem, males must carry a complete Y chromosome to be fertile. To ensure this, the males carry an additional Y on a compound *X.Y* chromosome. Both the female *X.X* and the male *X.Y* compound chromosomes (*C(1)RM, y* and *C(1;Y)In(1)EN, y*, respectively) carry the recessive marker *yellow*, *y*.

The problem of distinguishing particular segregant products was solved in a very elegant manner. The Y chromosome on which the *T(Y;A)* translocations were induced carried X chromosome markers near the tip of each arm, B^S on the long arm and y^+ on the short arm, $B^S Y y^+$. As a consequence, each segregant of the *T(Y;A)*s is genetically marked.

A cross to generate the reciprocal duplication and deletion for region 24C;24F using male *T(Y;2)H116*, with *T(YL;24F)* breakpoints, and female *T(Y;2)L126*, with *T(YS;24C)* breakpoints, is shown in *Figure 2*. The synthetic *Dp* 24C;24F will be recovered as a y^+ female, $T(Y;2)H116^{2DYP}L126^{2PYD}$, or $Ts(YSt;2Lt)H116 + Ts(YSt;2Rt)L126$; with the reciprocal deletion emerging as a y B^S B^S male $T(Y;2)L126^{2DYP}H116^{2PYD}$, or $Ts(YLt;2Lt)L126 + Ts(YLt;2Lt)H116$, *Figure 2a*. If the sex of the parental flies is reversed, then the sex of the *Dp* and *Df* segregants will be reversed. So, to establish a stock of the synthetic aneuploids, you should set-up the initial cross both ways, *Figure 2b*.

It is worth noting that B^S is an antimorphic mutation that gives a stronger phenotype with two copies. Similarly, the *Ts(YLt;2Rt)* segregants have a weak dominant Hairy wing phenotype which is stronger in two copies, resulting from duplication of the *achaete* gene adjacent to *y*. The Hairy wing phenotype

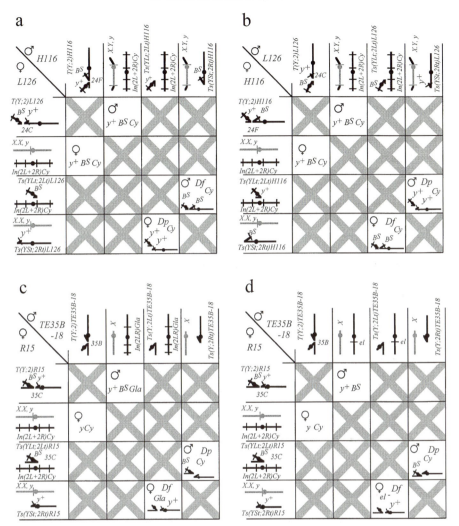

Figure 2. Segregation of the *T(Y;A)* translocation elements of Lindsley and Sandler *et al.* to generate reciprocal duplications and deletions. (a) Cross to generate a duplication and deletion of the 24C;24F segment. (b) Reciprocal cross to recover the duplication and deletion segregants in flies of the opposite sex to (a). (c) Use of differentially marked balancer chromosome to identify segregant products. (d) Use of recessive visible mutation in an F1 complementation test.

can be scored by the presence of ectopic microchaetae on the mesopleural plate just under the wing and is stronger in males than females. Strictly speaking the $B^S Y y^+$ chromosome is a complex rearrangement with the short arm of the Y capped with *1Lt*, y^+ and the long arm carrying an insertional duplication of B^S capped with *4Lt*. The FlyBase convention, fortunately, is to simplify this as above and regard the translocation segregants as carrying *YSt* marked with y^+ and *YLt* marked with *B*.

A limitation of this marking system is that the aneuploid products are only distinguishable from the parental classes when pairs of translocations are chosen that have the *Y* chromosome breakpoints on opposite arms. In many cases this is not a problem as the collection of translocations is large enough for alternative pairs to be selected. On occasions when this is not possible, a simple modification is to introduce either a differentially marked, or an unmarked autosome into the cross scheme, *Figure 2c*. This allows aneuploid segregant products to be recovered in combination with unmarked *T(Y;A)*s from conventional screens or between pairs of the Lindsley and Sandler *et al.* *T(Y;A)*s that carry breakpoints on the same arm of the *Y* chromosome.

Consider the example of *T(Y;2)TE35B-18*, this translocation was recovered from a conventional mutagenesis screen, it carries a complete *Y* and is male fertile. The stock is maintained with a 'floating balancer' system in which females are heterozygous for the second chromosome balancers *Gla* and *CyO*, while males carry the *T(Y;2)* and either *Gla* or *CyO*. In order to recover a deletion proximal to the translocation breakpoint, male *T(Y;2)TE35B-18/Gla* flies are crossed to *C(1)RM, y; T(Y;2)R15/In(2L+2R)Cy* females and the synthetic deletion, *Df* 35B1.2;35B-C, identified as *Gla* daughters, *C(1)RM, y; Ts(Y;2Lt)TE35B-18 + Ts(YLt;2Rt)R15/Gla, Figure 2c*. (The position of the *Y* chromosome breakpoint of *T(Y;2)TE35B-18* is unknown and omitted from the *Ts* designation, *Ts(Y;2Lt)TE35B-18*.) It can be useful to introduce visible mutants, in the cross that generates a synthetic deletion. This will test whether the deletion includes the visible locus and avoids the potential problems of male sterility that can result from deleting *Y* chromosome fertility factors. In this example, *T(Y;2)TE35B-18/el* males crossed to *C(1)RM, y; T(Y;2)R15* females will generate B^S *Cy* daughters showing the *el* phenotype, *Ts(YLt;2Lt)TE35B-18 + Ts(Y;2Rt)R15/el, Figure 2d*.

A rigorous explanation of the use of the Lindsley and Sandler *et al.* set of translocations is provided in the original paper and in the Grey Book (4). There is no substitute, however, for getting out a pencil and working specific examples through for yourself. In general the use of these stocks is quite straightforward although some of them are not very fertile.

3. The generation of aneuploids by recombination

3.1 Transpositions

It is sometimes possible to generate a useful duplication or deletion by recombination between a cytologically normal chromosome and an insertional transposition, although suitable stocks are rare. Consider the case of *Tp(2;2)TE35B-54, b*. This chromosome carries a three break insertional translocation with the 35B1.2;35C1 segment transposed to 38F giving the new chromosomal order: 2Lt—35B1.2│35C1—38F│35B1.2—35C1│38F—2cen——2Rt. A recombination event within the 35C1—38F interval will separate the 35B1.2│35C1 deletion from the 35B1.2—35C1 duplication. To recover the separate elements requires that the transposition chromosome and the cytologically normal chromosome carry different markers close to the breakpoints. Given that the transposition carries *b* (*black*, 34D4-6), if the normal chromosome were mutant for *pr* (*purple*, 38B5-C2) the deletion could be recovered as a *b pr* recombinant (in *Df(2L)TE35B-54, b/b pr* flies) and the duplication as a *b⁺ pr⁺* recombinant (in *Dp(2;2)TE35B-54/b pr* flies), *Figure 3a*.

3.2 Inversions

Similarly, recombination between the inverted segments of two similar inversions will generate aneuploid inversions carrying a duplication or deletion associated with each breakpoint. In the example of a recombinant between the two paracentric inversions *In(2L)TE35B-210, b* (28B12-D1;35B1.2) and *In(2L)C163.41* (27D1;35E1-2), the *In(2L)TE35B-210ᴸ C163.41ᴿ* product will carry *Dp* 27D1;28B12-D1 + *Df* 35B1.2; 35E1-2 (new order: 2Lt—28B12│35B1.2—27D1│35E1-2—2ccn——2Rt). The reciprocal product, *In(2L)C163.41ᴸ TE35B-210ᴿ*, will carry *Df* 27D1;28B12-D1 + *Dp* 35B1.2; 35E1-2 (new order: 2Lt—27D1│35E1-2—28B12│35B1.2—2cen——2Rt). In this example, markers are not useful in identifying the recombinant products. The only marker is *b*, within the inverted segment of the *In(2L)TE35B-210* chromosome. Depending whether the recombinant is left or right of *b*, either product could carry the marker and remain indistinguishable from non-recombinant *In(2L)TE35B-210, b* flies, *Figure 3b*. This example illustrates the general principle that you should never do a mutagenesis screen on unmarked chromosomes. It almost always leads to tears. In this situation, when the expected recombinant products can not be distinguished by markers, it may be possible to identify the aneuploid products either by failure to complement a visible mutation (e.g. *In(2L)TE35B-210ᴸ C163.41ᴿ/osp* flies are phenotypically outspread, as the deletion uncovers the *osp* locus in 35B2-3) or modification of a dominant phenotype (e.g. *In(2L)C163.41ᴸ TE35B-210ᴿ; H/+* flies have an enhanced Hairless (H) phenotype as the duplication carries the *Su(H)* locus in 35B9-10).

Analogous duplications can be produced by recombination between pericentric inversions, which span the centromere. In the example of the recombinant products between *In(2LR)TE35B-15, al dp b pr l(2)pwn cn sp* (35B1.2;44DE) and *In(2LR)Scoʳᵛ¹* (35D1-2;44C4-5), the *In(2LR)TE35B-15ᴸ Scoʳᵛ¹ᴿ* product (carrying *Df*35B1.2;35D1-2 and *Dp* 44C4;44DE) will be marked with *al* and *dp*; while the *In(2LR)Scoʳᵛ¹ᴸ TE35B-15ᴿ* (*Df* 35B1.2;35D1-2 and *Df* 44C4;44DE) will be marked with *sp*, *Figure 3c*. Either product might carry the markers from within the inverted segment of the original *In(2LR)TE35B-15* chromosome depending on the position of the

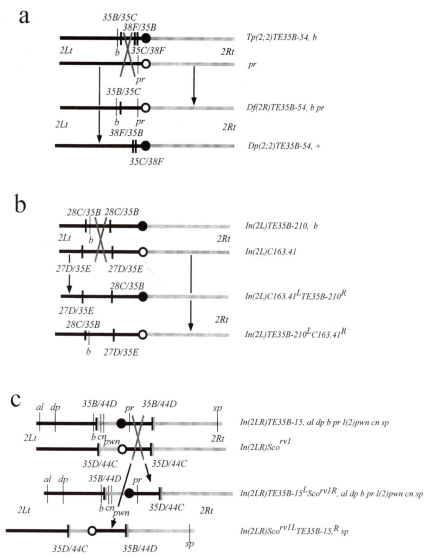

Figure 3. The generation of aneuploids by recombination. Note that the *L* arm is drawn in black and *R* arm in grey. (a) Transposition, recombination between the breakpoints of a three break insertional translocation, *Tp(2;2)TE35B-54, b* and a cytologically wild-type chromosome marked with *pr*. (b) Paracentric inversions, recombination between the inverted segments of two paracentric inversions with similar breakpoints, *In(2L)TE35B-210, b* and *In(2L)C163.41*, to generate the aneuploid recombinants *In(2L)TE35B-210^{L}C163.41^R, b (Dp* 27D1-28B12-D1 + *Df 35B1.2;E1-2)* and *In(2L)C163.41^{L}TE35B-210^R (Df* 27D1-28B12-D1 + *Dp 35B1.2;E1-2)*. (c). Pericentric inversions, recombination between the inverted segments of *In(2LR)TE35B-15, al dp b pr l(2)pwn cn sp* and *In(2LR)Sco^{rv1}* to generate the aneuploid products *In(2LR)TE35B-15^{L}Sco^{rv1R}, al dp b {pr l(2)pwn cn} (Dp* 35B1.2;35D1-2 + *Dp* 44C4;44DE) and *In(2LR)Sco^{rv1L}TE35B-15^R, {pr l(2)pwn cn} sp(Df* 35B1.2;35D1-2 + *Df* 44C4;44DE).

recombination with respect to these markers. In this case the aneuploid inversions can not be recovered by this scheme as the *noc* mutation at the 35B1.2 breakpoint of *In(2LR)TE35B-15* is lethal when heterozygous with *In(2LR)Sco^{rv1}*. In order to recover these aneuploids it is necessary to use a method devised by Loring Craymer.

4. Craymer's autosynaptic method

As defined by Loring Craymer in 1981 (5), autosynaptic chromosomes are the aneuploid products of recombination between a pericentric inversion and a cytologically wild-type chromosome. The name derives from the fact that homologous chromosome arms are carried on a single centromere and will therefore pair, or synapse, with each other (*Figure 4*). Craymer had the eccentric idea that genetics was really about what happens when two flies meet in a tube and that if you thought about it long enough it was bound to give you ideas. Like his Mentor, James Branch Cabell (6), Craymer was sick to death of the Southern mentality and makes few concessions to the unsophisticated reader. This can make him difficult to follow.

Briefly, the products of recombination within the inverted segment of a pericentric inversion and a cytologically normal chromosome are grossly aneuploid, *Figure 4*. In the normal course of events, these aneuploid products will segregate from each other to give gametes with lethal chromosomal elements. If, however, both complementary products can be recovered in the same fly it will be euploid. Craymer lists a number of methods to recover autosynaptic elements (4, 5). No method is easy, until you have a pre-existing autosynaptic stock. Then it becomes trivial. If an autosynaptic male is crossed to a female carrying the corresponding inversion, only the recombinant aneuploid gametes will survive. On Craymer's terminology, the left- and right-hand aneuploid recombinant products are the 'laevosynaptic'(*LS*) and

Paired Configuration   Unpaired Configuration

Figure 4. Recombination within the pericentric inversion *In(2LR)TE35B-4* (35B1.2;43A1.2) and a wild-type chromosome to generate the autosynaptic elements *LS(2)TE35B-4* and *DS(2)TE35B-4*. The chromosomes are drawn both synapsed along their whole length (paired configuration) and synapsed only in the pericentric region (unpaired configuration). In the rest of this chapter chromosomes will be drawn in the unpaired configuration.

117

'dextrosynaptic' (*DS*) elements, *Figure 4.* The *LS* element carries both copies of the left arm distal to the inversion breakpoint, while the *DS* element carries both copies of the right arm distal to the breakpoint. It is as if the left- and right-hand breakpoints of a standard 'heterosynaptic' inversion are separated onto homologous centromeres, leaving the breakpoint-distal regions to 'autosynapse' with each other. This process can be designated as female *In(nLR)A/+*, *LS(n)A//DS(n)A*, with the // indicating that the autosynaptic elements are carried on homologous centromeres. In the example shown in *Figure 4*, recombination within the *In(2LR)TE35B-4* inversion (35B1.2; 43A) gives the complementary products *LS(2)TE35B-4* (2Lt—35B|43A—2cen——2Lt) and *DS(2)35B-4* (2Rt—43A|35B—2cen——2Rt).

4.1 The recovery of aneuploid autosynaptic stocks

Given one autosynaptic stock, inversions with similar breakpoints can be re-covered as aneuploid autosynaptic stocks by crossing heterosynaptic females to autosynaptic males (5). For example, the *DS* element of *In(2LR)TE35B-226* (35B1;47B10-14) can be recovered crossing female *In(2LR)TE35B-226/+* to male *LS(2)DTD128//DS(2)DTD128* flies (35B2-3;48C6-8) and recovering the aneuploid *LS(2)DTD128//DS(2)TE35B-226* autosynaptic stock carrying *Df* 35B1;35B2-3 and *Dp* 47B10;48C, *Figure 5a.* The reciprocal constellation, *LS(2)TE35B-226//DS(2)DTD128* is not recovered as the 47B10;48C segment is too large to survive as a deletion. Males from the aneuploid *LS(2) DTD128//DS(2)TE35B-226* stock can then be crossed back to female *In(2LR) TE35B-226/+* to recover the *LS* element in a euploid *LS(2)TE35B-226//DS (2)TE35B-226* stock. This method of progressing along the chromosome to produce a set of autosynaptic stocks was termed 'inchworming' by Loring Craymer. The critical point is that the lack of recombination in male *Droso-phila* can be used to force the recovery of recombinants in the female line.

There are two particularly useful applications of this approach. First, aneuploid autosynaptic stocks can be 'resolved' to give the corresponding aneuploid inversion stocks, which are fertile with normal laboratory strains (5). Secondly, individual autosynaptic stocks can be used as the basis of selec-tive genetic screens to recover novel breakpoints in a given genetic region (7). Before going on to consider specific examples, it is necessary to consider the general problem of differential marking of autosynaptic elements and to introduce a nomenclature system that will cope with this.

When two euploid autosynaptic stocks with similar breakpoints are crossed together there are two classes of progeny that may be viable, depending on the relative positions of the breakpoints: *LS(n)A//DS(n)A* × *LS(n)B//DS (n)B*, *LS(n)A//DS(n)B* + *LS(n)B//DS(n)A*. (The *LS(n)A//LS(n)B* and*DS(n) A//DS(n)B* segregants will be grossly aneuploid and embryonic lethal.) The problem with this cross, as it stands, is that the products are unmarked and can not be distinguished from the parental genotypes. Even when only one

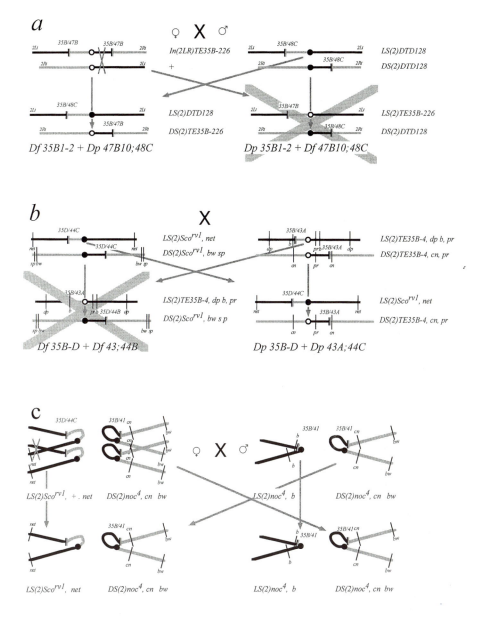

Figure 5. (a) Recovery of a novel *DS* element by crossing females carrying a hetero-synaptic inversion, *In(2LR)TE35B-226/+*, to males of an autosynaptic stock with similar breakpoints, *LS(2)DTD128//DS(2)DTD128*. (b) Recovery of hyperploid autosynaptic constellation by crossing two autosynaptic stocks with similar breakpoints, *LS(2)Sco^{rv1}, net//DS(2)Sco^{rv1}, net × LS(2)TE35B-4, dp, pr //DS(2)TE35B-4, cn, pr→LS(2)Sco^{rv1}, net//DS(2)TE35B-4, cn, pr*. (c) recombination between non-sister *2L* chromatids attached to different centromeres to generate a homozygously marked *LS(2)Sco^{rv1}, net* element.

surviving class is expected this is bad practice. It is, however, simple to construct uniquely marked autosynaptic elements. Given that the breakpoint-distal regions of each autosynaptic element are derived from the same parent, any viable recessive mutations distal to the inversion breakpoints can be used to mark an element. Markers within the inversion loop behave as in normal stocks. Dominant mutations will tend to be lost by recombination, unless they map close to the inversion breakpoint. An exception is with dominant mutations associated with an additional paracentric inversion. In these cases the position of the centromere is indicated with a full stop (e.g. *LS(2) DTD111.In(2L)Cy*; cytological order: *2Lt—29F|2cen—34A|22D1—34A4| 22D1—2Lt*).

A duplication for the 43A1.2;44C4-5 region could be synthesized by crossing *LS(2)Sco^{rv1}//DS(2)Sco^{rv1}* *(35D1.2;44C4-5)* flies to *LS(2)TE35B-4//DS(2) TE35B-4 (35B1.2;43A1.2)* flies, *Figure 5b*. The *Sco^{rv1}* stock carries the breakpoint-distal markers *net* on the *LS* and *bw sp* on the *DS*; *LS(2)Sco^{rv1}, net// DS(2)Sco^{rv1}, bw sp*. The *TE35B-4* stock carries *dp* and *b* distal and *pr* proximal to the breakpoint on the *LS* element (*LS(2)TE35B-4, dp b, pr*) and *cn* distal and *pr* proximal to the breakpoint on the *DS* element (*DS(2) TE35B-4, cn, pr*). The second comma indicates the position of the inversion breakpoint relative to the markers, the use of this nomenclature is illustrated in *Figure 6*. In the above cross the parental autosynaptic flies are *net bw sp* and *dp b pr cn* in phenotype. The surviving F1 flies (*LS(2)Sco^{rv1}, net//DS(2)TE35B-4, cn, pr*) are net in phenotype and carry *Dp* 43A1.2;44C4-5 + *Dp* 35B1.2;35D1-2. (Neither cn, nor pr phenotypes will be visible as *LS(2)Sco^{rv1}, net* is homozygous for *pr^+* and hemizygous for *cn^+*.)

4.2 Construction of marked autosynaptic elements

In order to recover novel autosynaptic elements and segmental aneuploids, it is essential to produce differentially marked elements. The *In(2LR)TE35B-4* inversion, illustrated in *Figure 4*, was spontaneous on a chromosome that carried *dp b TE35B pr cn*. These recessive markers, carried on the inversion chromosome, allow marked autosynaptic elements to be recovered directly from phenotypically dp b pr cn females. The *DS* element, marked with *pr* and *cn*, can be obtained by crossing females heterozygous for *In(2LR)TE35B-4, dp b pr cn* and a cytologically wild-type *dp b pr cn* chromosome to autosynaptic *LS(2)Sco^{rv1}//DS(2)Sco^{rv1}* males. The only surviving progeny of this cross will be *LS(2)Sco^{rv1}, +//DS(2)TE35B-4, cn, pr* flies, which will be phenotypically wild-type as *LS(2)Sco^{rv1}, +* is hemizygous for both *pr^+* and *cn^+*. The *LS* element can then be recovered by crossing the same heterozygous females (*In(2LRTE35B-4, dp b pr cn/dp b pr cn*) to autosynaptic *LS(2)Sco^{rv1}, +//DS(2)TE35B-4, cn, pr* males. This cross will give two classes of progeny: *LS(2)TE35B-4, dp b, pr//DS(2)TE35B-4, cn, pr* (which are phenotypically dp b pr cn) and *LS(2)Sco^{rv1}, +//DS(2)TE35B-4, cn, pr* (phenotypically wild-type).

The marked $LS(2)Sco^{rvl}$, $net//DS(2)Sco^{rvl}$, bw sp stock could not be constructed in this way as the original inversion was induced on an unmarked chromosome. Instead, heterozygous recessive markers were introduced into the autosynaptic stock and then selected to become homozygous in subsequent generations. A breeding scheme which combines this with Craymer's inchworm approach is as follows; female $In(2LR)Sco^{rvl}/net$ × male $LS(2)noc^4$, $b//DS(2)noc^4$, cn bw , $LS(2)Sco^{rvl}$, $(net)//DS(2)noc^4$, cn bw. Giving an unmarked LS (heterozygous for net, *Figure 5c*, female parent) in a hyperploid autosynaptic constellation (Dp 41;44C4-5 + Df 35B1.2;D1-2). Flies of this genotype are backcrossed to $LS(2)noc^4$, $b//DS(2)noc^4$, bw, cn to establish the hyperploid ($LS(2)Sco^{rvl}(net)//DS(2)noc^4$, bw, cn) stock, from which recombinant flies homozygous for net are selected. The recombination event necessary to generate the homozygous element occurs at the four-strand stage between non-sister chromatids attached to separate centromeres, *Figure 5c*. The marked DS can be recovered by crossing male $LS(2)Sco^{rvl}$, $net//DS(2)$ noc^4, cn bw to female $In(2LR)Sco^{rvl}/bw$ sp to give a $LS(2)Sco^{rvl}$, $net//DS(2)$ Sco^{rvl}, $(bw$ $sp)$ fly, from which a $LS(2)Sco^{rvl}$, $net//DS(2)Sco^{rvl}$, bw sp stock is selected.

4.3 The construction of synthetic aneuploids using $h;A$ autosynaptic stocks

As with autosomal translocation segregants, autosynaptic stocks can only be used to construct reciprocal duplication and deletion products when aberrations with similar breakpoints in both arms are available. This problem can be avoided, as in the Lindsley and Sandler *et al.* study, by using aberrations with one heterochromatic (h) and one euchromatic (A) breakpoint. The autosomal centromeres are surrounded by large blocks of centric heterochromatin which represent large targets for inversion breakpoints and many pericentric inversions are available with $h;A$ breakpoints. $h;A$ autosynaptic stocks can be used to construct segmental aneuploids analogous to those of Lindsley and Sandler *et al.*, but with the heterochromatic breakpoint neatly tucked into the autosomal centromere.

A set of autosynaptic stocks on the second chromosome with one breakpoint at, or close, to the centric heterochromatin has been constructed by Gubb and Roote *et al.* (8) and is available through the Bloomington and Umeå Stock Centres. The marker combinations have been chosen to facilitate identification of segmental aneuploids (*Table 1*, *Figure 6*). The *2L* and *2R* blocks of centric heterochromatin are designated h35-h38L and h38R-h46, respectively. These regions are under-replicated in polytene chromosomes and are simplified to *2Lh* and *2Rh* in *Table 1*.

A number of $h;3$ autosynaptic stocks generated by Craymer (5) are also available through Bloomington, *Table 2*.

To construct synthetic duplications and deletions, autosynaptic flies with

Table 1. The *h;2* autosynaptic stocks of Gubb and Roote *et al.*

LS	DS	A breakpoint	h breakpoint
S^{325},+. $In(2L)Cy$*	S^{325},, Sco	21F	2Rh
DTD18, ho^2	DTD18, cn bw	23A4-7	2Rh
DTD21, ho^2	DTD21, cn, {dp}	23A1.2	2Rh
DTD16,, dp	DTD16, bw, dp	23C	2Rh
DTD8, +.{net}	DTD8, sp	23C-D1	2Rh
DTD42, +.{net}	DTD42, bw sp	23E3-6	2Rh
DTD52	DTD52, vg	24D1.2	2Rh
DTD124,, b	DTD124, cn, b	24D2-3	2Rh
DTD109, +.{fy^2}	DTD109, bw	25E2-3	2Rh
DTD116, +.{net}	DTD116, sp	26A4-6	2Rh
DTD24, +.{net}	DTD24, bw sp	26C1.2	41A
DTD51, +.$In(2L)Cy$	DTD51, cn bw, {b}	27D1	2Rh
DTD11	DTD11, sp	28A	2Rh
DTD111, {al^2 pr}.$In(2L)Cy$,	DTD111, vg	29F	2Rh
DTD125, net	DTD125, bw sp	31E	2Rh
DTD107, $In(2L){TE23C ho^2}$.{dp}	DTD107, bw sp	32F + 28D;32E-F	2Rh
DTD4, fy^2	DTD4, bw	32F	41AB
DTD86, Bl	DTD86, {cn}	33B1.2	2Rh
b81a2	b81a2	34D4 + Df34D4;E3	41DE
D20, {net dp b}	P9, {pk cn sp}, pr	34E4-F2//34B7-12	2Rh//41D
DTD43, ho^2	DTD43, bw sp	35B1.2	2Rh
D9, dp b	D6	35B1-3//35D5-7	2Rh
Sco^{rv9}	D2, cn bw	35D1-2//34D4.5	2Rh//41B3-9
D5	CH25	36C1//36C	42F//2Rh
P3,,{pr}	noc^4, cn bw	37B.2//35B1.2	41CD//2Rh
C(2L)C3, b pr;	**C(2R)C1, sple**	**2Lh;2Rh**	**2Lh;Rh**
f6, net	f6, bw sp	48F6-49A1	39D3-E1
Rev-B, dp b, cn	Rev-B, bw, cn	52D10-E1	2Lh
Pu^L	Pu^L, {or} lf	57C4-6	2Lh
bw^{v32g}, net bw^v	bw^{v32g}, bw, lt^v {sp}	59E	2Lh
lt^{G10}, b el, {cn} bw	lt^{G10}, lt ,{cn}	59F3-4	2Lh
lt^{G16}, lt, stw^3 cn bw	lt^{G16}, cn bw	60E5-8	2Lh

* Constructed by Loring Craymer.

the required breakpoints are crossed to each other to generate aneuploid constellations. These aneuploid stocks remain stable and can be maintained in autosynaptic form before being resolved to give the corresponding heterosynaptic inversion. In general, duplications of several numbered divisions survive, although with reduced fertility. If nursed through a few generations, however, even severely aneuploid stocks tend to become reasonably healthy and can be amplified.

4.4 Conversion of *T(Y;A)* translocations to autosynaptic form

The similarity between *T(Y;A)* translocations and *h;A* autosynaptic stocks suggests that it might be possible to convert one form to another. In a sense

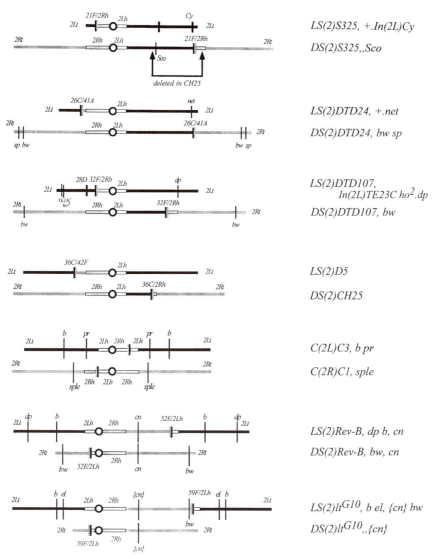

Figure 6. Representative *h;A* autosynaptic stocks from the Gubb and Roote *et al.* collection to show chromosomal configuration and marker mutations. Note that the heterochromatic regions *h35-h38L* and *h38R-h46, 2Lh* and *2Rh* respectively, are drawn as hollow boxes.

pericentric inversions can be regarded as reciprocal translocations between two chromosome arms that happen to be joined to the same centromere. Given that a *T(Y;A)* translocation has two large blocks of heterochromatin, one at the breakpoint and the other surrounding the centromere, it should be possible to induce a translocation between these two targets. This idea has

Table 2. Craymer's *h;3* autosynaptic stocks

LS	DS	A breakpoint	h breakpoint
LD3	*LD3, p^p*	61F	3Rh
P13,+. R^a	*P13, +. Ki*	63A	3Rh
P93,,st	*P93, +. Ki, h st*	64C-E	3Rh
P10, +. {R}	*P10, +. {Ki}*	65C5-9	3Rh
P91	*P91, +. {Ki}*	67C10-D1	3Rh
P42,, +. th st cp in ri	*P42,, st. In(3L)H27*	70F1-2	3Rh
C(3L)P3, ri^b	*C(3R)P3, sr*	*3LH*	*3Lh*
A114,, Sb Ubx. +	*A114*	92A-B1	3Lh

[a] When Loring Craymer departed the world of fly genetics, like some latter day Dom Manuel (6), his autosynaptic stocks were deposited with the Bloomington Stock Centre. Although these stocks carry marker mutations, Craymer, characteristically, neglected to explain his notation. This table shows my best guess at the actual marker configurations, using the notation explained in this chapter. Concerning the finer points of Loring's nomenclature the 'problems perhaps involved are relinquished to those really thoroughgoing scholars whom erudition qualifies to deal with such topics, and tedium does not deter' (6).
[b] Constructed by D. G. Holm.

been tested by irradiating female *C(1)RM, y; T(Y;2)A80, B^S y^+* and crossing them to male *LS(2)noc^4, b//DS(2)noc^4, cn bw* and recovering the novel *DS* derivative of *T(Y;2)A80* as a *LS(2)noc^4, b//DS(2)A80* fly lacking both *B^S* and *y^+*, *Figure 7* (J. Roote and D. Gubb, unpublished data). The novel *DS* element was crossed to *LS(2)b^{81a2}//DS(2)b^{81a2}* and resolved to give *In(2LR)b-81a2^L A80^R* (*Df* 34D4;35B1). This synthetic deletion gives a fertile stock which was mapped genetically to confirm the limit of the *A80* breakpoint. The initial cross required setting-up 20 tubes of 40 female *C(1)RM, y; T(Y;2)A80, B^S y^+* and 20 male *LS(2)noc^4, b//DS(2)noc^4, cn bw* flies. This cross was particularly difficult as the original *C(1)RM, y; T(Y;2)A80, B^S y^+* stock is less fertile than most of the Lindsley and Sandler *et al.* translocations. The same approach could be used to generate *h;A* autosynaptic stocks in regions that lack appropriate pericentric inversions.

4.5 Resolution of autosynaptic chromosomes

In order to recover heterosynaptic inversions, autosynaptic females are crossed to males from a normal laboratory stock. The cross needs to be set-up with a large number of flies, particularly if the inversion breakpoint is close to the centromere, as only recombinants within the inversion will survive. As an example, *LS(2)DTD11, +//DS(2)DTD11, sp × LS(2)DTD51 .In(2L)Cy//DS (2)DTD51, cn bw, {b}* gives two classes of progeny: *LS(2)DTD11, +//DS(2)DTD51, cn bw, {b}* (*Df* 27D1;28A) and *LS(2)DTD51 .In(2L) Cy//DS(2)DTD11, sp* (*Dp* 27D1;28A). This is a fertile cross and can be made on a small scale although the *Df* class emerges at 18% of the frequency of the

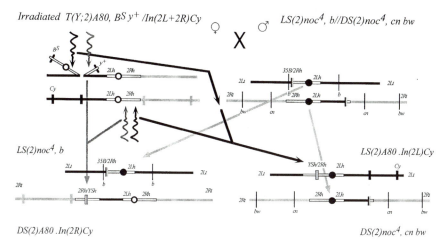

Figure 7. Conversion of a *T(Y;A)* translocation to *h;A* autosynaptic element. Chromosomal breakpoints are induced in the *YS* and *2R* heterochromatin by irradiation. The breakpoints can generate an *LS* (black thunderbolts) or *DS* (grey thunderbolts) element as a translocation event.

Dp. To resolve the autosynaptics, and generate the heterosynaptic inversion forms of the deficiences and duplications, several vials should be set-up with about 20 pairs of flies. The genotype of the males is generally not critical, a double balancer stock such as *In(2LR)Gla/CyO* can be used, e.g. female *LS(2)DTD51 .In(2L)Cy//DS(2)DTD11, sp* × male *In(2LR)Gla/CyO, Figure 8*. There are two viable heterosynaptic recombinants from this cross, the synthetic duplication *In(2LR)DTD51^{L}DTD11^R, sp* and the euploid chromosome *In(2L)Cy, Cy sp*, which can be distinguished as Cy Gla$^+$, Cy$^+$Gla and Cy Gla flies (corresponding to *In(2LR)DTD51^{L}DTD11^R, sp /CyO, In(2LR) DTD51^{L}DTD11^R, sp/Gla,* and *In(2L)Cy, Cy sp/Gla,* respectively) and backcrossed to *In(2LR)Gla/CyO* flies to recover a stock. Resolution of the *LS(2)DTD11, +//DS(2)DTD51, cn bw, {b}* stock similarly produces two heterosynaptic products, but in this case the markers are indistinguishable (either the *Df* chromosome, *In(2LR)DTD11^{L}DTD51^R*, or the normal sequence chromosome will carry *cn bw,* and possibly, *b.* The two classes could be distinguished genetically either by crossing to a visible mutation in the 27D1;28A interval, such as *wg* (27F1-2), or by recovering the inversion over a normal sequence chromosome that carries a dominant mutation within the cytological boundaries of the inverted region, such as *Sco.* When backcrossed to *Sco/CyO* flies, the *Df* will give a stable stock (*In(2LR)DTD11^{L}DTD51^R, {b cn bw }/Sco*) that can be confirmed cytologically. *Sco* will not be maintained when heterozygous with a cytologically normal chromosome, unless this were to carry a recessive lethal mutation mapping close to *Sco.*

In the particular case of autosynaptic stocks that carry markers at the

David Gubb

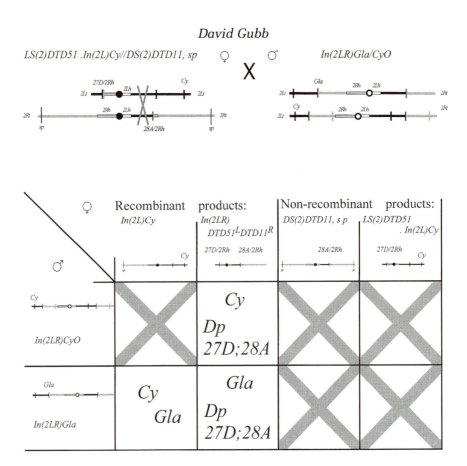

Figure 8. Resolution of an autosynaptic stock to give a heterosynaptic inversion carrying a duplication.

pericentric inversion breakpoints, identification of the heterosynaptic inversion products is easy. A number of stocks carrying w^+ at the breakpoint of both the *LS* and *DS* elements have been constructed from inversions with one breakpoint within *TE35B* (9), which carries a tandem duplication ($w^+ w^+$) inserted in the *noc* locus at 35B1.2. The resolved inversion products of such stocks will carry the w^+ marker, e.g. female $LS(2)b^{81a2}//DS(2)TE35B-222,, w^+,$ $In(2LR)b-81a2^L$ $TE35B-222^R$, w^+ (*Df* 34D4; 35B1.2), which can be distinguished from the unmarked cytologically normal chromosome in a w^- background.

The *LS* elements of several of the Gubb and Roote *et al.* stocks were constructed from unmarked inversions and carry distal recessive markers that were selected to become homozygous, but failed to do so, such as *DTD24*, *+.{net}, Figure 6*. Unless these markers recombine during long-term maintenance of these stocks, the resolved heterosynaptic inversion product will be unmarked, while the euploid product will carry the recessive marker.

5. Selective genetic screens

The complete sterility of autosynaptic males when crossed to females from a normal stock can be used as the basis of genetic screens for novel breakpoints (7). The idea is to induce novel autosynaptic elements directly as translocations between homologous chromosomes, in the female pre-meiotic germline. In practice, large numbers of recessively marked females can be irradiated and crossed to males from an appropriate autosynaptic stock. Crosses can either be made in crowded vials containing 40 females and 20 males and transferred to fresh food every few days, or in population cages. If the first method is used, vials should be examined about a week after transfer of adults. The majority of vials can be discarded at this stage retaining only those which contain a surviving larva or pupa. Lone pupae should be transferred to fresh vials as medium that is not actively worked by larvae develops a sticky surface that can trap emerging adults. Pupae can be eased from the surface of the vial with a wet paintbrush. (Oddly enough the salivary gland secretion that sticks pupae down is not tanned and remains water soluble. As a survival strategy this seems daft.) It is critical that novel elements should be marked uniquely as the frequency with which non-disjunctional sperm, carrying both paternal autosynaptic elements, fertilize nullosomic eggs, from the irradiated females, becomes significant under these conditions (7). Novel autosynaptic elements are generally recovered in hyperploid flies (7) suggesting that the recovery could be increased by crossing to an aneuploid autosynaptic stock. For example, if irradiated females were crossed to *LS(2)DTD107//DS(2)DTD111* novel *DS* elements would be recovered with the *LS(2)DTD107* element, which has a breakpoint in 32F. Novel *LS* elements, on the other hand, would be recovered with *DS(2)DTD111* having a 29F breakpoint. As a consequence of choosing these breakpoints the majority of novel *DS* elements would have a breakpoint distal to 32F, while the majority of novel *LS* elements would have a breakpoint proximal to 29F. In practice, this screen would be expected to give a strongly biased recovery of novel elements of the smaller class (7, 10); in this case the *DS* element. It would be useful, however, to bias the region in which breakpoints of novel *LS* elements were recovered to give alternative left-hand breakpoints for use during resolution of the novel aneuploid autosynaptic stocks to give heterosynaptic duplications and deletions.

5.1 Second generation selective screens

Given a set of *h;A* autosynaptic stocks it is possible to design more powerful selective screens based on deletion of large segments in the male line. As can be seen from *Figure 6*, a compound arm stock can be regarded as the limiting case of an autosynaptic with both breakpoints in centric heterochromatin. The other limiting case is with a euchromatic breakpoint near the tip of one arm, for example *LS(2)S325//DS(2)S325*, in which the *LS* is essentially a free 2L and the *DS* is essentially 2R|2L.2R, and *LS(2)lt-G16//DS(2)lt-G16*, which approximates to *2L|2R.2L//2R*. Between these limits, the *h;A* stocks form a series of 'hyper-compound' arms, with increasingly large inserts, balanced with the corresponding 'hypo-compound arms', carrying the reciprocal dele-tions, *Figure 6*. Although this is an odd way to consider the *h;A* stocks, it emphasizes their similarities and implies that it should be possible to derive some *h;A* elements as deletion derivatives from larger elements. On this view, *DS(2)CH25* (chromosome order, *2Rt——2Rh|36C—2cen——2Rt*) corre-sponds to a hyperploid *C(2R)* carrying the duplicated segment 36C;2Lh. This element was derived from *DS(2)S325,, Sco* (chromosome order, *2Rt——2Rh|21F—2cen——2Rt*) by deleting the *2Rh|21F—36C* segment including the *Sco* mutation, *Figure 6* (C. Henchcliffe and D. Gubb, unpublished data). 200 male *LS(2)S325, +.In(2L)Cy//DS(2)S325,, Sco* flies were irradiated with 4.5 kR X-rays and crossed to 400 female *C(2L), b pr//C(2R)Cl, sple*. The cross gave two *DS* elements with breakpoints distal to *pr* as *b pr⁺* flies, in addition to four putative *C(2R)* elements as *b pr* flies, *C(2L), b pr//C(2R)*. The same idea occurred to Bruce Reed independently, who generated a set of 44 *DS* 'carve-down' elements with breakpoints in the 56F3 to 60F interval (10). Deletions were induced either on a *C(2R)* element or a hyper-compound element. (Again, the principle can be seen by reference to *Figure 6*. The hypo-compound *DS(2)Rev-B*, for example, differs from a *C(2R)* element by the deletion of the 41A;52E segment. Deletions from the centric heterochromatin to around the 52 region will be a relatively common product in irradiated sperm from a *C(2R)* stock as this requires a simple two break event with one breakpoint within centric heterochromatin, which can occur on either arm of the *C(2R)* stock.) In Reed's experiment, irradiated *C(2L)C3, b pr;C(2R)Cl, sple* males were crossed to *LS(2)lt^{G16}, lt, stw³ cn bw//DS(2)bw^{v32g}, bw* females and the occasional surviving *lt* fly was backcrossed to *LS(2)lt^{G16}, b //DS(2)lt^{G10}, If* flies to establish a stock. It is hard to estimate the recovery fre-quency with autosynaptic screens as only very rare progeny survive. Reed recovered new *DS* stocks at a frequency of one for every three bottles set-up (each bottle containing about 25 irradiated *C(2L)C3, b pr;C(2R)Cl, sple* males and 50 *LS(2)lt^{G16}, lt, stw³ cn bw//DS(2)bw^{v32g}, bw* females), which is a significant improvement on previous screens.

In most euchromatic regions, the hyper-compound element could be re-covered by deletion of a segment from hyper-compound with a more distal breakpoint, while the hypo-compound element could be recovered by deleting a segment from a compound arm.

6. The use of aneuploid stocks to identify enhancers and suppressors

A powerful approach to identify the components of a genetic pathway is to screen for mutations which modify a given mutant phenotype. The general approach is to screen newly-induced mutations in a genetic background that gives a weak mutant phenotype. For example EMS mutations that modify the *glass* phenotype were recovered by Ma *et al.* (11). Mutations recovered in this type of screen generally correspond to recessive lack of function alleles of genes related to the original mutation. Deletions of such genes will also modify the mutant phenotype and this could be used as the basis of a screen. In general, however, constructing a set of deletions in the correct mutant background is too much work, unless there is reason to suspect that a particular chromosomal region contains a related genetic function. The deletion kit at the Bloomington Stock Centre currently consists of 184 stocks that cover about 75% of the genome (K. Matthews, personal communication). This could be used to screen for modifiers of dominant antimorphic mutations in an F1 cross, but even so represents a lot of work. Alternatively, large duplications could be screened initially and any regions in which modifiers are identified can be subdivided using smaller duplications and deletions.

This approach is dependent on having a dominant antimorphic mutation of the gene of interest. Such mutations are rare in conventional genetic screens as they result from alterations either in the substrate specificity of the gene product, or in its temporal or spatial patterns of expression. These types of change are easy to generate in transgenic constructs, however, and will become increasingly available. An additional application of a set of large duplications would be to screen for chromosomal regions containing modifiers of the expression of transgenic reporter gene constructs.

6.1 The CAM duplication set

In order to facilitate this type of genetic manipulation, a set of large duplications on the second chromosome has been constructed using Craymer's method, *Table 3* (Roote, Johnson, Hermann, and Gubb, unpublished data)

Table 3. The CAM duplication set

Duplicated segment	Abbreviation	Full genetic designation
21E2;23D1.2	*Dp(2;2)Cam1*	*In(2LR)DTD8^L S325^R*
23D1.2;26C1.2	*Dp(2;2)Cam2*	*In(2LR)DTD24^L DTD8^R*
26C1.2;29F	*Dp(2;2)Cam3*	*In(2LR)DTD111^L DTD24^R, bw sp*
29F;32F	*Dp(2;2)Cam4*	*In(2LR)DTD107^L DTD111^R, In(2L)DTD107, TE23CD vg*
32F;35B1.2	*Dp(2;2)Cam5*	*In(2LR)DTDnoc^{4L} DTD107^R, b*
35B1.2;36C	*Dp(2;2)Cam6*	*In(2LR)S1^{L}noc^{4R} cn bw*
36C;2Lh	*Dp(2;2)Cam8*	*In(2LR)C3^L CH9-25^R, b*
2Rh; 43A1.2	*Dp(2;2)Cam10*	*In(2LR)TE35B-4^L noc^{4R}, {al} dp b cn bw*
43A;47B10-14	*Dp(2;2)Cam11*	*In(2LR)TE35B-226^L TE35B-4^R, b pr l(2)pwn*
47B10-14;48C	*Dp(2;2)Cam13*	*In(2LR)DTD128^L TE35B-226^R, net ho^2 TE23CD*
49A;51EF	*Dp(2;3)Cam14T*	*Ts(2LR;3Rt)6r23 + Ts(2Rt;3Lt)H24, cn*
52D10-E1;57C4-6	*Dp(2;3)Cam15*	*In(2LR)lt-G16^L Pu-L^R, b bw*
57C4-6;60E5-8	*Dp(2;2)Cam16*	*In(2LR)lPu-L^L Rev-B^R, cn bw*
90C;93C	*Dp(3;2)Cam30*	*Ts(2Lt;3Lt)el-24+ Ts(2Rt;3Rt)TE35B-28, pr l(2)pwn cn*

---

and translocation segregants. These stocks are available through the Bloomington and Umeå Stock Centres. As pointed out by Craymer (5), such *Duplication + Inversion* stocks are useful as balancers for the inverted segment. This makes them ideal for use in screens for haplo-lethal, or haplo-sterile, deletions.

Acknowledgements

For critical reading of the manuscript and advice: John Roote, Bruce Reed, Pascal Heitzler, and Elanore Whitfield.

References

1. Muller, H. J. (1932). In *Proceedings of the 6th International Congress of Genetics*, Vol. 1, pp. 213–55.
2. Lefevre, D. L. (1976). In *The genetics and biology of Drosophila* (ed. M. Ashburner and E. Novitski), Vol. 1a, pp. 31–6. Academic Press, New York.
3. Lindsley, D. L., Sandler, L., Baker, B. S., Carpenter, A. T. C., Denell, R. E., Hall, J. C., *et al.* (1972). *Genetics*, **71**, 157.
4. Ashburner, M. (1998). *Drosophila: a laboratory handbook*. Cold Spring Harbor Laboratory Press.
5. Craymer, L. (1981). *Genetics*, **99**, 75.
6. Cabell, J. B. (1919). *Jurgen*. Dover, New York.
7. Gubb, D., McGill, S., and Ashburner, M. (1988). *Genetics*, **119**, 377.
8. Gubb, D., Roote, J., Coulson, D., Henchcliffe, C., Lyttle, T., and Reed, B. (1992). *Drosophila Inf. Service*, **71**, 142.
9. Gubb, D., Roote, J., Trenear, J., Coulson, D., and Ashburner, M. (1997). *Genetics*, **146**, 919.
10. Reed, B. H. (1992). Ph. D. Thesis. University of Cambridge.
11. Ma, C., Liu, H., Zhou, Y., and Moses, K. (1996). *Genetics*, **142**, 1199.

5

Enhancer traps

C. J. O'KANE

1. Enhancer trapping—principles and uses

Enhancer trapping is widely used in *Drosophila* for generation of cell type markers that are often exquisitely specific, for identification of novel genes on the basis of expression patterns, and more recently, for the manipulation of defined cell types by allowing targeted expression of transgenes. While its application must of course be dictated by rigorous scientific considerations, there can be few experimental exercises that rival the fun of seeing new expression patterns that are natural works of art, and occasionally discovering a pattern that labels exactly those cells you always wanted to work with.

1.1 What are enhancer traps?

An enhancer trap construct carries a reporter gene fused to a weak or basal promoter, which on its own is insufficient to drive detectable expression of the reporter gene, but can respond to transcriptional enhancers that lie outside the construct. When a transgenic line is generated using such a construct, expression of the reporter gene is regulated by genomic enhancers in the vicinity of the integration site, and the spatial and temporal expression pattern in that line is characteristic of that integration site (*Figure 1*). Therefore, generation of many new transgenic lines produces a wide variety of new expression patterns which are easily visualized. The first enhancer trap experiments depended on injection of a *P* element construct into embryos (1), thus limiting the number of inserts that could conveniently be generated. However, the arrival of stable genomic sources of transposase (2) has made possible tens of thousands of new insertions of a single construct in a co-ordinated effort, simply by using standard crossing schemes (3–5).

The most widely used reporter gene is the *lacZ* gene from *Escherichia coli*, which encodes β-galactosidase, an enzyme that can be detected using either a chromogenic substrate, X-gal (5-bromo-4-chloro-3-indolyl-β-D-galactoside), or immunohistochemistry. The most widely used basal promoter has been that of the *P* transposable element. The sequences necessary for its function have not been precisely defined, but are likely to lie around the 5' end of the *P* element and the major transcription start site at +87 (6).

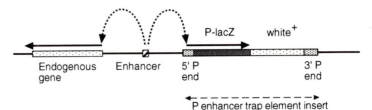

Figure 1. Principle of enhancer trapping. A *lacZ* reporter gene expressed from a minimal promoter on a *P* element can be controlled by enhancers that lie near to its insertion point. In most cases there is an endogenous *Drosophila* gene nearby which is also controlled by the same enhancers. In the scenario shown, insertion of the *P–lacZ* element does not inactivate the endogenous gene. However, insertion at other locations, e.g. in the endogenous gene, may lead to its inactivation.

1.2 Enhancer traps as cell markers

In *Drosophila*, most new enhancer trap insertions show some reporter gene expression. The expression patterns vary from being ubiquitous at a given stage of development, through patterns that are spatially regulated but widespread, to expression in a single cell type (*Figure 2*). Enhancer trap lines are therefore a useful way to generate cell markers, by screening new insertion lines to identify ones that express in the cells that one is interested in. Such markers are useful for several reasons:

(a) They allow cells to be followed throughout development, even when morphogenetic movements and changes in cell morphology might otherwise obscure this.

(b) Specific expression of a *P–lacZ* fusion in a given cell type early in development indicates that it already has a unique identity—even before it can be distinguished morphologically.

(c) They may allow novel cell types to be identified and molecularly defined.

(d) They allow the fates of specific cells to be followed in different mutant backgrounds.

Enhancer trap inserts can therefore be used in similar ways to other molecular markers such as antibodies to cellular proteins. They have the advantages that a greater variety of patterns can be generated more easily than by raising antibodies, and they also offer an easier route to cloning and mutating any nearby gene that has a similar expression pattern. One disadvantage is that a minimum of two generations is required to cross an enhancer trap insert into a homozygous mutant background (and more if the enhancer trap insert and the mutation of interest are linked).

1.3 Targeted expression with GAL4

It is now possible to drive expression of any cloned gene of interest (referred to hereafter as an effector gene) in the same expression pattern as an en-

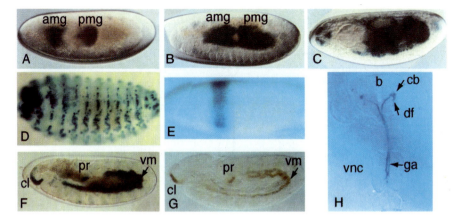

Figure 2. Examples of enhancer trap patterns and their uses. Except where otherwise noted, stainings were performed with X-gal. Except where otherwise noted, anterior is to the left, and dorsal on the top. (A, B, C) Insert *P{lArB}A490.2F3* (5), which expresses in a progression of stages in the embryonic midgut, allowing development of this tissue to be traced throughout development. (D) Insert *P{lac,ry⁺}A37* (79), which marks the embryonic peripheral nervous system. (E) Insert *P{lArB}A405.1M2* (5), which is inserted near the *spalt* gene and is expressed in a single band of cells at blastoderm stage; note the nuclear localization of the *P–lacZ* fusion product. (F) Insert *P{lArB}A183.1F2* (4), inserted in the 5' untranslated region of the *fasciclin III* gene. LacZ expression is seen at this stage in the visceral mesoderm, the clypeolabrum, and a small patch in the proctodeum. (G) For comparison, a wild-type embryo at about the same stage, stained with antibody against Fasciclin III protein. (H) A dissected adult nervous system, showing cytoplasmic β-galactosidase localization in the giant neurones, when expression of a UAS–*lacZ* fusion (see Section 4) is driven by P–GAL4 insert OK307 (31, 80). Staining can be seen in the cell body, the dendritic field, and the axon of each giant neurone. Anterior is to the top. Abbreviations: amg, anterior midgut; b, brain; cb, cell body of giant neurone; cl, clypeolabrum; df, dendritic field of giant neurone; ga, axon of giant neurone; pmg, posterior midgut; pr, proctodeum; vm, visceral mesoderm; vnc, ventral nerve cord.

hancer trap insertion. This approach uses an enhancer trap vector in which the reporter gene is not *lacZ*, but one encoding the yeast transcription factor GAL4 (7). A line showing tissue-specific expression of GAL4 can then be used to drive similar tissue-specific expression of any other effector gene that is fused to a promoter that carries a GAL4-dependent upstream activation sequence (UAS). This is usually done by crossing the appropriate *GAL4* line to a second line carrying the UAS–effector fusion, producing F1 individuals that carry both constructs and hence show the desired expression pattern of the effector (*Figure* 3). Hence, one *GAL4* line can easily be used to drive expression of any effector gene in the same cell type, with minimal effort. Furthermore, transgenic individuals with specific patterns of gene expression can be reproducibly generated, even when this expression pattern has a dominant lethal phenotype.

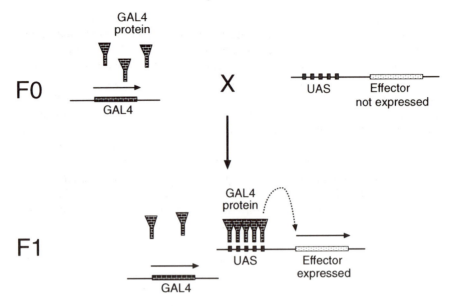

Figure 3. *GAL4* enhancer trapping. An effector gene and basal promoter are next to an upstream activation sequence (UAS) which contains several sites that can bind the yeast transcriptional activator GAL4. In a transgenic line that does not express GAL4 (F0, right-hand strain), there is no expression of the effector gene. If a line is available which expresses GAL4 in a tissue of interest (F0, left-hand strain), it can be crossed to the line carrying the UAS–effector gene fusion. In the F1 flies, which carry both constructs, GAL4 protein can bind to the UAS and activate expression of the effector gene.

1.4 Gene discovery using enhancer traps

As the enhancers that are detected by an enhancer trap insert are functional, there is often a nearby *Drosophila* gene with a similar expression pattern (*Figure 2*). Insertion of the enhancer trap element at many potential sites around the 5′ end of a gene may result in a *lacZ* expression pattern similar to that of the gene; *P* elements also show a strong preference for insertion near transcriptional regulatory elements, usually but not always near the 5′ end of the gene, and either upstream or downstream of the transcription start site. Hence, enhancer traps are not necessarily mutagenic, as structure or expression of the nearby gene is not always significantly disrupted. The *P–lacZ* fusion can be found in either orientation relative to that of the neighbouring gene (8–10).

What are the chances of finding a nearby transcript with an expression pattern similar to that of the insert? While the list of genes identified through enhancer trapping is long (see Section 2.1), it is also common to screen several kilobases around an insertion and not find such a transcript—or perhaps find a transcript with an entirely different expression pattern. One of the most systematic searches has been carried out by Ron Davis and colleagues (personal communication). Out of 15 enhancer trap lines that express *lacZ* in adult mushroom bodies, 14 had a nearby transcript that was also expressed there—although in about four cases the inserts were over 10 kb away (in one case clustered at 35 kb!) from the nearest exonic sequence, and so might have remained undetected in a more casual screen. In only one case was the hunt for a transcript unsuccessful (enhancer crap!). In short, increasing the effort expended in screening for nearby transcripts increases the chances of success —but also increases the stakes, in terms of lost effort, if one is unsuccessful!

An alternative approach to screening for transcripts is to screen for mutant phenotypes that affect the cells expressing the enhancer trap insert. While most insertions are not obviously mutagenic, deletions of flanking genes can be generated by imprecise excision of the insert (Section 7.3). Here, the success rate (albeit by a different criterion) is much lower. Of eight lines expressed in subsets of developing photoreceptors, only one gave a mutation with an obvious function in eye development (200 imprecise excisions were examined for each line; mitotic cell clones were made in the eye when the excision was homozygous lethal; M. Freeman, personal communication). However, the functional role of a nearby gene can be judged at an earlier stage, and it must be a decision for the investigator whether it is more important to screen with a lower success rate for an easily recognizable phenotype, or to screen with a higher success rate for a transcript whose function may be more difficult to ascertain.

While the expression pattern of an enhancer trap insert sometimes represents the expression pattern of a neighbouring gene accurately, e.g. *hedgehog* (11, 12), this is not always so. Expression of many genes is regulated by several separable enhancers, and the enhancer trap insert may bring the *P–lacZ* fusion under control of only some of these, e.g. an insertion near *Toll* which shows many features of its zygotic expression pattern but no maternal expression (5).

A further reason for caution is that some components of enhancer trap patterns may be due to regulatory elements within the vector itself. For example, *P–lacZ* insertions often show weak expression in the dorsal posterior epidermis of each embryonic segment, which is stronger in alternating segments. Most *P–GAL4* insertions are expressed to varying degrees in the salivary glands.

2. Generating new enhancer trap insertions

2.1 Target specificity

While a number of different promoters can act as enhancer traps to varying degrees, the *P* transposable element promoter has been the most widely used. Its major advantage is its proven worth in identifying many different expression patterns, cell types, and adjacent genes with expression patterns similar to that of the *P–lacZ* reporter. Its major limitation, as with any kind of *P* element mutagenesis, is that the target genes identified are clearly non-random. Loci with quite high insertion rates include *barren*, *couch potato*, *longitudinals lacking*, *mastermind*, *neuralized*, *rhomboid*, *rutabaga*, and *scabrous* (for allele data, see FlyBase and Encyclopaedia of *Drosophila*) (13, 14). Other target genes are hence less likely to be hit, and many genes are unlikely ever to be hit. The fraction of genes that can be identified by enhancer trapping is hard to guess, but as many enhancer trap insertions are not obviously mutagenic even when they are adjacent to essential genes, it is probably more than the estimated 50% (15) or so of genes that are sensitive to conventional *P* mutagenesis.

More recently, an enhancer trap vector based on the *hobo* transposable element has been developed, using the *hsp70* basal promoter (16). Like *P–lacZ* enhancer traps, this inserts at a wide range of sites and generates a wide variety of expression patterns, albeit with a lower frequency of novel tissue-specific staining patterns; some of these patterns also resemble the expression patterns of nearby genes. However, based on a sample size of 612 insertions, the *hobo* enhancer trap has a different distribution of insertion hot spots from that of *P* elements, making it a viable alternative for recovering patterns or target genes that occur only rarely, or never, using a *P–lacZ* vector. However, many laboratory strains may make *hobo* transposase, or have *hobo* elements that can be mobilized by it. Hence, it is necessary to use strains developed for the purpose of *hobo* mobilization (16), or to use molecular and genetic approaches to ensure that any strains used do not carry unmarked complete or defective *hobo* elements.

2.2 Choices of *P–lacZ* vector

Some of the commonest enhancer trap vectors are listed in *Table 1*. The vectors of choice now contain a replication origin and a selectable marker that are functional in bacteria, allowing rapid cloning of flanking sequences from many different insertion lines with minimal effort (see Section 6). In contrast, with the first published vector, *P[lac,ry$^+$]A*, DNA flanking the insertion site could only be cloned by either making a genomic library or by inverse PCR.

The two markers commonly used to follow enhancer trap vectors in *Drosophila* are the eye colour markers w^+ and ry^+. While both these markers are straightforward to use (even for a red-green colour blind person such as the author), the w^+ marker is easier to score. In addition, the w^+ gene fragment used in most vectors (a mini-*white* gene which lacks some intron sequences) usually leads to only low levels of eye pigment, sufficient for yellow, orange, or light red eyes; the darker eye pigmentation that results from multiple copies of mini-*white* allows more versatility when identifying flies carrying more than one copy of the element, e.g. when homozygosing novel insertions, or performing local transposition experiments. Curiously, mini-

Table 1. Some of the most widely used enhancer trap vectors[a]

Vector	Synonym	Fly marker	3′ rescue[b]	5′ rescue[c]	Bacterial marker	Ref.
P{A92}	P{lac,ry⁺]A	ry⁺	–	–	–	1
P{lArB}		ry⁺	2.9 kb	–	Amp	4, 5
P{lacW}		w⁺	1.8 kb	10.3 kb	Amp	3
P{PZ}		ry⁺	–	6.8 kb	Kan	13[d]
P{GawB}		w⁺	2.9 kb	11 kb	Amp	7
H{Lw2}	H{pHLw2]	w⁺	[e]	[e]	Kan	16

[a] The ability to perform plasmid rescue at either the 3′ or 5′ end of the insertion, and the bacterial marker used, are given. For sequences, restriction maps, and suitable enzymes for plasmid rescue, see the Transposons and Vectors section of FlyBase (13). Nomenclature is from FlyBase.
[b] Size of the plasmid sequence that lies between the two polylinkers after rescue. Another 233 bp of sequence from the 3′ end of the P element lies between the two polylinkers after rescue. Another 233 bp of sequence from the 3′ end of the P element lies to the right of polylinker B (*Figure 5*).
[c] Distance from site of enzyme used for rescue to 5′ end of P element. The size of plasmid sequence that lies between the two polylinkers is the same as for 3′ rescue, when this is possible.
[d] Mlodzik and Hiromi.
[e] *Hobo* enhancer trap. Various enzymes can be used for 5′ and 3′ plasmid rescue, not all in polylinkers.

white expression itself can cause males to court other males (17), emphasizing the need to control for effects that may have nothing to do with either the insertion site, or indeed any other cloned gene carried on the enhancer trap vector.

2.3 Generating new insertions

Provided that some insertions of an enhancer trap vector have already been generated by embryo injection, it is feasible to generate up to tens of thousands of new insertions, by crossing a strain carrying an existing insertion to a second strain that carries a stable transposase source. The progeny of such a cross are known colloquially as 'jumpstarter' flies (F0 in *Figure 4*), as mobilization of the insertion will occur in them; individuals carrying a novel insertion (transposants) can then be recovered in the *progeny* of the jump-starter flies (F1 in *Figure 4*). A straightforward mobilization scheme is described in *Figure 4*, although many variations on such schemes are possible.

The most common stable P transposase source is the P[ry⁺,Δ2–3]99B insertion on the third chromosome, which allows straightforward mobilization of any insertion on the first, second, or fourth chromosomes. For mobilization of insertions on the third chromosome, sources of P transposase on other chromosomes that remain stable during P mobilization have been generated by inserting a P transposase gene on a *hobo*-based transformation vector (B. Sanicola, M. Calvi, and W. B. Gelbart, personal communication).

Typically, about 1 in 100 of the progeny of jumpstarter flies can be identified as new transposants that carry neither the original insertion nor the transposase source. This frequency can be increased by increasing the copy number of the original insertion, as in the multi-lac strain. Even for the same construct, mobilization frequency is dependent on insertion site (3). If a large scale mobilization is planned, and several starting insertions are available, it is probably worth using each of them for a pilot mobilization before using the most easily mobilized one for the large scale experiment.

Most transposition in jumpstarter flies will occur pre-meiotically. Hence, multiple germ cells, and hence several F1 transposants, may arise which carry the same new insertion. To eliminate this, one can use only one transposant from each vial, having used only one or two jumpstarter parents to generate each vial. However, it is more convenient, both for setting-up these crosses and for scoring the progeny, to use mass crosses of jumpstarter flies and then collect multiple transposants from the bottles in which their progeny hatch. In practice, one usually settles for a compromise between independence and convenience, as the number of independent insertions is typically over 80% of the total number of transposants (5), and likely non-independent insertions can be identified by identical staining patterns, if transposants that arose from the same bottle are labelled as such in all subsequent experiments.

For most purposes, a reasonable compromise between independence and convenience is to use up to 20 jumpstarter males and 20 virgins in a bottle; transferring the flies to a fresh bottle once every few days will prevent the progeny from becoming overcrowded. However, these figures should be an upper limit: dividing the parents into as many separate jumpstarter crosses as possible will reduce subsequent problems with non-independent insertions. As most jumpstarter males, when crossed to two or three virgins, typically give rise to one or more transposant progeny (3, 5), such a scaled up cross should typically give rise to 10–20 new transposants—the exact figure will vary with parameters such as the construct mobilized, its original insertion site, and its copy number.

A common problem in enhancer trap screens is the recovery of mutations that are not caused by the *P* insertions. In several early screens most of these were due to second-site mutations present on the target chromosomes before mobilization. This source should be eliminated, at least when screening for inserts on a single chromosome, by prior isogenization of that chromosome (see *Figure 4*, and Chapter 1). However, isogenization of more than one chromosome at a time can in practice be difficult, e.g. if the balancing is not perfect, or the isogenic stocks are too weak. In the absence of isogenization, second-site mutations can be identified by their failure to revert to wild-type in crosses that should detect precise excision of the insert (Section 7.1).

However, there is a more insidious source of second-site mutations. Even in screens for new insertions on isogenized chromosomes (18), it is frequent to recover mutations that cannot be caused by the marked *P* insert, e.g. because they are homozygous lethal, but not lethal over a deficiency that covers the insert, or because they can be separated from the marked *P* insert by recombination. In some cases, these second-site mutations can even be reverted to

5: Enhancer traps

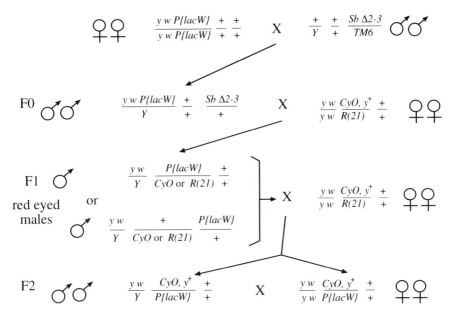

Figure 4. An example of a scheme, used by Török *et al.* (18) for generation of new enhancer trap inserts on the second chromosome, by using the *P{ry⁺,Δ2–3}*99B element (*Δ2–3*) to mobilize a *P{lacW}* element (or multiple elements, as in the Multilac strain) on the X chromosome. New insertions on any autosome can be identified as red eyed F1 males. Combinations of dominant markers and balancers other than those given can also be used. Use of a *CyO* balancer marked with *y⁺*, in a *y* background, simplifies identification of insertion homozygotes as late embryos and early larvae. *R(21)* is a dominant cold-sensitive allele of *Ketel* which is lethal at 18°C. While this marker is not essential, it simplifies establishment of stocks, as the only F2 survivors at 18°C are the ones shown, if the insert is on the second chromosome. It is advisable to have the same isogenized target chromosome in both the ammunition and the transposase-producing parental stocks before starting (see chapter on mutagenesis for examples of how to isogenize). *ry⁺* elements can also be mobilized by using a *ry* rather than a *w* background for all crosses. Simple variations on this scheme can generate inserts on other chromosomes. In the scheme shown, third or fourth chromosome inserts are also recovered in the F1, and can be inferred by the recovery of white eyed F2 flies (details of cross not shown). In this case, stable lines can be recovered if necessary by using standard balancer crosses, or identifying darker coloured eyes in the F3 if red eyed F2 flies are crossed to each other (discussed in Section 2.3), although these crosses are more laborious. Alternatively, dominant temperature-sensitive lethal mutations are also available on the third chromosome (*l(3)DTS* mutations), and could simplify the generation of third chromosome insertion stocks. If it is wished to screen for insertions on both main autosomes, these can be isogenized simultaneously by using a *white* stock with reciprocal translocation balancers such as *T(2;3)CyO-TM6*, although the resulting stocks may be weak. To generate insertions on all chromosome pairs, including the X, one can mobilize an ammunition element from a second chromosome balancer (5).

Table 2. The enhancer trapper's strain kit[a,b]

P transposase sources[c]

w; CyO/Sp; ry506 Sb P{ry+,Δ2-3}99B / TM6	Stable transposase insertion P{ry+,Δ2-3}99
w; CyO/Sp; ry506 Dr P{ry+,Δ2-3}99B / TM6	As above, but transposase tightly linked to Dr marker, which can be used to follow transposase even in non-balanced females
y w HoP6(w+)[d]	P transposase on a hobo element, X-linked Source: B. Calvi, M. Sanicola, and W. B. Gelbart
y w; Bc Elp / CyO, HoP1(w+)[d]	P transposase on a hobo element, on CyO Source: B. Calvi, M. Sanicola and W. B. Gelbart

Ammunition[e]

* P{lacW}	Many single P{lacW} inserts are in the Stocks section of FlyBase
*Multilac (y w, with four insertions of P{lacW})	A cytologically normal X chromosome. Source: D. Lindsley
* C(1)RM-Multilac	An attached-X version of the previous stock; doubles the copy number of P{lacW} (in females only); kept over / w1118 / Y or / 0 / C(1;Y); the compound XY chromosome should keep the stock more stable
* C(1;Y)-Multilac / C(1)RM	A compound XY version of the Multilac chromosome
* P{GawB}	Several P{GawB} inserts are listed in the Stocks section of FlyBase

Examples of w balancer stocks, for making homozygous or heterozygous balanced stocks

1st chromosome:	*FM7 series	Carry w[a]; Many FM7 stocks available over a lethal or an attached-X
2nd chromosome:	w; CyO/Sp	
	y w; CyO / R(21)	(18)
3rd chromosome:	w; TM3 / ry ftz e[s]	

2. Mobilizing a ry+ element

* CyO/Sp ; ry506 Sb P{ry+,Δ2-3}99B / TM6	A transposase source in a ry background
*P{lArB}, and other 'ammunition' elements	Many insertions of this and other ry+ elements are in Stock Centres

Examples of ry balancer stocks, for making homozygous or heterozygous balanced stocks

1st chromosome:	*FM6 series ; ry506	FM6 available over a lethal X chromosome or an attached-X
2nd chromosome:	CyO/Sp ; ry506	
3rd chromosome:	TM3, Sb ry / ftz ry e[s] (rf10)	

3. Fly stocks carrying GAL4-dependent reporter genes[f]

UAS–lacZ (5 GAL4 binding sites)	(7) Rabbit polyclonal from Cappel; monoclonal from Promega
UAS–lacZ (Fischer, 4 GAL4 binding sites)	(24)
UAS–nuclear–lacZ	(53)
UAS–btau	(51); Monoclonal available from Sigma
UAS–tau–lacZ	(50)
UAS–kinesin–lacZ	(48, 49)
UAS–IMPTNT (inactive tetanus toxin)	(47)
UAS–CD2	(27)
UAS–GFP	(54)
*UAS–GFP(S65T)	B. Dickson, documented in ref. 13
UAS–nuclearGFP–lacZ	(60)
UAS–tau–GFP	(55)
*UAS–yellow	(64)

4. Toxigenes

UAS–ricin	(41)
UAS–FRT–poly(A)–FRT–ricin	(31)
UAS–FRT–poly(A)–FRT–diphtheria toxin	(25)
UAS–reaper	(47a; A. Brand, unpublished; G. Halder, unpublished)
UAS–hid	(47a)
UAS–TNT (tetanus toxin)	(47)

5. Plasmids for use with the GAL4 system

pGaTB, pGaTN	For fusing promoters to GAL4, in BamHI and NotI sites respectively (55); Fusion must then be inserted in a P vector
pUAST	P vector, with polylinker for fusing genes to UAS element (55) Source: R. Kostriken (25)
pUAST derivative, 8 GAL4 binding sites	
pUFWT	P vector, with NotI site 3′ to a poly(A) site that can be removed by FLP recombination (31)

[a] Only strains for using P element constructs are listed. For mobilization of hobo (Section 2.1), special hobo-containing and hobo-free stocks must be used (16). Stocks marked with an asterisk are available at time of writing in Bloomington, Bowling Green, or Umeå Stock Centres.
[b] Balancer chromosomes are underlined; other chromosomes will not suppress recombination in females.
[c] Stocks given are representative examples that carry stable P{ry[+], Δ2–3}99B transposase. Other stocks available.
[d] Beware destabilizing these inserts with unwanted hobo transposase; see Section 2.1.
[e] Detailed Multilac genotypes can be found in the 'Stocks' section of FlyBase, by searching for 'ammunition'.
[f] Commercial sources of antibody listed, if known.

wild-type during crosses designed to detect precise excision (I. Kiss, personal communication)! Possible sources for such mutations are:

(a) Transposition followed by remobilization of the marked element, leaving a mutation at a site that no longer has an insert.

(b) An unmarked *P* element (e.g. a deletion derivative of the marked one).

(c) Parallel mobilization of one or more non-*P* transposable elements.

Recovery of second-site homozygous lethal and sterile mutations can be reduced by using the shade of eye pigmentation, rather than balancers, to identify flies homozygous for new inserts (if they are marked with w^+), as unrelated mutations on the same chromosome can be lost by recombination before the chromosome is homozygosed. However, some individuals in such a homozygous viable stock may still be heterozygous for a mutation linked to the *P* insertion; such mutations can be a nuisance if individual chromosomes are subsequently sampled from the stock, e.g. for imprecise excision experiments (Section 7.3).

3. LacZ enhancer traps as cell markers

Expression of a β-galactosidase reporter gene can be detected either by using the chromogenic substrate X-gal, or by immunohistochemistry. X-Gal staining is simple and rapid, and the low molecular weight of the substrate causes fewer complications with accessibility of cells in whole mount preparations. Antibody staining gives superior and more reproducible resolution at the cellular and subcellular levels, and avoids occasional complications caused by endogenous *Drosophila* β-galactosidase. The rest of this section is concerned with the methodology of X-gal staining. Procedures for immunohistochemistry are dealt with in Chapter 6; examples of β-galactosidase antibodies are given in *Table 2*.

Staining procedures usually require an initial fixation step. Formaldehyde and glutaraldehyde are acceptable fixatives, but overfixation reduces enzyme activity. Commercially available formaldehyde is adequate, but for maximum sensitivity it should be prepared free of methanol by dissolving paraformaldehyde. Methanol and ethanol should be avoided if possible, as they destroy LacZ activity; however, they may be used for a brief embryonic devitellinization step if required (*Protocol 1*).

After tissues are fixed and remaining fixative has been washed away, staining solution containing X-gal can be added. X-Gal must first be dissolved in an organic solvent such as dimethyl sulfoxide (DMSO) before being added to the aqueous staining solution; precautions must then be taken to minimize its precipitation, to avoid unsightly crystals adhering to the preparation. Staining solution also contains a mixture of potassium ferrocyanide and potassium ferricyanide as redox buffer. This ensures that the blue product of the β-galactosidase reaction remains insoluble and does not diffuse. The iron cyanides can be titrated out by tissue, thus reducing their effective concentration and allowing diffusion of stain. A concentration of 10 mM of each is adequate for most purposes and does not substantially reduce enzyme activity. If diffusion problems are still encountered they may be solved by using a larger volume of staining solution, or less tissue in the staining reaction.

Endogenous β-galactosidase activity is encountered principally in cells that form a stripe along the embryonic dorsal midline at stage 14 and then disperse throughout the interior of the embryo, in adepithelial cells that adhere to some imaginal disc preparations, and some of the gut at various stages of development. However, this enzyme has a pH optimum of about 5, compared to about pH 7 for the *E. coli* enzyme, and so staining at a higher pH can reduce its activity.

Fixative and staining solution must also have access to the tissues. In most embryonic stages this means removing the chorionic membrane, and either removing or permeabilizing the waxy vitelline membrane. From late embryonic stages into adult stages, the presence of a tough impermeable cuticle means that either tissues must be dissected, or holes made mechanically which are large enough to allow access of fix and stain to internal organs, e.g. with a needle or by sonication (19).

Mounting of stained tissue can be performed in several standard mountants. Glycerol-based mountants are straightforward to use, but preparations are more sensitive to wear and tear. Araldite is a more robust long-term mountant, although less simple to use. Initial clearing of whole mount preparations helps viewing of internal structures, whether stained or not, using Nomarski (DIC) optics.

An alternative to the whole mount procedures described here is to stain sectioned tissue. This may be useful if the cells or tissue of interest cannot easily be dissected, or stained using antibodies and examined by confocal microscopy. For examples of protocols, see refs 20 and 21.

Protocol 1. X-Gal staining of embryos[a]

Equipment and reagents

- P200 Gilson pipette
- Materials for embryo collection: healthy parents of the appropriate genotype, fresh yeasted apple or grape juice plates, laying cages to fit on plates, wash bottle, fine paintbrush, collection baskets
- 1.5 ml microcentrifuge tubes
- A rotating mixer which can hold 1.5 ml microcentrifuge tubes, e.g. blood tube rotator SB1 (Stuart Scientific BDH Cat. No. 403/0191/00)
- 70% ethanol and 100% ethanol, for ethanol series

- Microscope slides and coverslips
- Wash bottle with deionized or distilled water
- 20% formaldehyde stock solution: dissolve paraformaldehyde (Merck) in a beaker of water at 60°C on a hotplate in a fume-hood; when cool, store in 1 ml aliquots at –20°C
- Fix solution: 0.1 M Pipes (piperazine-N,N'-bis[2-ethanesulfonic acid]) pH 6.9, 2 mM EGTA, 1 mM $MgSO_4$, 8% formaldehyde (added by diluting formaldehyde stock solution)
- 80% ethanol, stored at –20°C

Protocol 1. *Continued*

- PBS: 10 mM sodium phosphate buffer pH 7.2, 150 mM NaCl—make up as a 10 × stock solution, and dilute to these concentrations before use
- PBT: PBS, 0.1% Triton X-100
- Staining buffer: 10 mM sodium phosphate buffer pH 7.2, 150 mM NaCl, 1 mM MgCl$_2$, 10 mM K$_4$[Fe^{II}(CN)$_6$], 10 mM K$_3$[Fe^{III}(CN)$_6$], 0.1% Triton X-100. Alternatively, use a higher pH (e.g. pH 7.8) if endogenous β-galactosidase activity needs to be minimized. This can keep for a few days in the dark at 4°C; longer-term storage should be in aliquots at –20°C. Immediately before use, pre-incubate 0.5 ml aliquots at 37°C.
- X-gal: stock solution should be 8% in dimethyl sulfoxide (DMSO), stored at –20°C in the dark—immediately before use, pre-incubate 12.5 μl aliquots at 37°C in the dark
- Gelvitol: 90 ml glycerol, 5 ml 1 M Tris–HCl pH 7.5, 5 ml water

- Chlorox/chlorine bleach solution: dilute commercial bleach to a concentration of 5% active chlorine immediately before use
- Heptane: immediately before fixation, prepare 1.5 ml microcentrifuge tubes containing 0.7 ml heptane and 0.35 ml of fix solution
- For mounting in gelvitol: flat slide holders
- For permanent mounting in Araldite: Histoclear (National Diagnostics)
- Araldite mountant: mix 1 ml Durcupan Resin (Bio-Rad), 1 ml DDSA (dodecenyl succinic anhydride; Bio-Rad), 40 μl DMP-30 (Bio-Rad), and 50 μl dibutyl phthalate (Sigma) rigorously with a spatula. Transfer 1 ml aliquots into 1.5 ml microcentrifuge tubes, spin for 5 min at 10 000 *g* (13 000 r.p.m. in a standard microcentrifuge) to remove all bubbles, and store aliquots at –20°C.

Method

1. Collect embryos of the appropriate age range from yeasted apple or grape juice plates, and wash into a mesh-bottomed basket. Embryos should not be older than about 18 h.

2. Dechorionate by dipping into 5% chlorox for 1–2 min, occasionally swirling gently. Embryos float to the surface and stick together when dechorionated.

3. Wash embryos thoroughly with water, and blot mesh dry from the bottom with tissue paper.

4. Wash the embryos from the mesh into a 1.5 ml microcentrifuge tube containing the heptane:fix mixture, using a Gilson pipette and the heptane:fix contents of the tube. If the embryos can form a single layer between the lower aqueous phase and the upper heptane phase, do not add any more!

5. Fix embryos for 10–20 min at room temperature, agitating continuously on a rotating mixer.

6. Meanwhile, mix pre-warmed aliquots of staining buffer (0.5 ml) and X-gal (12.5 μl), to make staining solution, and maintain at 37°C. At this temperature X-gal should remain soluble, but if a precipitate does form it may be redissolved by heating briefly to 65°C, or removed by a brief centrifugation in a microcentrifuge.

7. After fixation, remove as much heptane and fix as possible from the tube with a Gilson pipette.

8. Add 1 ml of cold 80% ethanol, and vortex vigorously for up to 30 sec. Devitellinized embryos will sink to the bottom.[b] Remove the 80% ethanol as soon as possible. Embryos retaining the vitelline mem-

brane will not sink, and can be removed with the 80% ethanol and discarded, or the ethanol can be removed, and the addition of cold 80% ethanol and vortexing can be repeated.

9. Immediately wash the embryos twice with 800 µl of PBT.

10. Resuspend embryos in 500 µl of pre-warmed staining solution (or less if there are few embryos).

11. Incubate at 37°C for a period varying from 5 min to overnight, depending on the level of β-galactosidase activity.

12. After staining, remove the staining solution, wash embryos with PBT, and remove excess PBT.

13. For gelvitol mounting:[c] add gelvitol to the tube containing the embryos, and allow the embryos to settle.

14. Pipette the embryos in a few drops of gelvitol onto the centre of a microscope slide, flanked by two coverslips on either side.

15. Cover the liquid with a third coverslip which is then supported by the two other coverslips on either side. If the liquid is not sufficient to fill up the space under the third coverslip after it has filled all the other gaps, it is possible to add more at the top or bottom edges of the coverslip, using a Gilson micropipette and taking care to avoid smearing it on the slide and coverslips.

16. Store the preparations flat in a protective slide holder. Embryos may be rolled by moving the top coverslip.

17. For a robust Araldite mount, continue from the end of step 12: wash embryos with 70% ethanol, and twice with 100% ethanol. Replace ethanol with Histoclear and allow embryos to settle.

18. Using a Gilson micropipette, transfer embryos onto the centre of a microscope slide in as little Histoclear as possible. Remove most of the Histoclear, but do not allow to dry completely; cover with Araldite and mount as in steps 14 and 15. Preparations can be viewed immediately, but are best left to harden (overnight at 60°C, or several days at room temperature).

19. As an alternative to mounting on a slide, embryos may be mounted between two coverslips, or may be sucked into a capillary tube to be viewed from any orientation about its longitudinal axis (22).

[a] This procedure includes treatment with cold 80% ethanol which removes the vitelline membrane. The ethanol step must be kept as short as possible to avoid inactivating β-galactosidase activity.
[b] Embryos devitellinized and stained in this way can also usually be stained using antibodies to other embryonic antigens (after step 11).
[c] A convenient way to mount the embryos is in a glycerol-based mountant, e.g. gelvitol; these preparations are not very robust, but with careful handling can last for years. A more permanent mount can be made using Araldite. Another common mountant, GMM (Gary's magic mountant), is not recommended for X-gal stained preparations because the stain slowly fades.

Protocol 2. X-Gal staining of third instar imaginal discs and brains[a]

Equipment and reagents

- Dissecting microscope
- Two pairs of dissection forceps (Type: 5SA), finely sharpened with a sharpening stone
- Original Arkansas sharpening stone (from any good hardware store)
- Siliconized slides, preferably with a single depression (BDH): to siliconize, wipe the surface with a tissue that has been immersed in dimethylchlorosilane solution in trichloroethane (BDH); siliconize in a fume-hood
- Microscope slides and coverslips
- For mounting in gelvitol: gelvitol and flat slide holders (see *Protocol 1*)
- For mounting in Araldite: Histoclear and Araldite mountant (see *Protocol 1*)

- Alternatively, a watchglass
- Ringer's solution: per litre, 7.5 g NaCl, 0.35 g KCl, 0.21 g $CaCl_2.2H_2O$, plus 50 ml of 1 M Tricin buffer pH 7.0—adjust pH to 7.1 with NaOH
- Alternatively, balanced salts solution (BSS): per litre, 3.2 g NaCl, 3.0 g KCl, 1.8 g $MgSO_4.7H_2O$, 0.69 g $CaCl_2.2H_2O$, 1.79 g Tricin, 3.6 g glucose, 17.1 g sucrose, 1.0 g bovine serum albumin—adjust pH to 6.95 with NaOH
- Staining buffer, X-gal, PBS (see *Protocol 1*)
- Fixative: 1% glutaraldehyde (made using Sigma grade 1 50% glutaraldehyde) in PBS
- 70% ethanol and 100% ethanol, for ethanol series

Method

1. Wash the larva briefly in PBS, and place it in a drop of Ringer's solution (or BSS) on a siliconized slide or a watchglass. BSS may allow survival of the cells for a longer period before fixation.

2. Sever the front third of the larva with forceps, and either turn it inside out, or carefully cut it open along its length and peel it open. The CNS, most of the discs, and most other internal organs will remain attached to the cuticle. Carefully remove as much as possible of any unwanted tissues, to avoid obscuring the tissues of interest.

3. If you wish to dissect more larvae, use forceps to carefully transfer the preparation to a drop of Ringer's solution or BSS in a microcentrifuge tube on ice,[b] and store for up to 30 min.

4. Prepare staining solution (*Protocol 1*).

5. Carefully remove the Ringer's solution or BSS from each preparation, add 100 μl fixative, and leave for 10–15 min.

6. Carefully remove the fixative, and wash the preparations with 100 μl PBS.

7. Add 100 μl of staining solution, and leave to stain at 37°C.

8. After staining, tissues may be dissected further in a drop of PBS on a siliconized slide or a watchglass, if this is necessary to spread them out further.

9. Wash preparations in 50% and 70% glycerol, and finally add gelvitol. Use forceps to transfer to a slide and mount in gelvitol; mount on a

depression slide covered with a coverslip, or on a normal slide using three coverslips as in *Protocol 1*.

[a] Using this procedure, it is possible to collect up to 10–15 dissected preparations over 30 min, and then fix and stain them together.
[b] For larger numbers of preparations, it may be more convenient to use microtitre plate wells rather than microcentrifuge tubes for fixation and staining.

Protocol 3.   X-Gal staining of ovaries

Equipment and reagents

- Dissecting microscope, two pairs of dissection forceps, and a sharpening stone (see *Protocol 2*)
- Microscope slides and coverslips
- Watchglass
- 1.5 ml microcentrifuge tubes
- Ringer's solution (or BSS), and fixative (1% glutaraldehyde in PBS) (see *Protocol 2*)
- Heptane
- Staining solution, X-gal, gelvitol (see *Protocol 1*)
- Buffer B: 100 mM KH_2PO_4/K_2HPO_4 pH 6.8, 450 mM KCl, 150 mM NaCl, 20 mM $MgCl_2$
- Devitellinizing buffer: 1 vol. buffer B, 2 vol. 20% formaldehyde (prepared as in *Protocol 1*), 3 vol. H_2O
- 70% ethanol, 100% ethanol

Method

1. Anaesthetize healthy egg-laying females of the desired age (e.g. four days), wash in ethanol, and dissect the ovaries in Ringer's or BSS in a watchglass.

2. Transfer the ovaries to a microcentrifuge tube containing 100 μl devitellinizing buffer and 600 μl heptane, and agitate gently for 10 min.

3. Carefully remove the liquid, add 500 μl fixation buffer, and leave for 10 min.

4. While ovaries are in devitellinizing buffer and fixative, prepare staining solution (see *Protocol 1*).

5. Carefully remove the fixative, and wash once with Ringer's solution.

6. Remove the Ringer's solution, add 500 μl staining solution, and leave to stain at 37 °C.

7. When staining is complete, rinse in PBS, and dehydrate in 70% ethanol for 2 min, and absolute ethanol for 2 min.

8. Transfer the ovaries to a microscope slide, add a few drops of gelvitol, and spread out the ovarioles using fine forceps or a needle.

9. Place two coverslips on either side of the ovary, and a third coverslip on top, resting on the two side coverslips.

Protocol 4.   X-Gal staining of adult whole bodies[a]

Equipment and reagents

- Dissecting microscope, two pairs of dissection forceps, and a sharpening stone (see *Protocol 2*)
- Dissecting dish: a 55 mm Petri dish, with a thin (about 5 mm) layer of Sylgard 182 Silicone Elastomer Base (Dow Corning) gel matrix on inner base
- Extremely fine dissection pins, e.g. 0.1 mm minutien pins No. 26002–10 (Interfocus Ltd, UK; Fine Science Tools Inc., USA)
- Post-fix decolorizing solution: 4% paraformaldehyde, 15% hydrogen peroxide (BDH Analar) in PBS, prepared freshly from stock solutions, and pre-cooled to 4°C
- Fixative (4% paraformaldehyde in PBS), PBS, PBT, staining solution (see *Protocol 1*)

- Fragments of microscope slide glass, roughly rectangular and about 20 × 10 mm. Prepare them by dividing up a slide with marks made by a diamond pen, and breaking the slide by a gentle tap (e.g. with the end of a spatula) downwards onto the opposite side of the slide. Take care to use suitable skin and eye protection against glass splinters.
- 1.5 ml microcentrifuge tubes
- 96-well microtitre plate
- Parafilm
- Ethanol: 30%, 50%, 70% (in distilled water), 100%
- Histoclear, and Araldite mountant (see *Protocol 1*)

A. *Adult dissection and staining[b]*

1. Kill and wash the fly by dipping it in ethanol for a few seconds.

2. Pin the fly, ventral surface upwards, to the bottom of a dissection dish containing fixative, by sticking a fine dissection pin through the centre of its abdomen. Make sure that the following dissection (steps 3–8) is completed within 5–10 min, to avoid overfixation.

3. Remove wings and legs close to the thorax, to allow penetration of fix and stain.

4. Gently use the forceps to extend the proboscis, without gripping it. Break open the cuticle joining the back of the proboscis and the head, by pinching it shallowly and pulling.

5. Remove the *large* silvery air-filled tracheae inside the head capsule. Do not attempt to remove any smaller tubular tracheae.

6. Use forceps to cut the thoracic cuticle laterally along the sides of the sternal plates, approximately between the bases of the first and second legs; take care not to insert the forceps too far into the thorax.

7. Carefully cut the loose patch of cuticle transversely, at its anterior and posterior edges, and peel it off. It may be necessary to cut any muscles anchored to it, to stop large chunks of muscle from being pulled out along with it.

8. After internal tissues have had access to fixative for 5–10 min, remove fixative, and rinse the preparation in two or three changes of PBT for 5 min.

9. Place the preparation into 70–100 μl of pre-warmed (37°C) staining solution (prepared as in *Protocol 1*) in a microtitre plate well.

10. Cover the microtitre plate with Parafilm and incubate at 37 °C until staining is complete. It may be convenient to keep the plate on a heating block (set at 40 °C) and covered with a box, while further fixed preparations are added to other wells.

B. *Clearing and mounting*

1. Bleach the cuticle and eye pigment by transferring the preparations to 1.5 ml microcentrifuge tubes containing 1 ml of cold post-fix decolorizing solution; leave them until all colour is lost (a few days at 4 °C or room temperature; higher temperatures are faster, but lead to internal bubbles).

2. Dehydrate the preparations in an ethanol series of 30%, 50%, 70%, and three steps of 100% ethanol; leave them for at least an hour at each ethanol concentration.

3. Clear the preparations for 30 min in 1 ml of Histoclear.

4. Transfer the preparations to a microscope slide in Histoclear, between two fragments of slide glass. Remove as much Histoclear as possible, but do not allow to dry, and cover preparations with Araldite. Cover with a coverslip.

5. Preparations may be viewed with care immediately, but the Araldite will harden for the next week and become more robust. Preparations should be left at room temperature until hard; higher temperatures cause bubbles in the carcass to expand.

[a] The views of internal stained structures in whole bodies are not as clear as in dissected organs, but this is a straightforward way to obtain an overview of adult expression patterns.
[b] The aim of the dissection is to allow access of reagents to internal tissue. If the incisions damage regions that are of interest (e.g. bases of legs or wings), alternative incisions can be investigated. Alternatively, only those parts of interest (e.g. heads) may be dissected and stained. Penetration is adequate for most of the head and body, but is variable for wings and legs.

Protocol 5. X-Gal staining of adult whole mount brains

Equipment and reagents

- Dissecting microscope, two pairs of dissection forceps, and a sharpening stone (see *Protocol 2*)
- Dissecting dish: a 55 mm Petri dish, with a thin (about 5 mm) layer of Sylgard 182 Silicone Elastomer Base (Dow Corning) gel matrix on inner base
- Extremely fine dissection pins, e.g. 0.1 mm minutien pins No. 26002–10 (Interfocus Ltd, UK; Fine Science Tools Inc., USA)
- 1.5 ml microcentrifuge tubes
- 96-well microtitre plate
- Fixative (4% paraformaldehyde in PBS), PBS, PBT, staining solution (see *Protocol 1*)
- Ethanol: 30%, 50%, 70% (in distilled water), 100%
- Histoclear, and Araldite mountant (see *Protocol 1*)

Protocol 5. *Continued*

Method

1. Pin the fly down in fixative in a dissecting dish (see *Protocol 4*).

2. Pull off the proboscis, cutting the oesophagus before the rest of the alimentary tract is pulled through the oesophageal foramen of the brain.

3. Peel off the head capsule, including the eyes, as if peeling the skin off an orange! Carefully remove tracheae that adhere to the brain.

4. If only the brain is required, use forceps to sever it cleanly at the cervical connective, which links it to the thorax, and proceed to step 9. If the thoracic ganglion is also required, proceed to step 5.

5. Use the forceps to make an incision in the cuticle all the way around the ventral thorax, as if using a can-opener. Ensure that the forceps do not penetrate too far into the thorax.

6. Peel off the cuticle, if necessary severing attached muscles, tracheae, and nerves, so that the thoracic ganglion is exposed.

7. Use the forceps to sever large nerves as far away from the ganglion as possible; insert an arm of the forceps underneath the ganglion and push it out of the thoracic cavity. Take extreme care to minimize contact between the forceps and the ganglion, and not to damage the cervical connective which links the brain and the thoracic ganglion. *Never* grasp any part of the CNS with forceps!

8. Carefully extricate the oesophagus from the oesophageal foramen, and remove any remaining adhering tracheae.

9. Lift the freed CNS, by resting it on top of the forceps, and place in PBT in either a microcentrifuge tube or a microtitre well.

10. Replace with 150–200 μl staining solution (prepared as in *Protocol 1*), and leave to stain at 37 °C.

11. After staining, dehydrate in an ethanol series of 30%, 50%, 70%, and two steps of 100% (about 20 min per step, carried out in the same tube or well); clear with Histoclear for 15 min.

12. Transfer the preparation to a microscope slide by lifting carefully on top of forceps, remove excess Histoclear (but do not allow to dry); cover in Araldite, and then with a coverslip.

13. Harden the preparation overnight at 60 °C.

4. Manipulating cells with *P–GAL4* enhancer traps

4.1 Components of the system

The *P–GAL4* enhancer trap system allows any cloned gene of interest (an effector gene) to be expressed specifically in the cells in which a *GAL4* insertion is expressed. GAL4-dependent expression of an effector gene normally requires only a simple cross, provided that a suitable *GAL4* line is available, and a transgenic line carrying a construct which has the effector gene under UAS control (*Figure 3*).

The number of characterized cell type-specific *P–GAL4* lines is growing, and more patterns can be identified either by screening existing collections of lines for expression at different stages of development, or by mobilization of an existing *P–GAL4* insertion. Alternatively, if a specific promoter is available, it can be cloned upstream of a promoterless *GAL4* gene, and the fusion introduced into flies by germline transformation; vectors available for this include pGaTB and pGaTN (7).

For expression of effector genes, the most widely used vector is pUAST; this carries a UAS with five GAL4 binding sites, and a basal promoter and transcription start site, upstream of a multiple cloning site, which is followed by a polyadenylation site (7).

4.2 Practical considerations

(a) As expression of *GAL4* can further amplify expression of the effector gene, expression of the latter is usually stronger than that of a reporter gene in a single-component enhancer trap. However, both approaches give a wide range of expression levels, and these ranges have a large overlap. While the high expression levels of GAL4-driven effector genes are not always a problem, care may sometimes be needed, e.g. a large amount of protein may overwhelm the mechanisms that cause its correct subcellular localization. An easy way to vary expression levels is to use several different insertions of the effector gene construct, whose expression levels differ. Expression of effector genes is also greater at higher culture temperatures (23). If this variation is insufficient, then using effector vectors with fewer (24) or more (25) than five GAL4 binding sites can also give different levels of expression from pUAST.

(b) Because of the need to express sufficient levels of GAL4 before the effector gene can be expressed, there is a delay in expression of the latter, of 30–90 minutes (7, 25–27).

(c) The basal promoter component of pUAST could act as an enhancer trap, and respond to flanking enhancers even when no *GAL4* is expressed, or even drive low ubiquitous levels of expression of the effector gene. Position effects on levels of GAL4-dependent effector gene expression are common. It is also important to control for GAL4-independent effects of an effector gene insertion, even if the levels of expression are lower than can be detected biochemically. However, even with genes encoding potent toxins such as ricin A chain or tetanus toxin light chain, if a few independent insertions are available it is usual to find some that have no obvious toxic effects when stably maintained in the absence of a *GAL4* gene. A more systematic way to reduce position effects would be to use a UAS vector in which the UAS–effector gene fusion is insulated from flanking regulatory sequences, e.g. by multiple binding sites for Suppressor of *Hairy wing* (su(Hw)) protein (28).

(d) In contrast to stable transgenic lines, transient expression of effector genes cloned in pUAST can occur during embryo injection, sometimes leading to difficulties in generating transgenic lines. In the cases of *wingless*, diphtheria toxin, and ricin A chain, this has been overcome by introducing a polyadenylation site, flanked by two FRT sites, between the promoter and the effector gene. Transgenic flies carrying this silenced construct are generated, and then the polyadenylation site is removed by FLP-mediated recombination (25, 29–32).

(e) A major limiting factor with the *P–GAL4* system is the specificity of patterns available. If a line is available which expresses in cells of interest, then it should be considered whether additional expression in other cells (which almost always occurs, to varying degrees) is relevant to any phenotype observed when an effector gene is expressed. For phenotypes that can be analysed at a cellular level, this consideration is usually easier to deal with, e.g. hindgut expression is unlikely to have much effect on early axonal pathfinding in the ventral nerve cord. However, for phenotypes that are analysed at the level of organ or tissue development, or even at the level of the whole organism, e.g. behaviour, this consideration is important. One way to approach the problem is to use several *P–GAL4* lines that express in the cells of interest, and which will usually have different 'secondary' sites of expression. Another is to use a specific promoter to drive *GAL4* expression if this is available, or can be isolated. In principle, mosaic analysis can also be used, e.g. using FLP catalysed recombination to remove a polyadenylation site from between a UAS element and an effector (25, 29–32). However, this may require examination of an impracticably large number of individuals in some circumstances.

(f) Usually, expression of one effector gene, e.g. *lacZ*, is a good indicator of the expression pattern of another effector gene insertion which is driven by the same *P–GAL4* insertion. However, it is best to be able to detect and localize effector gene expression directly. While *in situ* hybridization or immunohistochemistry using antibodies that recognize the effector gene product are obvious ways to do this, inclusion of an epitope tag (Section 5.5) should be considered when cloning an effector gene in pUAST. Such a tag is useful if no antibodies are available for the effector gene product, or to recognize expression of the effector gene product in a background of endogenous gene expression. One must of course check that tagged and non-tagged gene products are functionally equivalent, e.g. by checking whether both products can cause the same phenotypes when expressed under GAL4 control. *In situ* hybridization, using a probe which recognizes the 5' untranslated region that lies between the UAS-dependent promoter of pUAST and the site where the effector gene is inserted can also detect expression of an effector gene independently of its endogenous copy. O'Dell *et al.* (33) amplified such a fragment using primers pA(186) TTCAATTCAAACAAGCAAAG and pB(388) TATTCAGAGTTCTCTTCTTG on a pUAST template, to generate a 203 bp fragment extending from the putative transcriptional start site (185) to the multiple cloning site (399) where the effector gene is inserted. Single-stranded PCR using this amplified fragment as a template and only pA(186) as a primer generates a control sense probe; using only pB(388) as a primer generates an antisense probe which detects the transcript. As this probe can also hybridize to endogenous *hsp70* mRNA, it is advisable to hybridize it to a control strain which does not carry a pUAST-derived plasmid.

(g) A variety of problems includes the observation that some tissues may not readily express *GAL4* or effector genes (e.g. the female germline). Until the reason for this is known there is no alternative to making conventional promoter fusions to a gene of interest to express it in such a tissue. Also, some *GAL4* lines may show variegated or unreproducible expression in some cell types; however, a screen of enough *GAL4* lines will usually yield some that do not show variegation in the tissue of interest. Furthermore, in some *GAL4* lines, GAL4 expression may cause a mutant phenotype even in the absence of a UAS construct, presumably by transcriptional squelching (34). Appropriate control experiments must always be included to detect such an effect, and usually a straightforward solution is to use a different *GAL4* line which expresses in the same tissue. If squelching occurs, there will be selection for the fly stock to accumulate suppressing mutations that reduce GAL4 expression or activity; it may therefore sometimes be worth outcrossing an insertion of interest into a new genetic background if inconsistencies are found over time in the phenotypes that it can cause when it is used to drive effector gene expression.

The *GAL4* system does not yet generally allow a cell type-specific expression pattern to be further regulated temporally. However, the temperature-dependence of effector gene expression (see Section 4.2(a)) can sometimes allow a temporal regulation by temperature shifts (23). Heat shock of a *hsp70–GAL4* fusion (35) can also induce ubiquitous expression. Two 15 minute heat shocks at 37°C, separated by 30 minutes at room temperature, are sufficient to induce UAS–*lacZ* expression after a 60 minute recovery; however, levels of UAS–*cut* expression in this procedure were patchy and variable, and showed *cut*-dependent transformations of sensory organs in only a subset of segments (K. Blochlinger, personal communication). Different *hsp70–GAL4* insertions that differ in both basal and induced levels of expression, and different UAS–effector insertions may solve such problems to some extent. However, the extra step of GAL4 activation makes temporal control more ham-fisted than when the gene of interest is expressed directly from a heat shock-inducible promoter.

Eventually, temporal regulation of a cell type-specific effector gene pattern should become possible by using one of the inducible systems that has been used in mammalian or cell culture expression systems (36). Initial results suggest that a tetracycline-repressible transcription factor that has been used successfully in transgenic mice (37) can also be used successfully in *Drosophila* (B. Bello, D. Resendez, and W. J. Gehring, personal communication).

4.3 Altering and disabling cells

One extreme type of change which one may wish to bring about in a cell is to kill or disable it, in order to draw conclusions about its normal function. In *Drosophila*, cell killing has been achieved largely by laser ablation (38), chemical ablation of dividing cells (39, 40), or expression of gene products that cause cell death. Such gene products include toxins that inhibit protein synthesis such as ricin (31, 41) and diphtheria toxin (25, 42) A chains, and proteins that activate apoptosis such as *reaper* (43, 44), *grim* (45), and *hid* (46) products (*Table 2*). In addition, tetanus toxin light chain has been used; this abolishes synaptic vesicle release, but does not kill cells, and appears to be non-toxic to non-neuronal cells in *Drosophila* (47).

The considerations of unwanted GAL4-independent toxicity (Sections 4.2(c) and 4.2(d)) are relevant, and solutions such as an upstream polyadenylation site, removable by FLP recombination, have been useful in generating flies that have ricin or diphtheria toxin under GAL4 control (25, 31, 41). Transformants carrying ricin A chain (41), *reaper* (47a; G. Halder, personal communication; A. Brand, personal communication) and *hid* (47a) fused to a UAS element can be stably maintained, and have been used for embryonic ablation. Post-embryonic ablation is more problematic, because of the need to avoid death before the cells of interest can be ablated—this has been achieved by using cell type-specific promoters to drive apoptosis gene expression directly (43–46). General use of the *P–GAL4* system for post-embryonic ablation is difficult without the development of a temporally-inducible element (see Section 4.2). For some purposes, tetanus toxin light chain is useful for disabling post-embryonic cells; its lack of toxicity for non-neuronal cells allows survival of many toxin-expressing *GAL4* lines to adulthood. However, its mode of action makes it applicable to neurobiological rather than to general developmental problems.

Timing of ablation must also be considered. Evidence for cell death (e.g. acridine orange staining or TUNEL in the case of apoptosis genes, disappearance of cell-specific markers in the case of wild-type toxins) can be seen within about an hour of induction of their expression. However, a single UAS–*reaper* insert gives weaker phenotypes that a single UAS–ricin insert, but when multiple UAS–*reaper* inserts are used the phenotypes are similar (A. Brand, personal communication). Induction of ricin expression in glial cells at different times during embryogenesis gives different phenotypes, presumably reflecting dynamic changes in glial cell function (41). Hence, given the delay between expression of the toxic gene product and ablation, it is advisable to track the time course of ablation at a cellular level.

5. Alternative reporter genes

5.1 LacZ nuclear and cytoplasmic localization

The original *P–lacZ* enhancer trap vectors offered a *lacZ* gene fused to a fragment of the *P* transposase gene which encoded a nuclear localization signal. This fusion was convenient to use, had no apparent effect on cell physiology, and its nuclear localization facilitated counting of labelled cells. There is now a greater choice of reporter genes for visualizing different aspects of cell structure, particularly for use with the *GAL4* system. The standard UAS–*lacZ* strains encode a cytoplasmic β-galactosidase, which labels the major features, but not the finer details, of cell shape and axon/dendrite projection (*Figure 2*).

5.2 Visualizing cellular structure

The most widely applicable reporter genes are under UAS control (see *Table 2* for list, and examples of antibodies that can be used to detect them). Details of some of these, and of a growing number of other genes under UAS control, can be found in the Transposons section of FlyBase (13).

(a) Fine detail of cellular projections can be seen by using microtubule-associated reporters, such as a kinesin–lacZ fusion (48, 49), a tau–lacZ fusion (using bovine tau) (50), bovine tau itself (51), or human tau (G. Adam, documented in FlyBase) (13). The lacZ fusions of kinesin (48) and tau (A. Brand, personal communication) can cause severe cellular abnormalities; tau–GFP (Section 5.3) does not usually cause morphological abnormalities in neurones, although expression at high levels may interfere with cell division and cause lethality (J. Raff, A. Brand, personal communication).

(b) Inactive variants of tetanus toxin light chain label much fine detail of axons and synaptic boutons (47). To date, no detrimental effects of the inactive toxin variants have been described.

(c) Mammalian CD2 is a membrane protein, which in *Drosophila* also inserts into the plasma membrane, and can be used to visualize cell outlines (27, 52). In the few lines examined so far, it has no obvious effect on viability or cell morphology (N. H. Brown, personal communication).

(d) A nuclear UAS–*lacZ* is now also available (53).

5.3 *In vivo* reporting—green fluorescent protein

Green fluorescent protein from jellyfish can fluoresce without a requirement for any added substrate or cofactor. The gene encoding it can therefore be used as a reporter gene in live animals, hence allowing identification of specific genotypes before rather than after any subsequent experimental observations or manipulations. The lack of requirement for a staining reaction makes visualization of the expression pattern rapid. In addition to the standard uses of a reporter gene, balancer chromosomes that express GFP zygotically from an early stage would make genotype identification (and even sorting) possible at any stage of development—but in spite of a steady clamour on `bionet.drosophila`, no GFP balancers are available at time of writing.

The gene encoding native GFP from *Aequorea victoria* is available in *Drosophila* under UAS control (54). With a major excitation peak of 396 nm, and a minor one of 475 nm, Zeiss filter sets 18 (excitation 390–420 nm, emission 450 nm and above), 10 (FITC; excitation 450–490 nm, emission 515–565 nm), and 2 (DAPI; excitation 365 nm, emission 420 nm and above) can be used, in decreasing order of sensitivity (55). However, embryonic yolk and larval cuticle autofluoresce indistinguishably from GFP using filter set 10 (FITC). Using filter sets 2 (DAPI) and 18, autofluorescence is blue whereas GFP fluorescence is green. The basic features of wild-type GFP have been improved in several ways.

(a) The S65T variant of GFP, which fluoresces several times more brightly than wild-type GFP, and has a major excitation peak of 490 nm that makes it more suitable for an FITC filter set and most confocal lasers (56), is available under UAS control (B. Dickson, documented in ref. 13). Sensitivity of GFP(S65T) is also aided by autoxidation and activation of the chromophore being 80% complete within an hour, compared to four hours with wild-type GFP (56). The brighter fluorescence makes autofluorescence less of a problem, and some possible further solutions are discussed elsewhere (57). Other bright variants are available (58, 59) but have not yet been documented in *Drosophila*.

(b) Fusions of GFP to other protein domains can direct specific subcellular localization of GFP, e.g. tau–GFP for visualizing cellular projections (55), a nuclear GFP–lacZ fusion (60), and n–synaptobrevin–GFP(S65T) (M. Ramaswami, personal communication). Some of these fusions should be improved by the use of brighter GFP variants (A. Brand, personal communication).

(c) Variants of GFP have been developed which have different emission wavelengths. These allow double labelling procedures, e.g. to visualize different compartments of the same cells (61, 62).

5.4 Other *in vivo* reporter genes

Some *Drosophila* genes have also been used as reporter genes, at least in tissues where their activity can be assayed. First, out of about 1000 P{*lacW*} insertions examined, some seven showed patterned expression of the mini-*white* gene within the eye; most of these seven were in genes known to be involved in pattern formation (63). Secondly, a UAS–*yellow* construct has been used to screen *GAL4* lines for patterns that identify small regions or compartments of the adult epidermis (64).

5.5 Epitope tags

Epitope tags have a slightly different function from reporter genes, whose products should be physiologically inert—they may be used to detect expression of a transgene product that may well have a physiological effect. An epitope tag is designed not to affect function of the protein to which it is fused. Hence, such a tag may consist of a small number of amino acid residues (typically 10–20), added at one end of the protein—although GFP can also be used (Section 5.3). Nevertheless, it cannot be taken for granted that such an addition will not affect protein function, and appropriate controls using non-tagged proteins are advisable. Epitope tags used successfully in *Drosophila* immunohistochemistry are listed in *Table 3*. However, tissues can differ in cross-reactivity and it is advisable to check that an available antibody gives no background in any tissue(s) of interest before embarking on an involved scheme of cloning and injection.

6. You've trapped your enhancer—can you trap your gene?

6.1 Initial steps

The first step in the strategy should be to identify whether the enhancer trap insertion of interest is in a gene that has previously been mutated or cloned. To do this, it is normally necessary to map the insertion cytologically, by *in situ* hybridization to polytene chromosomes (65). If possible, chromosome squashes should be prepared from larvae of a homozygous strain of the insertion. The probe can be made from any DNA that is homologous to enhancer trap vector sequences, e.g. the vector itself, or a *ry* or *w* fragment where appropriate, an appropriate bacterial plasmid, or a *lacZ* fragment. Using *P* sequences as a probe (especially the ends of the element that are necessary in *cis* for transposition) may help reduce subsequent confusion by uncovering some of the second-site mutations that can occur (see Section 2.3). There may also be additional sites of hybridization if the probe includes *Drosophila* DNA from the vector, e.g. the *w* or *ry* loci, the *hsp70* locus (e.g. if the vector carries the *hsp70* polyadenylation site), or short sequences that originally

Table 3. Epitope tags[a]

Tag name	Sequence	Position	Antibody	Supplier	Reference
myc	QGTEQKLISEEDLN	C term	9E10	BabCO	81
FLAG	DYKDDDDK	N term	M5	Kodak	82
FLAG	DYKDDDDK	C term	M2	Kodak	J. Kiger, pers. comm.
Haemagglutinin	(YPY)DVPDYA	Internal	12CA5	Boehringer, BabCO	83

[a]See also discussion of GFP (Section 5.3). Some of these tags and antibodies are discussed in a bionet.drosophila article by Gerard Manning (posted on 2 October 1996, archived on FlyBase). Position of the tag within the protein may affect its immunoreactivity, but this has not been tested exhaustively for all tags and antibodies. Antibodies listed have not been tested for background in all *Drosophila* tissues.

flanked the *P* element used in vector construction (*w* in most cases). Such sites may be useful as positive controls for the *in situ* experiment, if there is any doubt about whether a particular spread may contain an insertion. *In situ* hybridization using an enhancer trap vector probe should also determine whether the strain has one or more insertions; multiple insertions do occur occasionally, especially with the higher mobilization frequency of the Multilac strain.

If the insertion is homozygous lethal, the insertion stock may only contain larvae in which the insert is heterozygous with a balancer chromosome. These larvae are unsuitable for cytological mapping, because of the disruption of chromosomal pairing and the complex inversion loops caused by the heterozygous balancer. Hence, heterozygous balancer stocks should be outcrossed to a cytologically wild-type stock for making chromosome spreads. Half the larval progeny of this cross will carry one copy of the insert, be cytologically normal, and can be used for *in situ* hybridization; the other half will carry one copy of the balancer and can usually be recognized by their failure to produce more than three recognizable chromosome arms in any spread, before they are actually used for an abortive *in situ* hybridization.

Knowing the map position of an insert identifies a wealth of information and stocks that are often essential to identify the gene adjacent to or affected by the insert. First, an insert that appears homozygous lethal, but viable over deficiencies from the region, means that the lethality is due to a second-site mutation on the same chromosome. If the insert itself does have a homozygous phenotype, proximity to mutant genes from the same vicinity can easily be determined by complementation tests. If it does not cause a homozygous phenotype, this is less straightforward, although imprecise excision (Section 7.3) can generate mutations in flanking genes that can then be used for complementation tests with existing mutations. However, the length of time that this procedure takes limits its use as a larger scale initial screening procedure, and it is of use mainly where there is reason to believe that an insert is close to a mutant gene of special interest, or where it is necessary to generate novel alleles at a later stage in the analysis of a gene.

There are two main ways to test for proximity of the insert to a cloned gene of interest. First, one can use the cloned gene to probe Southern blots of genomic DNA from the insertion strain and a control strain (i.e. one which carries the same parental chromosome and balancer chromosome, as appropriate), and determine whether any restriction fragments are altered in the insertion strain. Alternatively, one can clone DNA flanking the insert, and then use Southern blotting to check for hybridization to clones of interest that are known to map to the vicinity of the insert. DNA flanking the insertion can be cloned in three different ways.

(a) By using PCR primer information and existing P1 or cosmid clones, if the insertion has been used as an STS (sequence-tagged site) in genome mapping (see Section 6.3).

(b) By plasmid rescue, if the enhancer trap vector allows this. Flanking DNA can be cloned by plasmid rescue rapidly (within a few days of starting to make DNA from the insertion strain), and from several different insertions simultaneously (see Section 6.4).

(c) Exceptionally, by inverse PCR, if none of the above is possible (see *Protocol 7*).

6.2. Enhancer trap insertions and *Drosophila* genome maps

6.2.1 Your insertion site may already be characterized

A few large collections of enhancer trap insertions are being used by the Berkeley *Drosophila* Genome Project (BDGP) and the European *Drosophila* Genome Project (EDGP); these are principally insertions that cause homozygous lethality or sterility. As inserts from these collections have been distributed to many users, it is often the case that one may want to investigate an insert that is being characterized by these consortia. Several thousand inserts are being used for:

- cytological mapping
- complementation tests with other inserts and existing mutations in the region
- mapping relative to deficiency breakpoints by complementation tests
- cloning by plasmid rescue (see Section 6.4)
- sequencing of a few hundred base pairs of flanking DNA
- design of primers to allow specific amplification of a portion of flanking sequence (an STS—sequence-tagged site)
- localization of STSs to EDGP cosmid and BDGP P1 genomic maps

159

This information is available electronically from FlyBase (13), the Encyclopedia of *Drosophila* (14), BDGP (65a) and EDGP (65b). It makes it possible to:

(a) Localize the clone relative to other candidate genes, clones, and insertions that have been mapped on either the P1 or cosmid maps.

(b) Access a large amount of molecular and genetic information on the vicinity of the insertion.

(c) Obtain an existing P1 or cosmid genomic walk flanking the insertion site, which is often a prerequisite for identifying transcripts near the insertion.

6.2.2 Cloning DNA next to a 'genome project' insertion

i. Options

In principle, it should be possible to obtain plasmid rescued clones of such insertions from the BDGP or EDGP. However, at the time of writing, the organizational infrastructure is not in place to allow general distribution of these clones on request, leaving two possibilities:

(a) Use the P1 and cosmid clones, and any other clones from the region, as sources of genomic DNA. Use PCR to generate the relevant STS to allow more precise molecular mapping of the insertion on these clones, and then use the relevant clones for further work.

(b) Plasmid rescue the insert yourself (Section 6.3 and *Protocol 6*)! While this is not absolutely necessary if STS information is available and the first strategy is followed, it is still a convenient way to generate convenient genomic fragments from at least one end of the insertion.

ii. Verifying that cloned DNA is next to the insertion

It is advisable to confirm that an STS, derived from electronic information, is actually present next to the *P* insertion in your fly strain. The easiest way to do this is by PCR. DNA can be prepared from a single fly; for suitable procedures, see Chapter 11.

(a) Using the two primers that define the STS, and template DNA from either wild-type flies or flies carrying the insertion, will give a fragment whose size is defined in the relevant genome project.

(b) Using the right-hand primer from the STS, and a primer from the end of the *P* element used for plasmid rescue (or from the inverted repeat) that reads towards the flanking genomic DNA, will give a fragment that is slightly larger than the STS when used on template DNA from the insertion strain; this fragment should hybridize with a probe made from the STS. No such PCR product should appear when either primer is used alone, or when both primers are used with wild-type DNA template. A *P* element primer that we have used successfully for this approach is GGGACCACCTTATGTTATTTC, from the inverted repeat. While it sometimes gives spurious amplification products, the above control experiments should prevent these from confusing the results. Alternative primers can easily be designed from the sequence of the *P* vector used, which is usually available in the Transposons and Vectors section of FlyBase (13).

If additional confirmation is required of the veracity of an insertion, the STS can also be used as a probe for genomic Southerns, using the logic described in Section 6.3.2(c).

6.3 Plasmid rescue

6.3.1 Basics

Most enhancer trap insertions now carry a bacterial selectable marker and plasmid replication origin, flanked by restriction sites that allow flanking DNA to be cloned by plasmid rescue (*Figure 5*). In most cases, the proximity of these elements to one end of the vector make it straightforward to rescue DNA flanking that end. The size of the flanking DNA fragment rescued will depend entirely on the distance from the end of the insertion to the next site in the genome that can be cut by the restriction enzyme used for plasmid rescue. Fragments of at least 15 kb can be rescued routinely. While difficulties may be encountered occasionally if a site is further away than this, use of a second or third alternative enzyme will almost always yield an easily clonable fragment. Detailed restriction maps of most vectors are available in the Transposons and Vectors section of FlyBase (13).

Protocol 6. Plasmid rescue

Equipment and reagents

- Standard materials and equipment for *E. coli* transformation: should include competent cells (capable of at least 10^6 transformants per μg of plasmid vector), and plates containing medium with an appropriate selective agent for the rescued clones
- Appropriate restriction enzymes, DNA ligase, and reaction buffers

- Standard reagents for DNA extraction: 3 M sodium acetate pH 6.3, phenol buffered at pH 7.8–8.0 (e.g. Cat. No. 1690Q825 from Biosolve), chloroform:amyl alcohol (24:1), ethanol, 70% ethanol, TE (10 mM Tris–HCl, 1 mM EDTA pH 8.0), 0.5 M Na EDTA pH 8.0, microcentrifuge (e.g. MSE Micro Centaur)
- *Drosophila* genomic DNA[a]

Method

1. Digest 3–4 μg of genomic DNA (or the equivalent of 10–15 flies if DNA concentration cannot be easily measured) with an appropriate restriction enzyme in a volume of up to 30 μl.

2. Add 2 μl of 0.5 M EDTA, make up to 200 μl with TE, phenol extract and chloroform extract the DNA. Precipitate the DNA with 0.1 vol. 3 M sodium acetate and 2.5 vol. of ethanol.

3. Leave the precipitation at –70°C (or on dry ice) until the preparation is frozen, and centrifuge for 30 min at 4°C (10000 *g*; 13000 r.p.m. in a standard microcentrifuge).

Protocol 6. *Continued*

4. Wash the pellet in cold 70% ethanol, centrifuge again for 2 min, carefully decant the supernatant, and allow the pellet to dry.

5. Redissolve the pellet in 50 μl of TE.

6. Save 12 μl of the digest for gel electrophoresis; ligate[b] the remainder in a volume of 200 μl for at least 3 h at 14°C.

7. Stop the ligation reaction (using 5 μl of EDTA), purify and precipitate the DNA as in steps 2–4; redissolve the pellet in 15 μl of TE.

8. Compare undigested DNA (0.5–1.0 μg, or two to three fly equivalents), the 12 μl sample of digested DNA, and 5 μl of the religated DNA by gel electrophoresis, to confirm that digestion, religation, and DNA purification have been successful.

9. Use 5 μl of the ligation mix to transform competent *E. coli* cells, using a standard transformation procedure that can yield at least 10^6 transformants per μg of plasmid DNA. If the restriction enzyme used to digest the genomic DNA is suitable to allow plasmid rescue, this should result in anything from one up to a couple of dozen transformant colonies.

---

[a] Before plasmid rescue, genomic DNA must be prepared; for detailed protocols, see Chapter 11. To allow enough DNA for a few genomic Southerns as well, at least 100–200 flies (100–200 mg) should be used. A DNA extraction from isolated nuclei using proteinase K and Sarkosyl, followed by phenol extraction, and two ethanol precipitations gives DNA of adequate quality. If Southerns are unnecessary, then 10–20 flies are sufficient for plasmid rescue with a single restriction enzyme; a dozen rapid DNA extractions from this number of flies can be performed in parallel using a kit such as the Puregene™ Multiple Fly DNA Isolation Kit (Gentra Systems Inc.).
[b] Ligations are performed in a large volume to favour intramolecular over intermolecular ligation, thus minimizing the risk of also cloning unlinked DNA fragments.

Protocol 7. Inverse PCR[a]

Equipment and reagents

- Oligonucleotide primers: the sequences (capital letters only) come from the 3' fragment of the *P* element that is at the end of most vectors (nucleotides 2686–2907). The first, rightward facing primer is from nucleotides 2831 to 2849, and the second, leftward facing primer is from the complementary strand of nucleotides 2771 to 2752: 5' cgctctagaaTTCACTCGCACTTATTGCA 3' (58 bp from 3' end of *P*) and 5' ccagaattc-TAACCCTTAGCATGTCCGTG 3' (70 bp approx. to the most 3' polylinker).The sequences in lower case include restriction sites to facilitate cloning: *Xba*l and *Eco*RI sites in the first primer, and an *Eco*RI site in the second primer.

- *Drosophila* genomic DNA from the insertion strain: ideally, 4–8 μg per restriction enzyme used, prepared by any standard method. 1–2 μg is the absolute minimum, but this leaves no spare DNA for controls or for monitoring of reactions by electrophoresis.

- Appropriate restriction enzymes, DNA ligase, reaction buffers, and standard materials for DNA extraction, as in *Protocol 6*. Restriction enzymes with 4 bp recognition sites are preferable, although enzymes with 5 bp and 6 bp recognition sites may sometimes work.

- Thermostable DNA polymerase (e.g. *Taq* DNA polymerase) and buffer, and a PCR machine

Method[b]

1. Save 1 µg of genomic DNA for gel electrophoresis later. Digest 4–8 µg with an appropriate restriction enzyme. If less DNA is available, then scale down subsequent steps and omit the gel electrophoresis. If more DNA is available, the procedure may be scaled up, to allow more PCR reactions (e.g. if it is necessary to test different Mg^{2+} concentrations in the PCR).

2. Ethanol precipitate the digest, as in *Protocol 6*, steps 2–5, except that DNA should be finally dissolved in 80 µl of sterile water (not TE!). DNA dissolved in water should be kept on ice during use, and at –20°C for long-term storage.

3. Save 40 µl of the digested DNA for later use. Ligate the remainder and ethanol precipitate it as in *Protocol 6*, steps 6–7, except that DNA should be finally dissolved in 40 µl sterile water (not TE!).

4. Check that digestion, ligation, and DNA purification have worked, by examining 10–20 µl of digested DNA and religated DNA, and about 1 µg of undigested DNA by gel electrophoresis.

5. Calculate the volumes of reagents required for a PCR reaction with 10 µl of ligated DNA, and a control reaction with 10 µl of digested DNA. The reactions should contain 20–30 pmol of each primer, 0.2 mM of each dNTP, a standard PCR buffer, and a final Mg^{2+} concentration of 1.5 mM, made up to 50 µl with sterile water.

6. Mix the DNA and water in a 0.5 ml PCR tube, heat at 95°C for 5 min, and quickly chill on ice. Pulse spin in a microcentrifuge.

7. Add the rest of the PCR components as a master mix, and cover with 70 µl mineral oil (Sigma, Cat. No. 400–5).

8. Amplify for 35–40 cycles with 1 min denaturation at 95°C, 1 min annealing at 57°C, and 2 min extension at 72°C (10 min extension time in the final cycle); store at 4°C.

9. When the reaction is complete, separate the reaction mix from the oil by transferring the mix to a new tube. The mix should be stored at –20°C.

10. Check for reaction products by electrophoresing 10 µl of the reaction mix on a 1.5% agarose gel.

11. The remainder of the amplified product may be gel purified and sub-cloned, directly sequenced, or used as a probe to isolate overlapping genomic clones.

[a] This procedure is an adaptation of one obtained from Chiranjib DasGupta. Further procedures and information have also been documented by Jay Rehn (65a).
[b] If it is necessary to perform any troubleshooting, this may include using a different restriction enzyme, adjusting the final Mg^{2+} concentration in the PCR (typically adding between 0–5 mM Mg^{2+} to the reaction), or using a lower PCR annealing temperature of 55°C.

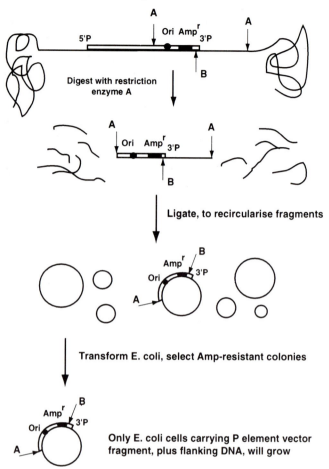

Figure 5. Cloning DNA flanking an insert, by plasmid rescue. Fly genomic DNA is digested with a restriction enzyme that cuts in polylinker A, but nowhere else between here and the 3' end of the insert. Many fragments are produced, including one containing the part of the *P* element with a plasmid replication origin (Ori), an antibiotic resistance gene (Amp^r), and some flanking genomic DNA (3' to the *P* insert in this case). Ligation of these restriction fragments produces circular molecules, and only the molecule that carries Ori, Amp^r, and flanking DNA will give rise to transformed *E. coli* cells that survive on selective antibiotic medium. If there is a second polylinker (B), sites in this are useful in restriction mapping the clones obtained (Section 6.3.2). Restriction enzymes with sites in polylinker B, but no other sites between here and the 5' end of the insert, may also be useful for plasmid rescue of DNA flanking the 5' end of the element.

6.3.2 Verifying plasmid rescued clones

Contaminants are of course a potential problem, especially when relatively few colonies are recovered. Several simple tests are available to verify any clones recovered. The first two tests are relatively rapid, but the latter two are more definitive, as they allow a direct comparison between the cloned DNA and the genomic DNA of the insertion strain; at least one of them should be performed before devoting extensive effort and resources to an insertion line.

(a) Simple restriction mapping. Single and double digests of the plasmid rescued clone with the restriction enzyme used for cloning from poly-linker A and an enzyme from polylinker B (see *Figure 5*) should give results that are partly predictable—a single fragment from the digest with the enzyme used for cloning, and two fragments (or more, if the genomic fragment also contains sites cleaved by the enzyme from polylinker B—this can be determined by a single digest with this enzyme) from the double digest, one of which should be the size of the plasmid vector that is bounded by polylinkers A and B.

(b) Hybridization to existing clones. The plasmid rescued clone can be hybridized to clones known to derive from the region. Obviously, these may include specific genes of interest, but cosmid and P1 clones from the fly genome projects are also of considerable use. If an enhancer trap insertion has been used as an STS, probing colony lifts of a few P1 or cos-mid clones from the same region can confirm the location of the plasmid rescued clone, if this has been used as an STS. Even with enhancer trap insertions that have not been used for STS generation, such hybridization will confirm that the plasmid rescued clone comes from the same cyto-logical location as the insertion, and open the door to all the materials and information generated by the fly genome projects.

(c) Genomic Southern blots of restriction digests of DNA from wild-type and insertion fly strains, probed with the rescued genomic fragment. The two strains should show differences in their restriction patterns which are to at least some extent predictable from the restriction map of the plasmid rescued clone. These digests should include the enzyme used for rescue, and one or two other enzymes that lie within the enhancer trap vector, close to the end which is cloned (e.g. those with sites in polylinker B, if an enzyme in polylinker A is used for rescue). If the two strains show no dif-ferences, then the putative plasmid rescued clone must be a contaminant.

(d) PCR of flanking DNA can be used if some sequence information is known. A single sequencing reaction on a plasmid rescued clone, using a primer from the relevant end of the *P* element, should give sufficient information to design a second primer specific for genomic sequence, which faces towards the *P* insert. Use of these two primers in PCR should give a specific amplification product when insertion strain genomic DNA is used as a template; this product should be absent when either primer alone is used, or when both primers are used on wild-type genomic DNA template, and should hybridize when probed with the cloned plasmid rescued fragment. The sequence data can also be used for homology searches, thus sometimes enabling a precise localization of the insertion site, or identification of a possible gene product.

Genomic Southerns that are probed with the plasmid component of the rescued clone are also advisable, as additional confirmation that the insertion strain carries only a single insertion. Genomic DNA should be digested with enzymes that can be used for plasmid rescue, as these give a unique plasmid-containing fragment for each insertion.

6.4 Detecting flanking transcripts

6.4.1 Isolating further flanking genomic sequence

Plasmid rescue normally allows cloning of DNA flanking one end of the enhancer trap insertion; however, a nearby transcript with a similar expression pattern can lie next to either end of the insertion and may not be detected when a plasmid rescued clone is used as a probe. Hence, it is necessary to isolate DNA from both ends of the insertion in order to screen the vicinity *systematically* for transcripts. This can be done by:

(a) Plasmid rescue of the other end of the *P* element (*sometimes*), if restriction sites in the vector and flanking DNA yield a fragment that contains genomic DNA and is small enough to clone.

(b) Identifying existing genomic clones, e.g. P1 or cosmid clones, by using the plasmid rescued clone as a probe in a library screen, colony hybridization, or Southern blot.

Restriction mapping of flanking DNA on both sides of the insert is normally a prerequisite for further analysis, and it is usually necessary to probe Southern blots of existing genomic clones to identify the restriction fragments that contain the plasmid rescued fragment. Lambda genomic clones may then be small enough to allow such restriction fragments to be used in some further experiments without the absolute necessity for further subcloning; however, cosmid and P1 restriction fragments will usually have to be subcloned to allow them to be produced in sufficient quantity and purity.

The high average gene density in *Drosophila* (usually one gene every 5–10 kb), and the bias to *P* element insertion in regulatory regions, means that genomic DNA that extends up to 10 kb in both directions is usually adequate to identify the closest transcript (but not always—see discussion of this issue in Section 1.4!). It is preferable to identify the closest transcript in each direction from the insert, unless other molecular and genetic data subsequently makes the evidence overwhelming that the single closest transcript is the one whose disruption is responsible for any mutant phenotype seen.

6.4.2 Detecting transcripts

When sufficient flanking genomic DNA has been isolated, several strategies are available to identify neighbouring transcripts. The fastest 'dry' approach is probably to sequence DNA directly flanking the P insert, on a plasmid rescued clone or inverse PCR product. This sequence can then be used to search any suitable site (e.g. 65a,65b) for *Drosophila* expressed sequence tags (ESTs) that overlap with the flanking genomic sequence. cDNA clones containing any EST identified can then be ordered from a standard supplier (65a). This approach is fast when it works, although coverage of expressed genes by the EST databases is not yet comprehensive enough to make it systematic, and it will not work when the P insert lies more than a few hundred base pairs from the transcript.

For an initial 'wet' screen of a genomic region, *in situ* hybridisation probably gives the best combination of convenience and reliability, alhough a reverse Northern may also be helpful initially if one wishes to identify or eliminate fragments from a large region.

i. Identifying repetitive DNA
It is necessary not to include repetitive sequences in any probe that will be used for transcriptional analysis. Such sequences are particularly common close to centromeric regions, and on the fourth chromosome. The quickest way to identify moderately or highly repetitive sequences in a genomic walk is to perform a 'reverse Southern', i.e. to label whole genomic DNA and use it to probe Southern blots of the cloned DNA. Cloned sequences that are over-represented in genomic DNA will hybridize most strongly. The more restriction enzymes that are used in the reverse Southern, the more precisely repetitive sequences can be defined.

ii. In situ hybridization
In situ hybridization to RNA is often a convenient way of screening for specific expression patterns of flanking genomic DNA. This has the advantage that one can immediately identify a transcript that has a similar expression pattern to that of the enhancer trap insertion, in the correct tissue or developmental stage. Protocols for *in situ* hybridization are discussed by Lehmann and Tautz (66).

Genomic fragments used for *in situ* hybridization should if possible not be larger than a couple of kilobases. Fragments larger than this risk either detecting expression of multiple, non-overlapping transcripts, or not including enough transcribed sequence to allow a good signal-to-noise ratio. Nevertheless, fewer than half a dozen overlapping restriction fragments allow 5–10 kb of genomic DNA to be scanned relatively rapidly for expressed sequences. It is also advisable to use single-strand probes, so that a complementary probe can act as an internal negative control if a putative signal is found.

iii. Reverse Northerns
Reverse Northern blots allow rapid screening of up to a few hundred kilobases of genomic DNA for transcribed restriction fragments in a single experiment, and may therefore be a good option for screening more than a few kilobases.

(a) Restriction fragments of genomic clones are separated by gel electrophoresis and transferred to a filter, as in a standard Southern blot procedure (67). Some fragments of DNA that are known to be transcribed at the stage(s) of interest should be included as a positive control.

(b) mRNA (i.e. poly(A) fraction) is prepared from the developmental stage(s) of interest. Many standard procedures and kits are available for this purpose, e.g. RNAgents total RNA isolation kit, followed by PolyA-Tract isolation of mRNA (both systems available from Promega).

(c) A labelled cDNA probe is prepared by reverse transcription of about 5 μg of whole mRNA, using random primers (67). Labelled nucleotides can contain either ^{32}P or digoxigenin (using the High Prime series of digoxigenin labelling and detection kits from Boehringer Mannheim).

(d) The filter is probed with labelled cDNA from the relevant developmental stage, to detect restriction fragments that are transcribed at that stage. Digoxigenin probe can be detected using CSPD as a chemiluminescent substrate.

Further analysis, e.g. *in situ* hybridization, Northern blotting, is then required to determine the number of transcripts and their expression patterns. Reverse Northerns may not detect low abundance transcripts, e.g. those expressed in a small number of cells. Hence, it is advisable to use for further analysis (e.g. *in situ* hybridization) not only the obviously transcribed fragments that lie closest to the insertion, but also any apparently untranscribed fragments that lie in between these and the insertion.

iv. Other molecular approaches

Northern blotting must be used for transcript analysis of flanking DNA at some stage. However, it is probably not the method of choice for initial screening. The spatial expression pattern is not detected directly, and if it is quite restricted, Northerns may not easily be made sensitive enough. Similar arguments on probe fragment size apply as with *in situ* hybridization (Section 6.4.2*ii*).

Again, it is usually necessary to use flanking DNA to screen libraries for cDNA clones at some stage. This can quickly yield information on the presence and structure of flanking transcripts. However, the fact that some clones can be under-represented in libraries because of their expression levels or their 'clonability', and that cDNA clones are not necessarily full-length, makes it unreliable as a sole or systematic way of screening for flanking transcripts.

v. Further analysis

Once a genomic region has been screened as systematically as possible, transcripts have been identified, and cDNA clones isolated, further molecular biological approaches must then be used to determine the precise structure of the gene and its transcript(s); these are beyond the scope of this chapter. In addition, it is necessary to determine whether a candidate gene identified is the one actually sought. An expression pattern resembling the enhancer trap is a good start, but if a mutant phenotype is being analysed, it may be necessary to put some work into determining whether the mutant phenotype is actually due to disruption of the gene that has been identified (see Section 7).

7. Generating a mutant

7.1 Has the insertion caused a mutation?

Typically, some 10% of insertions cause a phenotype that is homozygous lethal, sterile, or visible. As discussed in Section 2.3, it is however possible to recover phenotypes that are caused by second-site mutations on the same chromosome. It is vital to eliminate such mutations from further analysis.

(a) One can test for reversion of the phenotype to wild-type after precise excision of the element (*Figure 6*). Precise excisions require the insert to be heterozygous with a wild-type homologous chromosome in the jump-

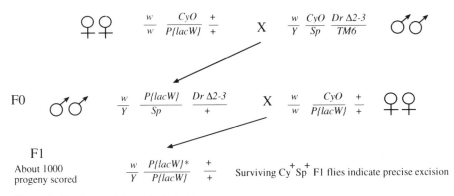

Figure 6. Precise excision of a lethal insertion on the second chromosome. In the F0, precise excision is favoured by picking P{lacW}/Sp rather than P{lacW}/CyO heterozygous balancer flies (Section 7.1). The excision chromosome is indicated by an asterisk in the F1 flies. If the insert gives a light red eye colour, most F1 flies that carry a putative precise excision should also have the light eye colour indicative of a single insert (the non-excised P{lacW} insert inherited from the female F0 parents). If the eye colour remains dark, then this indicates either that some rearrangements that leave the element inserted may no longer be lethal, that the element may have transposed elsewhere, or that the lethality was caused by an unmarked insert (either *P* or another transposon) on the second chromosome that has been fortuitously mobilized. Distinguishing between these possibilities requires more detailed molecular and genetic analysis of the revertants, which can easily be bred. An analogous scheme can be used for the third chromosome. However, precise excision of an X-linked insert is best performed in females, to provide a homologous template for correction of the excision (68). In this case, it is usually necessary to have the insertion heterozygous with an X chromosome balancer in the F0 jump-starter flies, to allow identification of excision chromosomes in the F1; if stable F1 revertants are required for further breeding, a third chromosome balancer is not usually required, given the low recombination between *Dr* and the Δ2–3 insert (about 1%).

starter fly. Having the insert heterozygous with a balancer reduces the frequency of precise excision; a homozygous or hemizygous insertion may not revert at all because of the lack of a wild-type template for correction of the gap that is left on excision (68). However, because of the reversion of some second-site mutations in these crosses (Section 2.3) this test alone is inadequate, although it is widespread in the literature. The rigour of this test can be increased by a perfect correlation between precise excision (assayed at the molecular level by PCR) and reversion of the phenotype, by checking for reversion by complementation with well mapped pre-existing alleles, or (in some cases) using eye colour as an indication of whether the insertion is still present (*Figure 6*).

(b) Deficiency mapping in as much detail as possible, or failure to complement other mutations that map nearby, can confirm that the lethality maps genetically to the expected cytological area.

(c) Second-site mutations will usually be separable from the marked insert by recombination.

7.2 Which gene is affected? Is the mutation a null?

If it can be shown satisfactorily that an insert has caused a mutation, it may be necessary to determine whether the mutation is a null allele. There are several approaches to this.

(a) A standard genetic test can be applied—null alleles (but also many partial loss-of-function alleles) have identical homozygous and hemizygous (e.g. heterozygous with a deficiency) phenotypes.

(b) The insertion site (e.g. in the middle of a conserved coding region) may make it very unlikely that the gene has any remaining function, although it is rare for the situation to be as clear-cut as this.

(c) *In situ* hybridization or immunohistochemistry may reveal whether detectable levels of gene product are expressed—although low levels that are still functional may escape detection, and non-functional gene products may still be detected.

(d) Most reliably, imprecise excision of flanking DNA (Section 7.3) can generate mutant lesions that cannot conceivably encode functional products, e.g. by lacking most of the coding region.

7.3 Generating (more) mutant alleles by imprecise excision

An initial insertion next to a gene of interest may well cause no detectable phenotype, or a phenotype which may not be null. In either case, imprecise excision of DNA that flanks the insertion can generate null alleles that can then be assayed for a phenotype (*Figure 7*). Excision events can be detected by loss of the w^+ or ry^+ marker of the insert. Typically, a few hundred excisions should be examined; this number should yield several lesions that affect flanking DNA within 2–3 kb of the insert; whether this is adequate will depend on the position of the insert relative to the flanking sequences that must be deleted. Imprecise excisions are favoured by the absence of a homologous wild-type chromosome that can pair with the insertion and act as a template for repair of the excision lesion, although they will also occur at an appreciable frequency even when a wild-type template is available (68–71). Disruption of pairing is therefore not absolutely essential, but it can be achieved conveniently by having the insert heterozygous over a balancer, or using only jumpstarter males for excision of an X-linked insert.

While it is common to screen excision chromosomes directly for a more extreme phenotype (e.g. for lethality, if the initial insert is viable), a PCR screening approach may be more informative. This makes no assumptions about the phenotype expected, and thus avoids the danger of only recovering larger lesions that include more than one gene, if mutations in the gene of interest turn out not to be lethal or not to have the phenotype expected. Direct sequencing of a novel PCR product also rapidly provides information on the molecular nature of the excision lesion. For deletions that remove the *P* element and some flanking DNA, one can use gene-specific primers that face each other and are separated by the *P* element. The original insert will usually not be amplified because of its large size; a wild-type chromosome will give an amplification product if the primers are not too far apart. Excisions that remove the *P* element and some flanking DNA (not including either of the primer binding sites) will give a new PCR product that is smaller than the wild-type product; usually the wild-type fragment will also be present, if the excision is heterozygous.

The presence of cryptic second-site mutations (Section 2.3) in some or all of the insert chromosomes can confuse the generation and interpretation of imprecise excision phenotypes. Hence, testing for a phenotype in flies that are heterozygous for the excisions and a deficiency of the region (or a relevant point mutation) is much preferable to testing homozygous excision flies. Also surprisingly, not only excisions of flanking DNA can cause a more extreme phenotype; some internal deletions of the *P* element can also do this (72).

A novel approach to selection of imprecise excisions has recently been developed (73). Male recombination between markers that flank a *P* insert is associated with a high frequency of deletions or duplications that extend in one direction from the insert (*Figure 8*). Usually, the *P* insert is retained, and deletions extend from the end of the element which has exchanged material with the homologous chromosome, ranging from a few nucleotides to cytologically visible deletions of over 100 kb. Screening of recombinants can be performed either by PCR (with one external primer and one *P* element primer) or genetically. The persistence of a *P* insert after deletion allows further deletions to be generated if a first round of deletions does not generate sufficient variety. Single male recombinants in the F1 are rare (< 1%), but

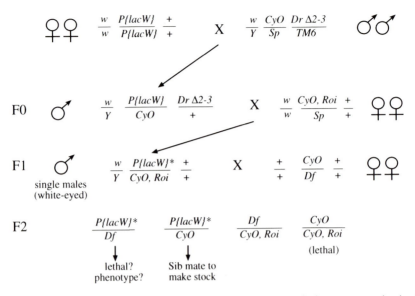

Figure 7. Imprecise excision of a viable insert on the second chromosome. In the F0 jumpstarter flies, imprecise excision is favoured by using P{lacW}/CyO balancer hetero-zygotes rather than P{lacW}/Sp flies. However, non-balancer flies also yield an acceptable frequency of imprecise excisions (Section 7.3). To recover independent excisions, it is best to set-up a couple of hundred individual F0 crosses, and to continue the scheme with one white eyed fly (which will carry some kind of excision event—denoted by an asterisk) from the F1 progeny of each cross. In the scheme shown, imprecise excision of a flanking gene is detected in the F2 by checking for a phenotype (e.g. lethality) in flies of genotype P{lacW}*/Df (i.e. Cy+ flies) that are hemizygous for the excision events. The heterozygous P{lacW}*/CyO F2 flies can then be used to establish a balanced stock; identification of these flies may be aided by use of a CyO, Roi balancer in the F0 female parents, if the deficiency chromosome is balanced over a simple CyO chromosome and has no other markers that can be followed. This scheme depends on availability of a deficiency that covers the insert (or of an existing allele of a flanking gene). If this is not available, then sib matings of P{lacW}*/CyO F2 flies can be used to generate homozygous excision flies in the F3 generation (not shown); however, this approach can inadvertently identify mutations caused by events other than simple imprecise excisions (see Sections 2.3 and 7.3). Alternatively, a PCR-based strategy can be used to screen for excisions. F1 flies may be screened by single fly PCR, but as this is destructive, it should only be done after they have successfully mated! F2 flies can be sacrificed for PCR if sibs of the appropriate genotype are still available for mating. Analogous schemes can be used for imprecise excisions of inserts on the X and third chromosomes. With third chromosome inserts, the jumpstarter flies will have the insert heterozygous with the Dr Δ2–3 chromosome, rather than with a balancer.

5: Enhancer traps

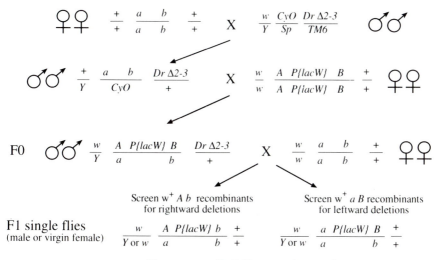

Cross to *w; CyO/Sp* to make stocks

Figure 8. Selecting unidirectional deletions flanking a second chromosome insert, by transposase-mediated male recombination (73). In the F0 jumpstarter flies in this scheme, the insert is shown flanked by two dominant markers *A* and *B* (usually wild-type alleles of *a* and *b* respectively), and over a chromosome that carries recessive mutant alleles *a* and *b* at the same flanking loci. The direction of the deletions can be chosen by selecting appropriate single recombinants in the F1. Female recombinants may show further homologous recombination between the insert and flanking markers, but this matters little as long as the insert can be followed. Establishment of stocks from single F1 recombinants is easiest if the F0 female parents are *w; a b*. Such flies can be made in parallel to the scheme shown, by crossing *a b* males to *w; CyO/Sp* females when setting-up the first cross, then crossing *w/Y; a b/Sp* male progeny and *w/+; a b/CyO* female progeny of this cross, to generate progeny which include *w; a b* females. If only *w+; a b* females are available as F0 parents, male F1 recombinants (which are *w+/Y*) can be crossed to *w; CyO/Sp* females, giving some red eyed male F2 progeny, which must carry the *P{lacW}* element and can therefore be used to make stocks that carry any flanking excision event. For an X-linked insert, F0 jumpstarter females must be used. It is then necessary to use tightly linked flanking markers, to minimize the proportion of recombinants that will arise from normal homologous recombination. For a third chromosome insert, the use of a third chromosome transposase source introduces some complications. First, it may be necessary to first recombine further markers onto either the insert chromosome or the *Dr Δ2-3* chromosome. Secondly, some classes of recombinant will still carry the *Δ2-3* insertion in the F1, allowing further unwanted mobilization events; the insert and the *Δ2-3* transposase source (or at least the *Dr* marker that is closely linked to it) must then be separated by recombination in F1 females. An alternative approach would be to use a transposase insertion on a different chromosome (e.g. *P* transposase on a *hobo* element; see *Table 2* and Section 2.3).

about a third carry a deletion, and so a relatively low number of individual lines must be bred to yield several deletions. In contrast, the standard imprecise excision approach (*Figure 7*) yields a high frequency of excisions in the F1; however flanking deletions are relatively rare and a few hundred lines must be bred to yield a few deletions, which probably involves more effort in the long run.

7.4 Which gene is responsible for the mutant phenotype?

In any molecular cloning, the source of much blood, sweat, tears, and sleepless nights, is the need to verify that a cloned gene indeed is the one responsible for a given mutant phenotype. This problem is reduced if the mutation is caused by a transposon insertion, but is not eliminated, particularly if there is one close candidate transcript on each side of the insertion. A detailed discussion of how this can be resolved in more difficult cases is beyond the scope of this chapter, but the following criteria are useful:

(a) The position of the insert, or of a deletion of flanking DNA, relative to each transcript, its exons and coding regions.

(b) The effect of the insert, or of a deletion of flanking DNA, on expression of each transcript.

(c) Other standard criteria, such as sequencing of additional point mutations, or rescue of the mutant phenotype in transgenic flies.

8. An enhancer trap miscellany

8.1 Modular misexpression

This approach is designed to identify and manipulate genes of interest, based not on their loss-of-function phenotypes, but on the phenotypes of their conditional over-expression in particular tissues (74). First, a 'pattern' line is chosen, which expresses GAL4 in a tissue of interest. Secondly, a *P* element construct, which carries a UAS-containing promoter which can read into adjacent genes, is mobilized to generate a large number of 'target' lines. Crossing the pattern line to a panel of target lines generates a panel of F1 flies that over-express a large set of genes in the tissue of interest. Genes whose over-expression in the GAL4-expressing cells causes a phenotype can thus easily be detected.

While the biological significance of dominant gain-of-function phenotypes must be treated with caution, they can be a useful way of assaying gene function, especially when the defect is specific and not generally deleterious. The modular misexpression screen can use any existing *GAL4* line, and target lines can be easily generated and used repeatedly for new screens.

8.2 Enhancer piracy

A developmentally or physiologically interesting gene is used instead of a relatively inert reporter gene in a vector that can function as an enhancer trap (75). In contrast to the modular misexpression approach, which detects the effects of a variety of genes in a single tissue, enhancer piracy identifies dominant effects of a single gene in a variety of tissues.

8.3 Stripe traps

Sometimes DNA within a *P* element vector can bias its specificity of insertion. Inclusion of a short fragment containing regulatory sequences from the *engrailed* gene causes one vector to act as an enhancer trap element with a preference for genes that are involved in segmentation or segment identity. This preference probably occurs because of 'pairing-sensitive' sites (which confer sensitivity to pairing on expression of nearby genes in the vector) in the *engrailed* regulatory sequence, which can direct the element to genomic locations where proteins that bind to such pairing-sensitive sites are already bound (76, 77). While the approach may not be widely applicable, it is certainly useful for identifying some groups of genes that have a common feature in their regulation.

8.4 Enhancer trap targeting

Any *P* element insertion can in principle be replaced by any other, when homology between a donor and a target insertion is used to insert a copy of the donor element upon excision of the target. This has been used to replace an unmarked *P* element in the *Broad-Complex* with a marked enhancer trap element at an identical location (78). It should thus be possible to introduce any other kind of sequence element (e.g. splice sites, reporter genes) at the location of any existing *P* element.

Acknowledgements

I am very grateful for help from Bruno Bello, Karen Blochlinger, Andrea Brand, Nick Brown, Ron Davis, Matt Freeman, Bill Gelbart, Georg Halder, Shigeo Hayashi, Istvan Kiss, Jordan Raff, and Mani Ramaswami, and for critical reading by several colleagues in my department.

References

1. O'Kane, C. J. and Gehring, W. J. (1987). *Proc. Natl. Acad. Sci. USA*, **84**, 9123.
2. Robertson, H. M., Preston, C. R., Phillis, R. W., Johnson-Schlitz, D. M., Benz, W. K., and Engels, W. R. (1988). *Genetics*, **118**, 461.
3. Bier, E., Vaessin, H., Shepherd, S., Lee, K., McCall, K., Barbel, S., et al. (1989). Genes Dev., 3, 1273.

4. Wilson, C., Pearson, R. K., Bellen, H. J., O'Kane, C. J., Grossniklaus, U., and Gehring, W. J. (1989). *Genes Dev.*, **3**, 1301.
5. Bellen, H. J., O'Kane, C. J., Wilson, C., Grossniklaus, U., Pearson, R. K., and Gehring, W. J. (1989). *Genes Dev.*, **3**, 1288.
6. Karess, R. and Rubin, G. M. (1984). *Cell*, **38**, 135.
7. Brand, A. H. and Perrimon, N. (1993). *Development*, **118**, 401.
8. Bellen, H. J., Vaessin, H., Bier, E., Kolodkin, A., D'Evelyn, D., Kooyer, S., et al. (1992). *Genetics*, **131**, 365.
9. Han, P.-L., Levin, L. R., Reed, R. R., and Davis, R. L. (1992). *Neuron*, **9**, 619.
10. Mlodzik, M., Baker, N. E., and Rubin, G. M. (1990). *Genes Dev.*, **4**, 1848.
11. Ma, C., Zhou, Y., Beachy, P. A., and Moses, K. (1993). *Cell*, **75**, 927.
12. Lee, J. J., von Kessler, D. P., Parks, S., and Beachy, P. A. (1992). *Cell*, **71**, 33.
13. FlyBase, http://flybase.bio.indiana.edu/ (1997).
14. Encyclopedia of Drosophila, http://shoofly.berkeley.edu/ (1996).
15. Kidwell, M. G. (1986). In Drosophila: a practical approach (ed. D. B. Roberts), p. 59. IRL, Oxford.
16. Smith, D., Wohlgemuth, J., Calvi, B. R., Franklin, I., and Gelbart, W. M. (1993). *Genetics*, **135**, 1063.
17. Zhang, S.-D. and Odenwald, W. F. (1995). *Proc. Natl. Acad. Sci. USA*, **92**, 5525.
18. Török, T., Tick, G., Alvarado, M., and Kiss, I. (1993). *Genetics*, **135**, 71.
19. Patel, N. H. (1994). In Drosophila melanogaster: practical uses in cell and molecular biology (ed. L. S. B. Goldstein and E. A. Fyrberg), p. 445. Academic Press, San Diego.
20. Han, P.-L., Meller, V., and Davis, R. L. (1996). *J. Neurobiol.*, **31**, 88.
21. Yang, M. Y., Armstrong, J. D., Vilinsky, I., Strausfeld, N. J., and Kaiser, K. (1995). *Neuron*, **15**, 45.
22. Prokop, A. and Technau, G. M. (1993). In Cellular interactions in development: a practical approach (ed. D. A. Hartley), p. 33. Oxford University Press, Oxford.
23. Speicher, S. A., Thomas, U., Hinz, U., and Knust, E. (1994). *Development*, **120**, 535.
24. Fischer, J. A., Giniger, E., Maniatis, T., and Ptashne, M. (1988). *Nature*, **332**, 853.
25. Lin, D. M., Auld, V. J., and Goodman, C. S. (1995). *Neuron*, **14**, 707.
26. Greig, S. and Akam, M. (1993). *Nature*, **362**, 630.
27. Dunin-Borkowski, O. and Brown, N. H. (1995). *Dev. Biol.*, **168**, 689.
28. Roseman, R. R., Pirrotta, V., and Geyer, P. K. (1993). *EMBO J.*, **12**, 435.
29. Wilder, E. L. and Perrimon, N. (1995). *Development*, **121**, 477.
30. Struhl, G. and Basler, K. (1993). *Cell*, **72**, 527.
31. Smith, H. K., Roberts, I. J. H., Allen, M. J., Connolly, J. B., Moffat, K. G., and O'Kane, C. J. (1996). *Dev. Genes Evol.*, **206**, 14.
32. Basler, K. and Struhl, G. (1994). *Nature*, **368**, 208.
33. O'Dell, K. M. C., Armstrong, J. D., Yang, M. Y., and Kaiser, K. (1995). *Neuron*, **15**, 55.
34. Gill, G. and Ptashne, M. (1988). *Nature*, **334**, 721.
35. Brand, A. H., Manoukian, A. S., and Perrimon, N. (1994). In Drosophila melanogaster: practical uses in cell and molecular biology (ed. L. S. B. Goldstein and E. A. Fyrberg), p. 635. Academic Press, San Diego.
36. Spencer, D. M. (1996). *Trends Genet.*, **12**, 181.
37. Gossen, M. and Bujard, H. (1992). *Proc. Natl. Acad. Sci. USA*, **89**, 5547.
38. Montell, D. J., Keshishian, H., and Spradling, A. C. (1991). *Science*, **254**, 290.

39. Broadie, K. S. and Bate, M. (1991). *Development*, **113**, 103.
40. DeBelle, J. S. and Heisenberg, M. (1994). *Science*, **263**, 692.
41. Hidalgo, A., Urban, J., and Brand, A. H. (1995). *Development*, **121**, 3703.
42. Kalb, J. M., DiBenedetto, A. J., and Wolfner, M. F. (1993). *Proc. Natl. Acad. Sci. USA*, **90**, 8093.
43. Hay, B. A., Wassarman, D. A., and Rubin, G. M. (1995). *Cell*, **83**, 1253.
44. White, K., Tahaoglu, E., and Steller, H. (1996). *Science*, **271**, 805.
45. Chen, P., Nordstrom, W., Gish, B., and Abrams, J. M. (1996). *Genes Dev.*, **10**, 1773.
46. Grether, M. E., Abrams, J. M., Agapite, J., White, K., and Steller, H. (1995). *Genes Dev.*, **9**, 1694.
47. Sweeney, S. T., Broadie, K., Keane, J., Niemann, H., and O'Kane, C. J. (1995). *Neuron*, **14**, 341.
47a. Zhou, L., Schnitzler, A., Agapite, J., Scwhartz, L. M., Steller, H., and Nambu, J. R. (1997). *Proc. Natl. Acad. Sci. USA*, **94**, 5131.
48. Ito, K., Urban, J., and Technau, G. M. (1995). *Roux's Arch. Dev. Biol.*, **204**, 284.
49. Giniger, E., Wells, W., Jan, L. Y., and Jan, Y. N. (1993). *Roux's Arch. Dev. Biol.*, **202**, 112.
50. Callahan, C. A. and Thomas, J. B. (1994). *Proc. Natl. Acad. Sci. USA*, **91**, 5972.
51. Ito, K., Sass, H., Urban, J., Hofbauer, A., and Schneuwly, S. (1997). *Cel Tissue Res.*, **290**, 1.
52. Dunin-Borkowski, O., Brown, N. H., and Bate, M. (1995). *Development*, **121**, 4183.
53. Doherty, D., Feger, G., Younger-Shepherd, S., Jan, L. Y., and Jan, Y. N. (1996). *Genes Dev.*, **10**, 421.
54. Yeh, E., Gustafson, K., and Boulianne, G. L. (1995). *Proc. Natl. Acad. Sci. USA*, **92**, 7036.
55. Brand, A. (1995). *Trends Genet.*, **11**, 324.
56. Heim, R., Cubitt, A. B., and Tsien, R. Y. (1995). *Nature*, **373**, 663.
57. Zylka, M. J. and Schnapp, B. J. (1996). *Biotechniques*, **21**, 220.
58. Crameri, A., Whitehorn, E. A., Tate, E., and Stemmer, W. P. C. (1996). *Nature Biotech.*, **14**, 315.
59. Cormack, B. P., Valdivia, R. H., and Falkow, S. (1996). *Gene*, **173**, 33.
60. Shiga, Y., Tanaka-Matakatsu, M., and Hayashi, S. (1996). *Dev. Growth Differ.*, **38**, 99.
61. Rizzuto, R., Brini, M., De Giorgi, F., Rossi, R., Heim, R., Tsien, R. Y., et al. (1996). *Curr. Biol.*, **6**, 183.
62. Heim, R. and Tsien, R. Y. (1996). *Curr. Biol.*, **6**, 178.
63. Bhojwani, J., Singh, A., Misquitta, L., Mishra, A., and Sinha, P. (1995). *Roux's Arch. Dev. Biol.*, **205**, 114.
64. Calleja, M., Moreno, E., Pelaz, S., and Morata, G. (1996). *Science*, **274**, 252.
65. Pardue, M.-L. (1994). In Drosophila melanogaster: practical uses in cell and molecular biology (ed. L. S. B. Goldstein and E. A. Fyrberg), p. 333. Academic Press, San Diego.
65a. Berkeley *Drosophila* Genome Project (personal communication). See http://fruitfly.berkeley.edu/ (1997).
65b. European *Drosophila* Genome Project. See http://croma.ebi.ac.uk/EDGP/ (1997).

66. Lehmann, R. and Tautz, D. (1994). In Drosophila melanogaster: practical uses in cell and molecular biology (ed. L. S. B. Goldstein and E. A. Fyrberg), p. 575. Academic Press, San Diego.
67. Sambrook, J., Fritsch, E. F., and Maniatis, T. (ed.) (1989). Molecular cloning: a laboratory manual, 2nd edn. Cold Spring Harbor Laboratory, Cold Spring Harbor, New York.
68. Engels, W. R., Johnson-Schlitz, D. M., Eggleston, W. B., and Sved, J. (1990). *Cell*, **62**, 515.
69. Klämbt, C., Glazer, L., and Shilo, B. Z. (1992). *Genes Dev.*, **6**, 1668.
70. Kretzschmar, D., Brunner, A., Wiersdorff, V., Pflugfelder, G. O., Heisenberg, M., and Schneuwly, S. (1992). *EMBO J.*, **11**, 2531.
71. Gailey, D. A., Taylor, B. J., and Hall, J. C. (1991). *Development*, **113**, 879.
72. Staveley, B. E., Heslip, T. R., Hodgetts, R. B., and Bell, J. B. (1995). *Genetics*, **139**, 1321.
73. Preston, C. R., Sved, J. A., and Engels, W. R. (1996). *Genetics*, **144**, 1623.
74. Rørth, P. (1996). *Proc. Natl. Acad. Sci. USA*, **93**, 12418.
75. Noll, R., Sturtevant, M. A., Gollapudi, R. R., and Bier, E. (1994). *Development*, **120**, 2329.
76. Kassis, J. A. (1994). *Genetics*, **136**, 1025.
77. Kassis, J. A., Noll, E., VanSickle, E. P., Odenwald, W., and Perrimon, N. (1992). *Proc. Natl. Acad. Sci. USA*, **89**, 1919.
78. Gonzy-Tréboul, G., Lepesant, J.-A., and Deutsch, J. (1995). *Genes Dev.*, **9**, 1137.
79. Ghysen, A. and O'Kane, C. (1989). *Development*, **105**, 35.
80. Phelan, P., Nakagawa, M., Wilkin, M. B., Moffat, K. G., O'Kane, C. J., Davies, J. A., et al. (1996). *J. Neurosci.*, **16**, 1101.
81. Xu, T. and Rubin, G. M. (1993). *Development*, **117**, 1223.
82. Afshar, K., Scholey, J., and Hawley, R. S. (1995). *J. Cell. Biol.*, **131**, 833.
83. Wolfgang, W. J., Roberts, I. J. H., Quan, F., O'Kane, C., and Forte, M. (1996). *Proc. Natl. Acad. Sci. USA*, **93**, 14542.

6

Looking at embryos

ERIC WIESCHAUS and CHRISTIANE NÜSSLEIN-VOLHARD

1. Introduction

In *Drosophila*, the pattern observed in mutant embryos often provides the first indication for the role of the wild-type gene in normal development. The loci which affect embryonic pattern, represent only a small fraction of the total genome and their phenotypes are often very discrete. With practice, new mutations in a given locus can be recognized strictly on abnormalities observed among the unhatched embryos from a mutagenized line. The easy recognizability of embryonic phenotypes greatly simplifies mutagenesis experiments and has proved very useful in subsequent analysis of the mutations obtained.

In this chapter we describe a number of very simple procedures for looking at *Drosophila* embryos. All these techniques are used on a regular basis in our laboratories and have been selected over the past several years because they provide the quickest, most direct characterization of mutant phenotypes. Probably the single most useful skill is the ability to recognize various stages and abnormalities directly in living embryos. In Sections 2–4 we describe techniques for collecting and looking at embryos, and provide a staging system which utilizes morphological features visible in eggs on agar plates under a stereomicroscope. In dealing with fixed material, our emphasis is on whole mount procedures for examining embryonic cuticle and/or internal structures. In addition to being relatively quick, whole mounts are extremely useful in that they provide the direct three-dimensional information necessary in analysing spatial patterns. At the opening of each section in the chapter, a brief description of the procedure is provided to orient the reader and indicate the goals of a particular technique. A more detailed discussion then follows.

2. Collecting embryos

Females are transferred to egg collection chambers which rest on Petri dishes containing an apple juice–agar egg medium. The type of egg collection chamber used in a given experiment depends on the number of flies and stocks from which eggs are to be collected. Change the plates at regular intervals and allow the eggs which have been laid on the plates to develop to the appropriate stage for examination.

2.1 Care of adults

Before transferring flies to egg collection chambers, keep females together with males in bottles with fresh yeast for at least one day. Alternatively, they can be mated and pre-fed in the collection container itself by adding excess yeast on the first day's collection and discarding the first plate.

Females begin producing eggs the second day after eclosion and when well fed and cared for will produce between 30 and 100 eggs a day over the two week period which follows. Females lay better in the dark and produce the largest number of eggs in the late afternoon and evening. When this makes it inconvenient to obtain sufficient eggs of the appropriate stages during normal working hours, it is possible to alter the laying preferences of females by keeping the flies on a shifted light-dark cycle.

2.2 Collecting large numbers of embryos from a single stock

When large numbers of embryos are needed from the same stock, we use disposable 100 ml plastic beakers (Tri-Pour® polypropylene) into which tiny holes have been burned with a hot needle. Alternatively, the bottom of any plastic beaker of the appropriate size can be replaced by a stainless steel screen, or a window covered with nylon mesh may be constructed. Shake the flies into the beaker using a funnel and place the Petri dish on top. Tri-Pour beakers have an inside diameter just the right size for a 60 ml Petri dish. A

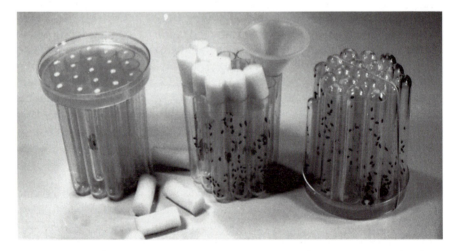

Figure 1. 'Blocks of plastic tubes used to collect eggs simultaneously from **19** different stocks. Prior to use, an apple juice–agar plate is yeasted in a pattern corresponding to the arrangement of the tubes in the block. Stocks to be tested are shaken into the appropriate tubes using a small funnel, and the tubes stoppered with foam plugs until all have been filled. After the stoppers have been removed and the Petri dish positioned over the tubes, the block is inverted. The flies in each tube lay their eggs in defined positions on the surface of agar.

slight ridge on the inside rim prevents the dish from falling into the beaker and also keeps the edge of the beaker out of contact with the yeasted apple juice plate. Use a rubber band or tape to hold the Petri dish in place, and invert the whole assembly so that the flies lay their eggs on a horizontal surface.

2.3 Collecting eggs from multiple stocks

When smaller numbers of embryos must be collected from a large number of different lines (as in a mutagenesis experiment), we use (*Protocol 2*) a 'block' of clear plastic tubes glued together in a defined pattern (1, 2) (*Figure 1* and *Protocol 1*).

Protocol 1. Construction of blocks[a]

Equipment an reagents

- Polystyrene tubes
- Rubber bands
- Sewing needle
- Needle holder
- Acetone

Method

1. Hold the appropriate number of tubes[b] together by rubber bands and invert on to a table-top so that the open ends of all tubes rest firmly on the surface.

2. Arrange the tubes in the correct pattern and check their alignment to make sure that all are exactly vertical.[c]

3. Glue the tubes together by dribbling one or two drops of acetone at each tube–tube interface. The acetone dissolves the surface of the adjacent polystyrene and fuses the tubes together.

4. Punch tiny holes[d] in the bottom of each tube after the block has solidified using a sewing needle mounted in a standard needle holder and heated with a Bunsen flame. Caution: carry out this part of the procedure in a fume-hood with maximum exhaust. The small amount of polystyrene adhering to the needle will burn and produce noxious fumes.[e]

[a] It is usually a good idea to make a dozen or so blocks at the same time. They can be washed and used over and over again. With care they will last four to five years.
[b] The size of the tubes used in making the block depends on the number of flies they are to contain. For egg collections from single females, we use 3 ml tubes from Sarstedt, Inc. which have a diameter of 10 mm and a height of 75 mm. 50 of these can be glued together in a block which fits on a 100 mm Petri dish. For collection of eggs from multiple (three to ten) females, nineteen 11.5 ml tubes are used, each with a diameter of 15.7 mm and a height of 100 mm.
[c] This is the most critical part of the procedure. If the base of the block is not perfectly flat before adding the acetone any irregularities will allow flies to escape (or move from one position to another) when the block is resting on the agar plate during egg collection.
[d] These holes should be large enough to allow moisture, but not flies, to escape during egg collection.
[e] A safer procedure for making holes is to have them drilled in the departmental workshop.

Protocol 2. Collecting eggs from multiple stocks

Equipment and reagents
- Collecting block (*Protocol 1*)
- Apple juice–agar plates
- Yeast
- Flies

Method

1. Shake flies from the different stocks into each tube using a small funnel and close the opening of the tube with foam rubber Figure 1.

2. Tap flies to the bottom of the tube when all the tubes have been filled and shake the flies down to the bottom by tapping the block on the table.

3. Remove the foam stoppers and place an apple juice–agar plate on the block. Prepare the plate in advance by dotting with yeast suspension in a pattern corresponding to the arrangement of the tubes in the block.

4. Position the plate over the block such that each yeast dot is centred over each tube.[a]

5. Hold the agar plate in place on the block with a rubber band.[b]

6. Invert the block with the plate and place in a tray or plastic box.

[a] Do not push the block into the agar or let it come in contact with the yeast. A dirty block encourages the flies to lay eggs on the side of the tube.
[b] If the tension of the rubber band is not uniform, the block will be pulled down into the agar on one side and flies will escape out of the other. A good way for testing the tension is to put the agar plate to your ear and pluck the rubber band on each side lightly (as though playing a guitar). The pitch should be as low as possible and the same on both sides.

2.4 Collecting eggs from single females

The procedures used to collect eggs from single females are essentially the same as for collecting eggs from stocks, except that the blocks consist of a larger number of tubes, each with a smaller diameter (see *Protocol 1*). For filling a large number of blocks with individual females, it is convenient to use ether rather than CO_2 to anaesthetize the flies.

2.5 Preparation of agar plates

Apple juice–agar plates are made according to the recipe described in Chapter 1. If the plates are kept at 4°C, warm them to room temperature before use, and dot or smear with live yeast suspension. The yeast is absolutely necessary as females stop laying eggs almost immediately when unyeasted plates are used. The plates may be used after the yeast has surface dried.

When collecting eggs from flies in blocks, yeast the plate in a pattern which corresponds to the position of the tubes in the block. One easy way of doing this is to place the Petri dish to be yeasted on an inverted block of the appropriate pattern (*Figure 1*). Because the plate is transparent, the pattern of the block can be seen through the agar and used to position the yeast drops. The yeast suspension should have the consistency of motor oil such that it forms stable drops about 2–4 mm in diameter.

The success of many experiments depends on the quality of the plates. If the agar is too soft, the flies will push their eggs into the agar. The plates should only be somewhat moist. If they are too wet, the flies will stick to the surface; if they are too dry, the eggs will be laid in the yeast. Dry plates may be moistened with 5% acetic acid. The amount of yeast added to the plate should be just enough that it will all be eaten by the time the plate is changed. Any yeast which remains on the plate obscures observation of the embryos. With experience one learns how much yeast to add for a given number of flies.

2.6 Collection intervals and maintaining plates

The maximum collection period should be shorter than the time required for embryonic development at that temperature (i.e. about 40 h at 18°C, about 22 h at 25°C, and about 18 h at 29°C). For more carefully timed collections, shorter intervals are used (1 h at 18°C, 30 min at 25°C, and 20 min at 29°C).

After the plate has been changed, add fresh yeast to the outer margins of the agar to attract the young hatched larvae from the central regions of the plate. This prevents them from carrying unhatched embryos from one position to another. In cases where large numbers of larvae are expected to hatch, remove the yeast and larvae from the margin of the plates using a spatula and add fresh yeast at 12–24 h intervals. Repeat this procedure until all normal embryos have hatched.

3. Looking at living embryos

Living embryos can be examined directly on the plates or transferred to a glass slide for viewing in the compound microscope. The embryos are enclosed in two protective coverings: the outer air-filled chorion and the inner vitelline membrane. Because the chorion is opaque, it must be removed or made transparent by covering with oil before the embryo can be observed.

3.1 Scoring late stages directly on plates under a stereomicroscope

After the hatched larvae have been removed, cover the embryos with paraffin oil and allow them to sit until the chorion becomes transparent (~ 30 sec). If the plates are too moist or the agar too soft, the oil will not penetrate the chorion. The problem can be partially remedied by allowing the plate to air dry uncovered in a 29°C incubator for several hours before adding the oil.

Most embryonic phenotypes can be recognized by viewing overall morphology in transmitted light at ×25 or ×40. The illumination should be well focused with little scattering, for example, the kind produced by the inexpensive Zeiss light boxes. Cuticle phenotypes and denticle bands are more visible in direct illumination or in transmitted light made more diffuse by putting a tissue under the Petri dish.

Estimate the percentage embryonic lethality from the fraction of the apparently developed eggs which fail to hatch. For more exact values count the empty chorions and classify the unhatched eggs according to phenotype. Decayed unfertilized eggs usually have a grainy, translucent appearance, as though the cytoplasm has separated into finely dispersed oil droplets and yolk. Based on this morphology, they can usually be distinguished from embryos which have died during early stages. Embryos blocked in pre-cellular stages often round up into a ball (e.g. the 'cannon-balls' obtained following massive irradiation of cleavage stages), cellularized and partially cellularized embryos usually retain some morphological indication of the stage when they died. Even after they have totally degenerated, the cytoplasm of early arrested embryos does not separate oil droplets and remains much more homogeneous (e.g. Nullo-X embryos).

3.2 Scoring early stages on plates under stereomicroscope; determining earliest phenotypic manifestations

Make egg collections for short intervals, usually of not more than 1 h, and for visibility the amount of yeast on the plate should be minimal. Scratch the surface of the plate with forceps to encourage flies to lay eggs. When the embryos have reached late syncytial stages (about 2 h at room temperature), cover the eggs with a thin, not too viscous, halocarbon oil (e.g. Voltalef 3S or Halocarbon Oil 27, Sigma) to make the chorion transparent. Do not use paraffin oil because it may cause abnormalities during early development.

Select 30–40 syncytial blastoderms and arrange them in rows of ten. Record the developmental stages of each embryo at regular intervals and note any developmental abnormalities. Record tentative designations of the genotype. When all embryos have been classified as either mutant or wild-type, return the plate to the incubator and allow the embryos to continue development for 24 h or until phenotypes can be scored unambiguously and the accuracy of the early designations assessed. Once the criteria for identifying mutant embryos has been established, it is usually no longer necessary to arrange embryos in rows or record data for each embryo. Instead, blastoderm embryos can be grouped together and as individuals approach the critical stage they can be examined in greater detail and classified as mutant or wild-type.

3.3 Photographing single embryos using the compound microscope

Eggs should be extremely clean and free of adherent yeast or agar. Such eggs can usually be found at the margins of the plate when egg collection intervals are kept very short. Once they have been immersed in Voltalef oil and their chorion is transparent, clean 'dirty' eggs by rolling them on a wet tissue paper and, after blotting to remove excess water, return them to oil. Alternatively, push them through a slit scratched into the agar. Transfer eggs to be photographed to a drop of halocarbon oil on a glass slide or 60 mm Petriperm plate. Petriperm® plates (Hereus) have bottoms made of an air-permeable membrane and thus allow better survival when using large coverslips. Although the plates are more expensive than glass slides, they can be used over and over again. One disadvantage is that the air-permeable membrane is birifringent and cannot be used with Nomarski optics.

Use two fragments of a no. 1 coverslip as supports on either side of the egg or embryo and cover with an 18 mm no. 1 coverslip or, when normal slides rather than Petriperm plates are used, cover the embryo with a coverslip strip of about 4 mm width. Nudging the coverslip causes the embryo to roll between it and the glass slide. Although this can be used to orient the embryo, excessive rolling should be avoided since the embryo slips inside the vitelline membrane and gastrulation can become crooked. It is better to orient the embryo first and then add the coverslip very carefully.

For somewhat better resolution remove the chorion by treating the eggs for 2–5 min with dilute bleach solution. The procedure also removes any adherent dirt or yeast. It has the major disadvantage that dechorionated eggs are more sensitive to damage and are difficult to handle. Removal of the chorion eliminates several useful landmarks for orienting the egg. We have found dechorionation useful only in eggs intended for surface views.

Although eggs thus prepared can be examined using phase-contrast or Nomarski optics, almost all our photographs are made using bright-field illumination and a Zeiss Planapo-chromat ×10 objective. Contrast is controlled by closing the iris diaphragm built into the condenser. If the light source also has a built-in diaphragm in the base of the microscope, better resolution is obtained if the diameter of the beam entering the condenser is reduced such that in the plane of focus it is only slightly larger than the embryo.

3.4 Videotaping living embryos

Essentially the same procedures are used to mount embryos for videotaping as for photomicrographs. The video camera should have a high resolution (~ 800 lines at the centre) and operate with low light intensities (e.g. E. R. Newvicon Cameras, No. WV-1850 from Panasonic). Unfortunately, the time lapse recorders which are commercially available have relatively low resolution (~ 300 lines at centre) and the play-back image is never as sharp as the original. A useful feature on any recorder is a 'date/time generator' which records the time directly on the tape while the embryos are being filmed. The time labels thus produced on each embryo are useful for bookkeeping, as well as in experiments where the duration of individual events must be determined. Other useful features included in many tape decks (e.g. Panasonic 6010 or 8050) are mechanisms to freeze individual frames for detailed analysis, a single frame advance, and a search function which allows viewing the tape at the 'fast forward' and 'rewind' speeds.

3.5 Klarsicht

The resolution of most post-blastoderm stages can be greatly improved by introducing the *klarsicht* mutation into the stocks from which films are being made. *klarsicht is* a maternal effect mutation on the third chromosome (3- 0, Nüsslein-Volhard, unpublished data) which has no effect on embryonic viability but enhances the contrast between yolk and cytoplasm in post-blastoderm stages. Although the mutation is not well understood, the phenotype seems to arise from the reduced number of lipid droplets which are incorporated into the basal side of blastoderm cells during cellularization (Wieschaus and Sweeton, unpublished observations, see also *Figures 4, 6*, and *8*).

4. Embryonic stages and morphological criteria for identifying approximate ages of living embryos

The staging system outlined below is primarily useful for work with living embryos in stages after blastoderm formation (stages 6–16). Approximate timings at room temperature (22 °C) are given in the parentheses following the stage designation. The numbering of the stages and the morphological criteria for their identification are the same as those suggested by Campos-Ortega and Hartenstein (3) in their monograph on *Drosophila* development, except that those authors add a stage 17 corresponding to the final differentiated embryo. The cleavage stages in our scheme (i.e. stages 1–4) are intended to supplement the more detailed system (1–14) developed by Zalokar and Erk (4) as modified by Foe and Alberts (5) for early embryos.

Photographs of wild-type and *klarsicht* embryos at each stage are provided in *Figures 2–8*. The individual stages are summarized in the schematic diagrams of *Figure 9*. Numerous descriptions of the general course of *Drosophila* development have been published (3, 6–9), as well as more detailed studies of the early cleavage stages (4, 5, 10–16), gastrulation (16–20), neurulation (21), and subsequent morphogenesis (22, 23).

4.1 Stage 1: freshly laid egg

(0–15 min) The cytoplasm of the embryo is homogeneous.

4.2 Stage 2: early cleavage

(15 min–1 h 20 min) At the posterior end of the embryo, the egg contracts away from the vitelline membrane. A cap of clear cytoplasm becomes visible

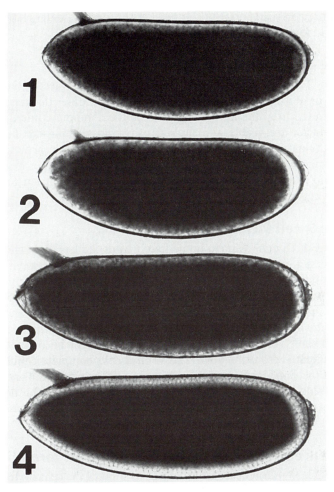

Figure 2. Wild-type embryos during cleavage stages of *Drosophila* development.
Stage 1: *freshly laid egg* (0–15 min).
Stage 2: *early cleavage* (15 min–1 h 20 min).
Stage 3: *pole cell formation* (1 h 20 min–1 h 30 min).
Stage 4: *syncytial blastoderm* (1 h 30 min–2 h 30 min).

at the posterior pole. This 'polar plasm' thins out during each cleavage cycle and forms again during interphase. The remainder of the embryo is covered by a very thin irregular layer of clear cytoplasm.

4.3 Stage 3: pole cell formation

(1 h 20 min–1 h 30 min) The layer of cytoplasm on the surface of the embryo becomes thicker and inhomogeneous due to the migration of the nuclei into the cortex. At the posterior pole, the cap of polar cytoplasm thins out and

becomes irregularly shaped. Buds appear at the surface, enlarge, and finally pinch off and form pole cells.

4.4 Stage 4: syncytial blastoderm

(1 h 30 min–2 h 30 min) The syncytial nuclei at the surface of the embryo divide four or more times. The cortical cytoplasm becomes thicker and more clearly delimited from the underlying yolk. In the compound microscope the individual nuclei are visible and their density and packing can be used to determine the particular cleavage cycle. As the nuclei enter each mitosis, the nuclear membrane breaks down, the cortical cytoplasm becomes more homogeneous, and then a wave-like contraction pulls the cytoplasm to the two poles of the egg and then back again.

4.5 Stage 5: cell formation

(2 h 30 min–3 h 15 min) Almost immediately after the last cleavage division, the nuclei begin elongation. Cell membranes move down between adjacent nuclei, partitioning them into individual cells. The advancing front of this infolding is visible as a line cutting through the nuclei and running parallel to the surface of the egg. Initially, the membrane advances slowly and reaches the base of the nuclei only after about 30 min. In the 10 min which follow, the cells double their volume and the blastoderm cell layer becomes thick and clearly demarcated from the yolk. The ventral cells complete cellularization earlier than the dorsal cells and the clear zone underlying the blastoderm at the ventral side disappears. The lack of synchrony between the dorsal and ventral sides means that there is never a uniform 'cellular blastoderm' stage in *Drosophila*.

4.6 Stage 6: early gastrulation, ventral furrow formation

(3 h 15 min–3 h 35 min) Gastrulation begins as soon as the cells on the ventral side of the embryo have completed cellularization. A ventral furrow forms as a longitudinal cleft along the ventral midline of embryo between 20–80% egg length. The cells immediately anterior and posterior to the furrow become thin. A posterior plate forms carrying pole cells towards the dorsal side of the embryo. The cephalic furrow becomes apparent as an oblique lateral infolding at two-thirds egg length. The dorsal cell layer is initially very thick and columnar at this stage. As the pole cells shift dorsally, the anterior and posterior dorsal folds begin to form.

4.7 Stage 7: midgut invaginations

(3 h 35 min–3 h 45 min) The cephalic furrow has deepened and is easily visible when viewed from the side. The posterior midgut plate is parallel to the long axis of the egg and thus has shifted the pole cells to the dorsal side of the embryo. The anterior midgut invagination can be identified at the anterior

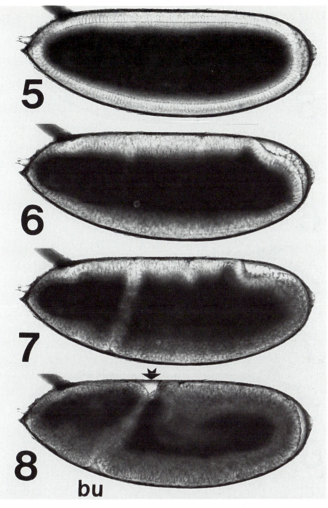

Figure 3. Wild-type embryos during cellularization and gastrulation.
mid-Stage 5: *cell formation* (2 h 30 min–3 h 15 min).
mid-Stage 6: *early gastrulation, ventral furrow formation* (3 h 15 min–3 h 35 min).
early Stage 7: *midgut invaginations* (3 h 35 min–3 h 45 min).
mid-Stage 8: *germ band extension* (3 h 45 min–4 h 30 min), note the characteristic buckle in the germ band on the ventral side immediately behind the head fold (bu). The indentation observed on the dorsal side of this embryo (—) is a variable feature and frequently totally obscured by the folds of the amnion and serosa (see *Figures 4* and *6*).

end of the ventral furrow. The end of this stage is characterized by the disappearance of the pole cells into the posterior midgut invagination at 30% egg length. The cephalic furrow is tilted to form a steeper angle with the longitudinal egg axis. The posterior dorsal fold is very deep and curves laterally towards the posterior tip of the embryo.

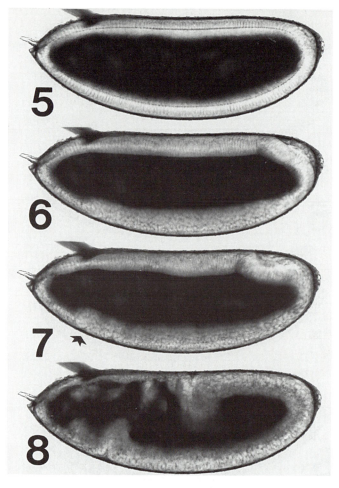

Figure 4. *klarsicht* embryos during cellularization and gastrulation.
mid-Stage 5: *cell formation* (2 h 30 min–3 h 15 min).
mid-Stage 6: *early gastrulation, ventral furrow formation* (3 h 15 min–3 h 35 min).
Stage 7: *midgut invaginations* (3 h 35 min–3 h 45 min), the *arrow* (→) indicates the anterior midgut invagination.
early Stage 8: *germ band extension* (3 h 45 min–4 h 30 min).

4.8 Stage 8: germ band extension

(3 h 45 min–4 h 30 min) The cell layers invaginated in the ventral furrow flatten along the ectoderm forming a multilayered band (= the germ band) which curves around the posterior tip of the egg, elongating along the dorsal side until the point of posterior midgut invagination reaches the head region at 65% egg length. The cells between the opening of the posterior midgut invagination and the cephalic furrow thin out such that the dorsal

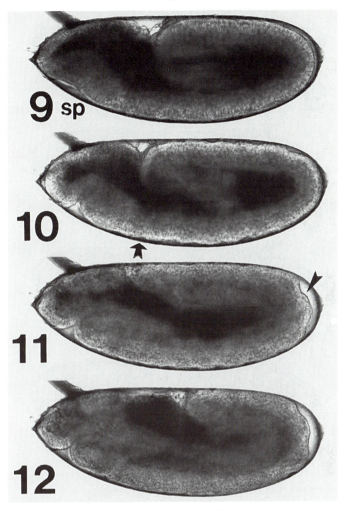

Figure 5. Wild-type embryos during extended germ band stages.

Stage 9: *stomodeal plate formation* (4 h 30 min–5 h 10 min), the shallow depression on the ventral side indicates the site of the future stomodeum (sp).

Stage 10: *stomodeal invagination* (5 h 10 min–6 h 50 min), the *arrow* (→) indicates the posterior-most extent of the anterior midgut anlage.

Stage 11: *three-layered germ band* (6 h 50 min–9 h), note the distinct space between the embryo and the vitelline membrane at the posterior pole (→).

Stage 12: *shortening of germ band* (9 h–10 h 30 min).

folds gradually disappear. The cephalic folds gradually disappear. The cephalic furrow is no longer easily visible. On the ventral side of the embryo, the germ band buckles into the interior of the embryo at the level of the cephalic furrow producing a transient gap between it and the vitelline membrane.

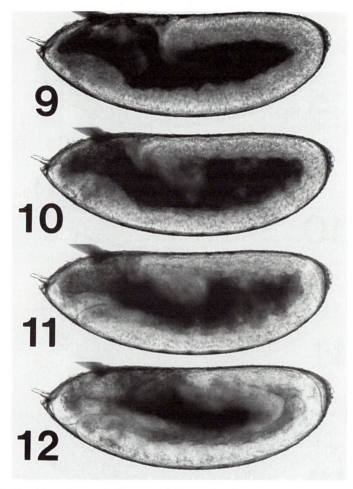

Figure 6. *klarsicht* embryos during extended germ band stages.
Stage 9: *stomodeal plate formation* (4 h 30 min–5 h 10 min).
Stage 10: *stomodeal invagination* (5 h 10 min–6 h 50 min).
Stage 11: *three-layered germ band* (6 h 50 min–9 h).
Stage 12: *shortening of germ band* (9 h–10 h 30 min).

4.9 Stage 9: stomodeal plate formation

(4 h 30 min–5 h 10 min) The gut opening has reached the head region at two-thirds egg length and the cephalic furrow is no longer visible. The dorsal head region is very thin, not clearly distinct from the yolk. The amnion and serosa are fully formed. The germ band appears continuous with the anterior midgut anlage, the ventral buckle and the anterior midgut opening observed in stage 8 are no longer distinct. The anterior midgut anlage appears as a character-istic broadening at the anterior–ventral part of the germ band. The future

192

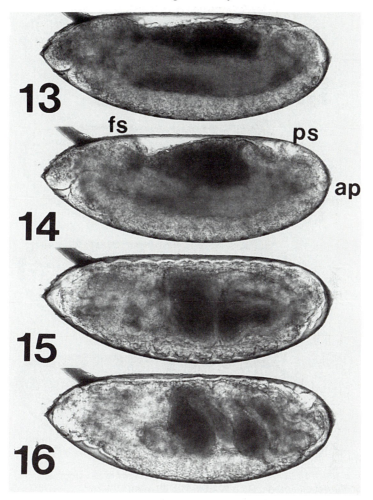

Figure 7. Wild-type post-shortening stages in *Drosophila* embryonic development.
Stage 13: *shortened embryo* (10 h 30 min–11 h 30 min).
Stage 14: *head involutiona and dorsal closure* (11 h 30 min–13 h), the frontal sac (fs), anal pads (ap), and posterior spiracls (ps) are apparent.
Stage 15: *dorsal closure complete* 913 h–15 h).
Stage 16: *condensation of CNS* (15 h to the completion of embryonic development).

stomodeal invagination is identifiable as a shallow gap at the level of the earlier anterior midgut invagination. The germ band appears compact and light, contrasting sharply with inner, dark yolk. The earliest indications of a segmental pattern can sometimes be observed as regularly spaced bulges of the dark interface between outer and inner cell layers of the germ band. The yolk sac shows a characteristic hook shape.

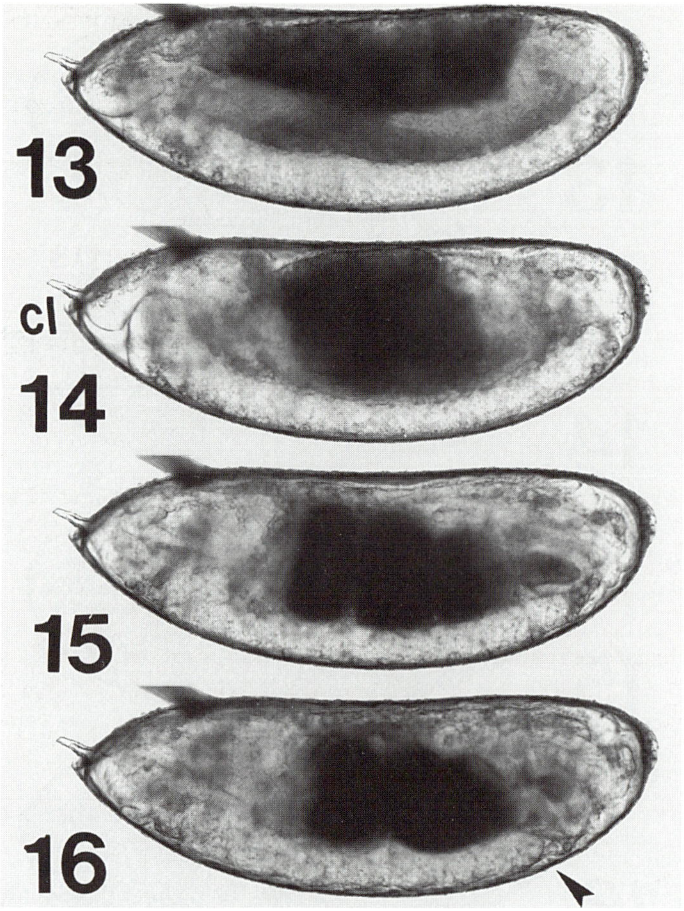

Figure 8. *klarsich* embryos during post-shortening stages of development.
Stage 13: *shortened embryo* (10 h 30 min–11 h 30 min).
Stage 14: *head involution and dorsal closure* (11 h 30 min–13 h), the clypeolabrum is prominent at the anterior end of the embryo.
Stage 15: *dorsal closure complete* (13 h–15 h).
Stage 16: *condensation of CNS* (15 h to the completion of embryonic development). The (→) indicates the posterior tip of the CNS.

4.10 Stage 10: stomodeal invagination

(5 h 10 min–6 h 50 min) The contrast between the cell layers and yolk is gradually lost and internal morphology is more difficult to follow. When the stomodeum forms, the anterior tip of the embryo bends ventrally and an invagination appears and deepens. The anterior midgut anlage moves posteriorly, reaching and passing the level of the posterior gut opening, at the same time flattening out along the ventral germ band. Ectodermal segmentation

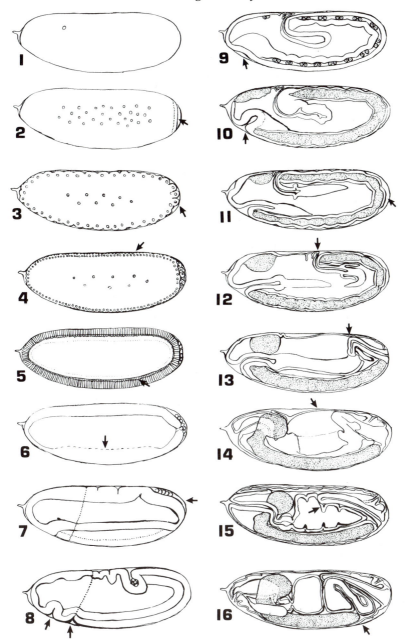

Figure 9. Schematic diagrams of 16 stages of *Drosophial* embryogenesis. The small *arrows* in each diagram indicate those morphological features which are most diagnostic of that particular stage (see text). In stages 9–16, the developing nervous system has been stippled. Stages 14 and 16 are depicted as slightly parasagital optical sections to emphasize the three-dimensional locations of the various organ systems.

becomes apparent as regularly spaced, shallow indentations formed as the germ band retracts slightly from the vitelline membrane.

4.11 Stage 11: three-layered germ band

(6 h 50 min–9 h) Segmentation is clearly visible. Three layers are discernible in the germ band. The embryo has retracted from the vitelline membrane at the posterior tip of the egg. The anterior midgut has flattened out and grown along the germ band, obscuring the distinction between head and trunk ventrally. The stomodeum is deep. The posterior midgut stretches along the dorsal side of the germ band and, due to neuroblast proliferation, the dense yolky regions gradually disappear from the head.

4.12 Stage 12: shortening of germ band

(9 h–10 h 30 min) The germ band contracts and becomes very compact, starting at the anterior end. During this compaction it gradually loses its three-layered appearance. The posterior gut opening moves posteriorly and the yolk sac, covered by the serosa, extends to the dorsal surface of the embryo. The gap between the embryo and the vitelline membrane at the posterior pole persists. Anterior and posterior midgut anlagen form finger-like projections, visible as light, pointed bands parallel to the germ band which gradually approach each other. The head region is distinct and devoid of dark yolk zones.

4.13 Stage 13: shortened embryo

(10 h 30 min–11 h 30 min) The germ band has now completely contracted and the gap at the posterior end of the embryo has disappeared. The anterior and posterior midgut anlagen have fused laterally, forming a continuous band at right angles to the hindgut and head region. The yolk sac is initially concave dorsally and changes to a convex shape at the end of this stage. The head bends dorsally and a gap at the anterior ventral region of the head enlarges. A ridge forms dorsally at the posterior margin of the head.

4.14 Stage 14: head involution and dorsal closure

(11 h 30 min–13 h) The convex protrusion of the yolk sac is covered by the serosa at the dorsal middle region of the embryo. The germ band stretches anteriorly and a head portion involutes into the interior of the embryo continuously with the stomodeal opening. The clypeolabrum is distinct. Frontal sac formation begins as the dorsal ridge moves anteriorly covering the head region. The posterior midgut broadens dorsally. Ventral layers become apparent and gradually flatten out along the yolk sac. The hindgut grows antero-dorsally forming a sharper angle with the midgut. The anal plates and posterior spiracles become distinct. Dorsal closure continues.

4.15 Stage 15: dorsal closure complete

(13 h–15 h) Head involution and dorsal closure of the ectoderm have been completed. During this stage, dorsal closure of the musculature and gut occurs. The midgut initially assumes a 'heart shape'. Subsequently constrictions divide the midgut into three regularly spaced subdivisions.

4.16 Stage 16: condensation of CNS

(15 h to the completion of embryonic development) The major remaining morphogenetic movements are the condensation of the nervous system and the conversion of the initially sac-like gut into a long convoluted tube. As the nervous system contracts along the ventral side of the embryo, the convoluting gut fills in the space between the posterior tip of the ventral nervous system and the anal pads. Muscular movements begin in the gut and are soon apparent in the somatic musculature as well.

5. Cuticle preparations

5.1 Embryo preparations

The following procedures for making cuticle preparations are all minor variants of the original technique described by van der Meer (24). The choice among procedures depends on a balance between the desired quality of the preparation and the number of different stocks and embryos to be examined. The first procedure (*Protocol 3*) may be used for determining rates of hatching, embryonic phenotypes, and so on.

Although in *Protocol 3* the embryo is still covered by the chorion and some adherent yeast, it clears quite well and the chorion pattern, while still visible, does not often interfere with the analysis of the cuticular phenotype. The major advantage of this procedure is that it is quick and no embryos are lost.

Protocol 3.   Quick preparations of undechorionated eggs

Equipment and reagents
- Slides
- Coverslips
- Hoyer's mountant[a]
- Lactic acid

Method
1. Collect embryos and allow them to develop until all normal individuals have hatched.
2. Determine the number of hatched eggs by counting the empty chorions on the agar surface.
3. Pick the unhatched eggs from the plate with a forceps and transfer to a drop of Hoyer's mountant (or 1:1 Hoyer's:lactic acid) on a glass slide.

Protocol 3. *Continued*

4. Cover the drop with a small (~ 10 mm) coverslip and incubate the preparation overnight at 60°C.

---

[a] Add 30 g of gum arabic (acacia powder) to 50 ml of distilled water in a 400 ml beaker, stir overnight until completely dissolved. Under continuous stirring, add 200 g of chloral hydrate slowly in small quantities to avoid clumping. Add 20 g of glycerol. Centrifuge the mixture until the mountant is clear and devoid of debris (generally 3 h to overnight at 12 000 g).

In *Protocol 4* it is possible to skip the glycerine:acetic acid fixation (steps 4–7) and transfer the eggs immediately after dechorionation to Hoyer's mountant or to the Hoyer's:lactic acid mixture.

To obtain better resolution 'pop' the vitelline membrane by pressing on the coverslip after the embryos have cleared but before the Hoyer's mountant becomes too hard. This is best done under a stereomicroscope. Return the preparations to the oven to harden or flatten by wrapping them in aluminium foil and placing two or three 200 g weights on the coverslip for three days. This procedure yields preparations of very high quality, but it is crucial that the preparation has cleared before the weights are added.

Protocol 4. Standard preparations in the vitelline membrane

Equipment and reagents

- Nets and baskets[a]
- Petri dish
- Glycerine:acetic acid (1:4)

- Sodium hypochlorite (commercial bleach undiluted)
- Hoyer's mountant (see *Protocol 3*)

Method

1. Remove the unhatched embryos from the plate with a small spatula or brush and transfer to a wire basket.[b]

2. Dechorionate the eggs by transferring the net or basket to sodium in a 35 mm plastic Petri dish.[c]

3. Remove the net from the bleach, dry briefly on a paper towel, wash with a few drops of water, and return to the emptied 35 mm Petri dish.

4. Add sufficient glycerine:acetic acid (1:4) fixative to the dish to cover the eggs but not enough to cause them to float up out of the basket.

5. Cover the Petri dish, place it inside a larger Petri dish to reduce evaporation, and incubate for 1 h at 60°C.

6. Remove the dish from the oven and allow it to cool to room temperature.

7. Transfer the basket to a paper towel to remove excess fixative.

---

8. Scoop the eggs from the individual positions using a fine-tipped brush and transfer them to a drop of Hoyer's mountant on a glass slide.

9. Cover them with a 10 mm or 18 mm coverslip and incubate at 60°C until the internal tissue clears (generally overnight).

[a] A simple basket can be made from a 1.0 mm square of stainless steel mesh formed into a basket shape by pressing and folding it over the end of a pencil or needle holder. In larger scale experiments where many stocks must be tested, nets which are subdivided to provide several positions for different stocks are useful (*Figure 10*). They can be made by welding polypropylene rings (made by cutting 3–5 mm sections off polypropylene tubing of appropriate diameter) onto a disc of stainless steel mesh cut to fit into a 35 mm Petri dish. Heat the steel mesh on a hotplate covered with aluminium foil and press the plastic rings gently onto the hot steel mesh. Take care that the rings are welded onto the net on their entire circumference to prevent the embryos from slipping from one ring to the next. The rings must be high enough to hold some drops of liquid without spilling, but not so high that it becomes difficult to get all the embryos out.

[b] When embryos from a number of stocks are to be processed at the same time, 'nets' with positions for seven stocks are convenient and time-saving.

[c] The chorion has been removed when the shiny surface of the vitelline membrane is visible and the eggs have become hydrophobic (i.e. when removed and returned to the bleach, they float on the surface due to surface tension).

The greatest resolution of detail and the most photogenic preparations are obtained when the embryos are removed from the vitelline membrane prior to embedding. The procedure described in *Protocol 5* is particularly useful when all of the progeny from a cross must be examined regardless of whether or not they hatch.

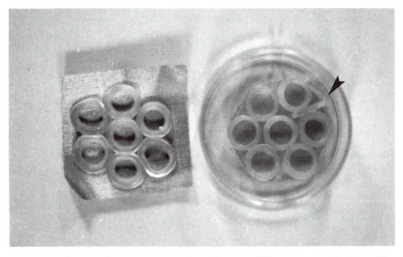

Figure 10. Nets for simultaneous processing of seven different egg collections. The nets are made by welding polypropylene rings onto stainless steel mesh and then trimming the mesh so that the net fits into a 35 mm plastic Petri dish. A small strip of plastic (→) inserted between two of the rings serves as a marker to unambiguously identify the ring at position one.

Protocol 5. Embryos removed from their vitelline membranes

Equipment and reagents

- Fine tungsten needles
- Nets and baskets (*Protocol 4*)
- Slides
- Coverslips
- Embryos

- Sodium hypochlorite (undiluted commercial bleach)
- Glycerine:acetic acid (1:4)
- Hoyer's solution (*Protocol 3*)

Method

1. Collect embryos at 12 h intervals and allow them to develop an additional 6 h at 25 °C.

2. Dechorionate in a stainless steel net in bleach and transfer to a plastic Petri dish by inverting the net over the dish and gently squirting water through the net.[a]

3. Return the Petri dish with the embryos still under water to 25 °C for another 18–24 h to allow all embryos to complete embryonic development. First instar larvae that hatch remain in the water and eventually drown.

4. After 24 h, remove the vitelline membrane from embryos that do not hatch using a fine tungsten or glass needle.

5. Incubate larvae and embryos in water for 30 min at 60 °C.

6. Remove the water and add 1:4 glycerine:acetic acid and fix the larvae in this solution for 15–60 min at 60 °C.[b]

7. Transfer the larvae and embryos individually to Hoyer's solution on a slide using a drawn-out mouth pipette or brush.

8. Add coverslips of the desired size and allow the preparations to clear at 40 °C, typically for 24 h although they can usually be scored much earlier.

[a] If the chorion has been completely removed the vitelline membrane will adhere to the plastic and cause the embryos to stick to the bottom of the Petri dish.
[b] Best results are obtained when preparations are allowed to continue fixation at room temperature for at least 24 h in the glycerine:acetic acid.

5.2 Other tricks for especially flat preparations of high contrast

When the coverslip is added, capillary action pulls the Hoyer's solution out over the entire area covered. If exactly enough Hoyer's is used, this causes the embryo to flatten gently under the coverslip without bursting. The Hoyer's mountant should not be too viscous. Mount the embryos rapidly to prevent the Hoyer's from drying out on the slide before the coverslip is added.

Hoyer's which is too thick can be diluted with a few drops of water. Peeled embryos can also be flattened with weights. It is crucial that the embryos first clear in the mountant for several hours at 60 °C before adding the weights.

For higher contrast, dilute the Hoyer's mountant 1:1 with lactic acid. Clearing time is then reduced to 1 h at 60 °C. This procedure yields high contrast preparations particularly useful when the dorsal hair patterns must be analysed in detail. Unfortunately, the preparations tend to develop crystals.

5.3 Analysis of cuticle patterns

Examine cuticle preparations using a compound microscope equipped with objectives for both dark-field and phase-contrast optics. As a simple substitute for true dark-field optics use low power objectives with the condenser in the phase-contrast position (6.3 objectives with the condenser set at Ph 2; ×10 or ×16 with the condenser in the Ph 3 position). Useful landmarks and sensilla in the external cuticle (25) and head (26, 27) are shown in *Figures 11* and *12*.

5.4 Photographing cuticle preparations

A high contrast film such as Kodak Technical Pan Film is most useful for photographing details of the dorsal hair and denticle pattern as observed in phase-contrast optics. For dark-field pictures of whole embryos, a lower contrast film like Kodak Panatomic-X gives the best detail, although when using an automatic exposure meter the DIN setting should be set higher than recommended, i.e. 23 rather than 16.

6. Whole mount procedure for looking at embryos at early stages

6.1 Fixation of embryos

All these procedures for fixing embryos are intended to preserve cellular morphology. They contrast with those used in cuticle preparations where the major goal in clearing embryos is to dissolve the cells or make them as transparent as possible.

Because the vitelline membrane is impermeable to water-based fixatives, the embryo is treated with fixative dissolved in heptane and the vitelline membrane is then removed. Two alternative procedures are used to devitellinize embryos, the first by hand-peeling (28) and the second by shaking with methanol (29). While the hand-peeling procedure is tedious, it is applicable in all experiments. We have not found fixation obtained in the methanol procedure suitable for semi-thin or electron microscopy (EM) sections. The first (A) and second (B) fixatives vary for different staining procedures; our preferences for each procedure are indicated in the protocols.

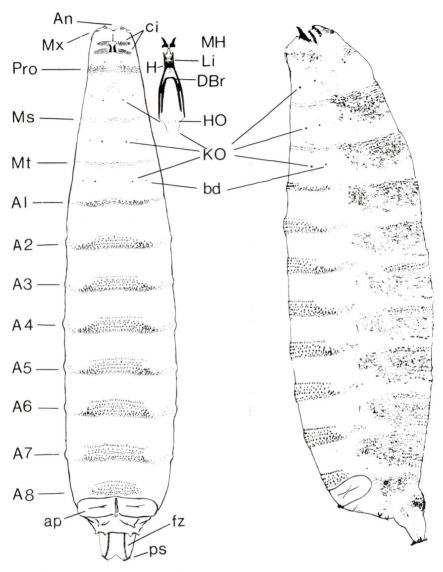

Figure 11. Landmarks for analysis of cuticle patterns in *Drosophila* at the end of embryogenesis. The two embryos are shown from the ventral and lateral aspect. In each case anterior is to the top, and in the lateral view, ventral is to the left. The small insert shows a dorsal view of the internalized mouth parts. Pro, prothoracic denticle band; Ms, mesothoracic denticle band; Mt, metathoracic denticle band. A1–A9, abdominal denticle bands 1–8; ps, posterior spiracles; fz, filzkörper; ap, anal pads; ci, cirri; An, antennal sense organ; Mx, maxillary sense organ; MH, mouth hooks; Li, labial sense organ, also known as hypophyseal sense organ; H, H piece. DBr, dorsal bridge; HO, hypopharyngeal organ, also known as organ X; KO, Keilin's sense organ; bd, black dot sense organs. Designations follow those of Lohs-Schardin *et al.* (25) and Jürgens *et al.* (27).

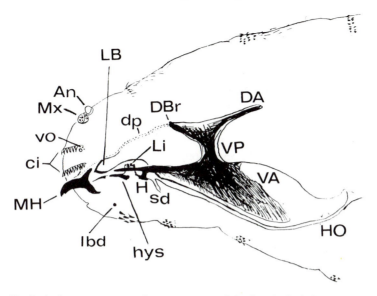

Figure 12. Cuticular structures and sense organs of the head. ci, cirri; vo, ventral organ; An, antennal sense organ; Mx, maxillary sense organ; LB, labrum; dp, dorsal pouch; DBr, dorsal bridge; DA, dorsal arm; VP, vertical piece; VA, ventral arm; HO, hypopharyngeal organ, also known as organ X; MH, mouth hooks; Li, labial sense organ, also known as hypophyseal sense organ; H, H piece; sd, salivary duct. Designations follow Schoeller (26) and Jürgens *et al.* (27).

Protocol 6. Fixation procedure for hand-peeled embryos

Equipment and reagents

- Apple juice plates (Chapter 1)
- Steel mesh baskets (*Protocol 4*)
- Screw-capped vials
- Double sided sticky tape
- Halocarbon oil
- Heptane
- Sodium hypochlorite (undiluted commercial bleach)
- Fixative A[a] and B[a]
- Phosphate-buffered saline (PBS)
- Embryos

Method

1. Select embryos of the appropriate stage directly from the apple juice plate under halocarbon oil and transfer to a small steel mesh basket in a drop of water.

2. Dechorionate in sodium hypochlorite and wash in water.

3. Shake 5 ml of heptane and 5 ml of fixative A vigorously in a 20 ml screw-cap vial. Add the basket with embryos, shake the contents again to get the embryos out of the basket, and remove the basket.

4. Allow the embryos to remain at the heptane/fixative interface for 20

Protocol 6. *Continued*

min, depending on the fixative and the subsequent staining procedures employed.

5. Place a small square of double stick tape on the bottom of a 35 mm Petri dish and fill the dish with enough fixative B to cover the bottom.

6. Transfer the embryos, with as little fixative as possible, from the vial on to the tape in the Petri dish using a short Pasteur pipette. One way of doing this is to pipette the embryos onto the plastic surface of the Petri dish near the tape and gently roll them onto the tape.

7. Nudge the stuck embryos with a tungsten needle, and those which have adhered to the tape will pop out of the vitelline membrane.

8. Transfer the peeled embryos, after fixation in fixative B, to an Eppendorf vial using a short Pasteur pipette and remove as much of the fixative as possible using a drawn-out pipette.

9. Wash the embryos in PBS and rinse twice in PBS.

10. Store overnight at 4°C for future use.

Alternative to steps 2–6

1. Blot the basket of dechorionated embryos on a paper towel and place in a depression slide.

2. Cover the basket with the upper heptane layer from a 1:1 mixture of heptane and fixative A which has been shaken vigorously before use.

3. Fix the embryos directly in the fixative-saturated heptane for 10 min, with occasional jiggling of the basket to prevent the embryos from adhering to each other. Pre-blastoderm stages may be fixed longer.

4. Blot the embryos quickly on a paper towel to remove the heptane, taking care not to allow the embryos to dry out in the basket.

5. Transfer them to double stick tape on the bottom of a Petri dish containing fixative B and peel as described above.

[a]Fixative A: 25% glutaraldehyde. Fixative B: 2.5% glutaraldehyde in PBS.

Protocol 7. Methanol procedure for fixation of large numbers of embryos[a,b]

Equipment and reagents

- Apple juice plates (Chapter 1)
- Steel mesh baskets (*Protocol 4*)
- Screw-capped vials
- Double sided sticky tape
- Halocarbon oil
- Heptane
- Sodium hypochlorite (undiluted commercial bleach)
- Fixative A[c] and B[c]
- Methanol
- Ethanol
- Embryos

Method

1. Dechorionate embryos in small stainless steel nets (*Protocol 6*).

2. Add heptane and fixative A to a 15 ml screw-cap tube and shake vigorously for 5 min. Add the dechorionated embryos to the tube, remove the net, and shake the contents of the tube for 20 min.

3. Remove the lower (aqueous) phase along with about 5 ml of the heptane using a Pasteur pipette.

4. Add 10 ml of methanol and shake the contents vigorously. At this point the vitelline membranes surrounding each embryo rupture and the embryos fall into the methanol layer.

5. Collect the embryos from the 100% methanol and rehydrate in a decreasing alcohol series for staining or treatment with antibodies.

[a] In the original procedure of Mitchison and Sedat (19) after fixation the embryos were transferred to pre-chilled heptane:methanol at –70 °C for 10 min. When the tube was warmed by swirling it under warm running tap-water, the embryos devitellinized with very high efficiency. For certain antisera, this is still the preferred procedure and the only one which gives reliable staining.

[b] With certain fixatives, it is also possible to keep the embryos in the net throughout the entire procedure. The embryos are fixed directly in heptane saturated with fixative as above and transferred to fresh heptane with methanol. This procedure works especially well when followed by the fuchsin staining protocol.

[c] Fixative A: 25% glutaraldehyde. Fixative B: 2.5% glutaraldehyde in PBS.

6.2 Heat methanol fixation

Because this protocol does not use formaldehyde or glutaraldehyde, it is especially useful in embryos to be stained for antigens that are sensitive to those fixatives. Morphology is well preserved at the level of whole mounts, but the procedure is not useful for sectioned material nor for transmission electron microscopy. The relatively mild fixation does not appear to preserve cytoplasmic proteins as well as it does proteins associated with the cell surface or cytoskeleton (30).

Protocol 8. Heat methanol fixation

Equipment and reagents

- Scintillation vials
- Mesh basket
- Eggs
- Heptane

- Triton salt solution: 0.4% NaCl in 0.03% Triton X-100
- Methanol

Method

1. Chill a small bottle of Triton salt solution on ice.

2. Add 1–2 ml Triton salt solution to a scintillation vial, screw cap on loosely, and heat in boiling water.

3. Collect and dechorionate eggs in mesh basket.

Protocol 8. *Continued*

4. Pull the hot vial out of the boiling water by its cap and unscrew the cap.

5. Immediately drop the basket of eggs into the hot solution, recap the vial, and shake once.

6. Stand for 20 sec.

7. Add 10–15 ml of the ice-cold solution and allow the vial to cool on ice for 1 min.

8. Remove the basket with tweezers.

9. Pour off all the Triton salt solution once the embryos have settled to the bottom of the vial.

10. Add heptane, then add methanol, and shake to remove the vitelline membrane as in *Protocol 7*.

11. Proceed as usual for antibody staining (see Chapter 7).

6.3 Staining and analysis of whole mounts
6.3.1 Nomarski optics

Protocol 9. Staining and analysis of whole mounts

Equipment and reagents

- Slides
- Coverslips
- Fixatives A[a] and B[a]
- Ethanol

- Phosphate-buffered saline pH 7.2, 0.8% NaCl (PBS)
- Faure's mountant[b]

Method

1. Dehydrate embryos fixed and peeled using of the procedures described in *Protocols 6–8* and store in 70% ethanol.

2. Rehydrate the embryos and/or mount directly in Faure's mountant under a no. 1 coverslip using two additional coverslips as spacers, for analysis of surface morphology in Nomarski optics.[c]

3. As an alternative to step 2, dehydrate through 100% ethanol and transfer into thick cedar wood oil (the kind formerly used with oil immersion lenses) or into Canada balsam/methyl salicylate, to obtain better internal morphology.

[a] Fixative A: 25% glutaraldehyde. Fixative B: 2.5% glutaraldehyde in PBS.
[b] Mix 30 ml of glycerine, 30 g of gum arabic (acacia powder), 50 g of chloral hydrate, and 50 g of distilled water. Stir overnight until dissolved and centrifuge as in the procedure for Hoyer's mountant until the solution has cleared. Faure's mountant is used to mount adult cuticle in cell lineage studies involving mitotic recombination of X chromosome loss, as well as in the Nomarski preparations.
[c] Such preparations have the general appearance of low power scanning EM (*Figure 13*) but offer the advantage that it is also possible to focus into the interior of the preparation.

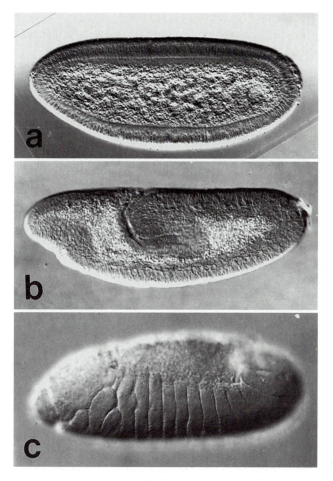

Figure 13. Nomarski whole mounts of *Drosophila* embryos showing various internal structures. (a) Optical cross-section of embryo late during the process of cellularization. The ventral nuclei are already somewhat distorted in preparation for forming a ventral furrow. (b) Optical cross-section of embryo at stage 9 showing indentation at site of future stomodeal invagination and posterior midgut primordium. (c) Surface view of embryo at the completion of germ band shortening, stage 10.

6.3.2 Rhodamine labelled phalloidin/Hoechst protocol

The specificity of phalloidin for F-actin allows rapid visualization of the cytoskeleton during cleavage and cellularization (14). In later stages, the staining highlights cell shape and certain internal structures (e.g. the fibre tracts in the central nervous system) (*Figure 14*). Hoechst 33258 binds to DNA and can be used to identify mitotic figures and to count nuclei.

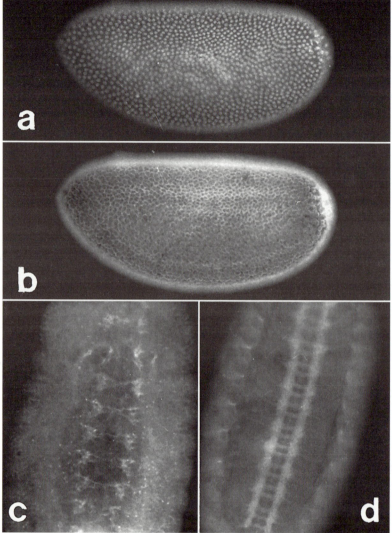

Figure 14. *Drosophila* embryos stained with rhodamine labelled phalloidin and Hoechst 33258. (a) Nuclear distribution in late cleavage stage 4 embryo (Hoechst 33258). (b) F-actin localization in same embryo as (a) viewed using rhodamine labelled phalloidin. (c) Early development of fibre tracks in the ventral nervous system of stage 13 embryo, stained with rhodamine labelled phalloidin. (d) Fibre tracks of the ventral nervous system at the completion of dorsal closure (stage 15) stained with rhodamine labelled phalloidin.

Protocol 10. Rhodamine labelled phalloidin and Hoechst labelling

Equipment and reagents

- Heptane
- Formaldehyde
- PBS
- Hoechst 33258

- Rhodamine labelled phalloidin (Molecular Probes Inc.)[a]
- Fixatives A[b] and B[b]

Method

1. Remove the vitelline membrane by hand peeling (*Protocol 6*).[c]
2. Pre-fix for 4 min in fixative A.[c]
3. Fix for 4 min in fixative B.[c]
4. Wash the embryos twice in PBS for at least 10 min to remove the formaldehyde.
5. Fixed embryos can be stored overnight at 4°C prior to staining.
6. Remove the PBS wash solution from the embryos with a thin Pasteur pipette and replace with Rh–phalloidin solution.
7. Stain the embryos for 20 min only, in the dark, inverting the tube occasionally to reduce sticking and aid uniform staining.
8. Remove the Rh–phalloidin, wash the embryos with PBS, and allow to stand for at least 10 min.
9. In an Eppendorf vial, mix 5 μl of Hoechst 33258 stock solution with 500 μl of PBS, to a final concentration of 1 μg/ml.
10. Replace the PBS wash solution with the Hoechst 33258 solution and allow the embryos to stain for 4 min only.
11. Replace the Hoechst solution with PBS wash solution and rinse the embryos twice for at least 10 min. At this point the embryos can be stored in sealed Eppendorf tubes, wrapped in foil, at 4°C.
12. Mount the embryos on glass slides in 1:1 glycerine:PBS under a coverslip with no spacers. View the preparations in a fluorescence microscope equipped with excitation filters for UV and rhodamine wavelengths.
13. Although the slides are not permanent, they can be stored for several days in the dark at 4°C.

[a] Upon receipt, aliquot 10 μl volumes into Eppendorf tubes, seal, and store in a freezer for future use. Immediately before staining, remove an aliquot of the phalloidin stock solution from the freezer and wrap in foil to reduce bleaching by room lights. Allow the liquid to evaporate under a vacuum until all of the methanol has been removed then make up the phalloidin to 200 μl with PBS.
[b] Fixative A: 5 ml of heptane and 5 ml of 8% formaldehyde. Fixative B: 8% formaldehyde.
[c] Phalloidin binding is lost following the methanol fixation procedure.

One very useful way to record data from fluorescently stained preparations is by using the same video recorder set-up described in Section 3.4, set in the non-time lapse mode. The patterns in each individual embryo can be recorded at ×16 and ×100 and analysed later. A second advantage of video recording is that the increased red sensitivity of most light-sensitive video cameras greatly enhances the image, reducing the exposure time, and consequently the photo-bleaching.

6.3.3 Basic fuchsin stained preparations

Fuchsin is a nuclear stain similar in usefulness to Hoechst, but simpler to use since it does not require a fluorescence microscope for viewing.

Protocol 11. Basic fuchsin stained preparations

Equipment and reagents

- Stainless steel basket
- Fuchsin
- Ethanol
- Glacial acetic acid
- Paraformaldehyde
- Heptane
- HCl
- Fixatives A[a] and B[a]

Method

1. Remove the vitelline membrane either by hand or by treatment with methanol (*Protocols 6* and *7*).

2. Fix embryos for 4 min in fixative A.

3. Transfer to 4 ml of fixative B for 1 h or overnight.

4. Wash four times for 20 min in 70% ethanol.

5. Heat the fixed embryos in a stainless steel net or basket in 2 M HCl for 10 min at 60°C, wash with water, and then with 5% acetic acid.

6. Transfer the basket to a drop of 1% fuchsin in 2.5% acetic acid. Allow early embryos to stain for 30 min, later post-gastrulation stages for 15 min.

7. Destain the embryos in 5% acetic acid.

8. Dehydrate in two changes each of 70%, 95%, and 100% alcohol.

9. Embed in Permount or cedar wood oil or, after removal of the alcohol using acetone, in Araldite or other plastics.

[a] Fixative A: shake 4 ml of 95% alcohol, 0.5 ml of glacial acetic acid, and 1.6 ml of 20% paraformaldehyde with 4 ml of heptane. Fixative B: 95% alcohol, 0.5 ml of glacial acetic acid, 0.4 ml of paraformaldehyde.

6.3.4 Sectioned material

Although the detailed analysis of sectioned material is beyond the intended scope of this chapter, most standard fixatives and embedding procedures

(12, 16, 19) can be used on *Drosophila* embryos once they have been removed from their vitelline membranes by hand-peeling (*Protocol 6*). The following are appropriate fixatives:

(a) Fixative A: 0.5 ml of phosphate buffer pH 7.2, 0.5 ml of 25% glutaraldehyde, two to three drops of acrolein.

(b) Fixative B: 2 ml of 25% glutaraldehyde, 5 ml of phosphate buffer pH 6.8, 3 ml of water.

For peeling fixative B can be diluted 1:1 in water, which facilitates removing the embryos from the vitelline membrane. After the vitelline membranes have been removed, incubate the embryos in full-strength fixative B for 1 h or overnight at 4°C. Wash in 0.05 M cacodylate buffer pH 7.2 (PBS), and post-fix in: 4% OsO$_4$ in cacodylate buffer, 2 h at 0°C.

6.4 Protocols for scanning electron microscopy

Examination of embryos with the scanning electron microscope is a useful procedure for determining the overall morphology and the cell surface detail. In *Protocol 12* methanol 'popping' is used to remove the vitelline membrane. This procedure is quick and easy and gives good overall morphology. In *Protocol 13* we describe 'hand-peeling' the vitelline membrane. This gives better surface morphology. It also facilitates breaking embryos in cross-section to display internal morphology.

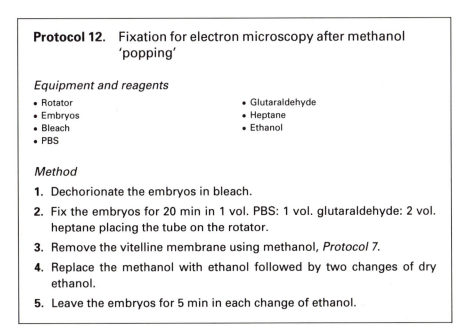

Protocol 12. Fixation for electron microscopy after methanol 'popping'

Equipment and reagents
- Rotator
- Embryos
- Bleach
- PBS
- Glutaraldehyde
- Heptane
- Ethanol

Method

1. Dechorionate the embryos in bleach.

2. Fix the embryos for 20 min in 1 vol. PBS: 1 vol. glutaraldehyde: 2 vol. heptane placing the tube on the rotator.

3. Remove the vitelline membrane using methanol, *Protocol 7*.

4. Replace the methanol with ethanol followed by two changes of dry ethanol.

5. Leave the embryos for 5 min in each change of ethanol.

Protocol 13. Hand-peeling to remove the vitelline membrane

Equipment and reagents
- Double sided tape
- Petri dish
- Glass needles
- Tungsten needles
- Heptane
- Glutaraldehyde
- 0.1 M cacodylate buffer: 2.14 g sodium cacodylate in 50 ml H_2O
- Osmium oxide
- Ethanol

Method

1. Dechorionate embryos with bleach in a wire mesh basket in a deep depression slide.
2. Shake heptane with 25% glutaraldehyde in 0.1 M cacodylate buffer.
3. Fix embryos for 15 min in this heptane preparation.
4. Transfer the fixed embryos to double sided sticky tape in a small Petri dish.
5. Immerse in 0.1 M cacodylate buffer and hand-peel membranes with a glass needle.
6. Slice or break open the embryos with a bent tungsten needle after they have been removed from the vitelline membrane to reveal cross-sections.
7. Collect embryos or fragments into an Eppendorf tube and post-fix in a fume-hood in 1% osmium in cacodylate buffer for 30 min (2 ml 4% OsO_4 in 6 ml cacodylate buffer).
8. Rinse three times in cacodylate buffer and dehydrate for 10 min in each of the ethanol series 50%, 25%, 100%, 100%.

Many of theses procedures are simple modifications of standard protocols. We have found that the use of Peldri (Ted Pella Inc.) (*Protocol 14*) provides a simple procedure for handling fly embryos that avoids critical point drying.

Protocol 14. Preparation of embryos for SEM using Peldri

Equipment and reagents
- Peldri[a]
- Ethanol

Method

1. Place Peldri in a water-bath at 37°C in a fume-hood until it has completely melted.
2. Add 0.4 vol. of melted Peldri II to an Eppendorf tube containing fixed embryos in 1 ml 100% ethanol.

3. Mix well by sucking back and forth with a Pasteur pipette.[b]

4. Incubate for 1 h in a 37 °C water-bath.[c]

5. Replace the Peldri:ethanol mixture with 100% Peldri.

6. Repeat several times to be sure that all the ethanol is removed.

7. Incubate for 1 h in 100% Peldri.

8. Cool on ice to solidify the Peldri.[d]

9. Sublimate the solid Peldri in a fume-hood overnight or in a vacuum chamber for several hours.

10. Remove the embryos from the Eppendorf tube to a clean Petri dish.[e]

11. Mount the embryos in the desired orientation on to a metal SEM stub using double sided sticky tape.[f]

12. Coat with gold palladium twice for 60 sec each time, using a sputter coater, and changing the orientation of the mount between each coating to ensure an even distribution.

[a] If Peldri is not available, equally suitable results can be obtained using TetraMethylSilane (TMS from Ted Pella Inc.) after incubating the dehydrated embryo in acetate for 5 min.
[b] At this ratio of Peldri:ethanol the embryos should sink to the bottom of the tube.
[c] The tube must be kept warm until the final step in order to prevent the Peldri from solidifying.
[d] If it is not solid after 5 min there are probably traces of ethanol left in the Peldri.
[e] The embryos at this stage should be white and very hard.
[f] If static electricity makes it difficult to handle the embryos wait for 1–2 h and try again.

References

1. Nüsslein-Volhard, C. (1977). *Drosophila Inf. Serv.*, **52**, 166.
2. Nüsslein-Volhard, C., Wieschaus, E., and Kluding, H. (1984). *Wilhelm Roux Arch. Dev. Biol.*, **193**, 267.
3. Campos-Ortega, J. A. and Hartenstein, V. (1985). *The embryonic development of Drosophila melanogaster.* Springer–Verlag, Berlin and Heidelberg.
4. Zalokar, M. and Erk, I. (1976). *J. Microsc. Biol. Cell.*, **25**, 97.
5. Foe, V. E. and Alberts, B. M. (1983). *J. Cell Sci.*, **61**, 31.
6. Poulson, D. F. (1950). In *Biology of Drosophila* (ed. M. Demerec), p. 168. Wiley, New York.
7. Bownes, M. (1975). *J. Embryol. Exp. Morphol.*, **33**, 789.
8. Fullilove, S. L. and Jacobson, A. G. (1978). In *The genetics and biology of Drosophila* (ed. M. Ashburner and T. R. F. Wright), Vol. 2c, p. 106. Academic Press, New York.
9. Hartenstein, V., Technau, C. M., and Campos-Ortega, J. A. (1985). *Wilhelm Roux Arch. Dev. Biol.*, **194**, 213.
10. Rabinowitz, M. (1941). *J. Morphol.*, **69**, 1.
11. Sonnenblick, B. P. (1950). In *Biology of Drosophila* (ed. M. Demerec), p. 62. Wiley, New York.

12. Turner, R. F. and Mahowald, A. P. (1976). *Dev. Biol.*, **50**, 95.
13. Warn, R. F. and Macgrath, R. (1982). *Dev. Biol.*, **89**, 540.
14. Warn, R. M., Macgrath, R., and Webb, S. (1984). *J. Cell Biol.*, **98**, 156.
15. Karr, T. L. and Alberts, B. M. (1985). *J. Cell Biol.*, **102**, 1494.
16. Rickoll, W. L. (1976). *Dev. Biol.*, **49**, 304.
17. Turner, T. R. and Mahowaid, A. (1977). *Dev. Biol.*, **57**, 403.
18. Rickoll, W. J. and Counce, S. J. (1980). *Wilhelm Roux Arch. Dev. Biol.*, **188**, 163.
19. Hartenstein, V. and Campos-Ortega, J. A. (1985). *Wilhelm Roux Arch. Dev. Biol.*, **194**, 181.
20. Technau, G. M. and Campos-Ortega, J. A. (1985). *Wilhelm Roux Arch. Dev. Biol.*, **194**, 196.
21. Hartenstein, V. and Campos-Ortega, J. A. (1984). *Wilhelm Roux Arch. Dev. Biol.*, **193**, 308.
22. Madhavan, M. M. and Schneiderman, H. A. (1977). *Wilhelm Roux Arch. Dev. Biol.*, **183**, 269.
23. Turner, T. R. and Mahowald, A. P. (1979). *Dev. Biol.*, **68**, 96.
24. Van der Meer, J. (1977). *Drosophila Inf. Serv.*, **52**, 160.
25. Lohs-Schardin, M., Cremer, C., and Nüsslein-Volhard, C. (1979). *Dev. Biol.*, **73**, 239.
26. Schoeller, J. (1964). *Arch. Zool. Exp. Gen.*, **103**, 1.
27. Jürgens, G., Lehmann, R., Schardin, M., and Nüsslein-Volhard, C. (1986). *Wilhelm Roux Arch. Dev. Biol.*, **195**, 359.
28. Zalokar, M. and Erk, I. (1977). *Stain Technol.*, **52**, 89.
29. Mitchison, T. J. and Sedat, J. (1983). *Dev. Biol.*, **99**, 261.
30. Peifer, M., Sweeton, D., Casey, M., and Wieschaus, E. (1994). *Development*, **120**, 369.

Immunolabelling of *Drosophila*

ROBERT A. H. WHITE

1. Introduction

Specific antibodies provide the easiest way, in general, of studying the distribution and subcellular localization of proteins. Such knowledge can give important clues to their role in development and cell biology. Using an antibody to detect the protein product of a specific gene is a convenient way of following gene expression patterns in development and allows the analysis of both transcriptional and post-transcriptional gene regulation. Antibodies are also widely used as markers for individual tissue and cell types enabling these to be followed during morphogenesis (see ref. 1). Multiple labelling can be used to correlate the expression patterns of different proteins. The use of immunolabelling is now so established that many research papers merely say 'Antibody labelling was by standard methods'. However, it is clear that appropriate fixation, labelling, and detection can make a huge difference to the sensitivity and clarity of the result, and indeed whether the labelling actually gives a tolerably faithful representation of the *in vivo* protein distribution.

2. Antibodies

2.1 Types of antibody

The production of antibodies is well covered elsewhere (2, 3). Here we are dealing with the use of antibodies and there are two types to consider:

(a) Polyclonal antibodies or antisera are collections of antibodies taken from the blood of immunized animals. They are complex, irreproducible mixtures available in limited quantity.

(b) Monoclonal antibodies are the product of a single antibody-producing cell immortalized by fusion with a myeloma cell partner. The tissue culture supernatant taken from these hybridoma cells growing *in vitro* contains a single antibody species. Monoclonal antibodies are reproducible reagents available in unlimited quantity.

Polyclonal antibodies can be used as crude sera but usually purification is required to reduce unwanted background labelling. Purification of the immunoglubulin fraction or isolation of the IgG subfraction, thus removing low affinity IgM antibodies, can be useful, but the most important way of improving an antiserum is affinity purification. This is the purification of antibodies specific to the antigen of interest by absorption and elution from the antigen on a solid support.

For most applications, monoclonal antibodies can be used as tissue culture supernatant which can be concentrated, if necessary, by ultrafiltration (Centricon-100, Amicon). The antibody can also be purified, either from the hybridoma supernatant or from ascites fluid produced by growing the hybridoma cells *in vivo*. While monoclonal antibodies have the advantages mentioned above, individual monoclonal antibodies vary widely in affinity and, as they are directed against an individual antigenic determinant, their binding of the antigen may be more sensitive to fixation conditions. If the protein exists in a variety of post-translationally modified isoforms, monoclonal antibodies may be isoform-specific. As monoclonal antibodies recognize a single antigenic determinant, panels of monoclonal antibodies against different determinants on the same protein can provide information about different parts of the antigen. A useful source of monoclonal antibodies is the Developmental Studies Hybridoma Bank, Department of Biological Studies, The University of Iowa, 436 Biology Building, Iowa City, IA 52242; e-mail: dshb@uiowa.edu.

The following is a guideline to some of the properties of the different antibody preparations.

(a) Crude antiserum:
   (i) Total protein concentration is about 80 mg/ml.
   (ii) Total immunoglobulin concentration is about 10 mg/ml.
   (iii) Specific antibody concentration for a good serum is about 200 µg/ml.

(b) Affinity purified antiserum:
   (i) Usually prepared at a concentration of 1–10 mg/ml.

(c) Hybridoma supernatant:
   (i) Total protein concentration depends on culture medium, e.g. 10% FCS will be about 10 mg/ml protein.
   (ii) Specific antibody concentration 1–50 µg/ml.

(d) Ascites fluid:
   (i) Specific antibody represents about 90% of total protein at a concentration of 2–10 mg/ml.

2.2 Storage

Antibodies are relatively stable molecules but can easily be damaged by repeated freezing and thawing. The recommended storage conditions are:

(a) Antiserum: store aliquoted at –20°C for long-term or with 10 mM sodium azide for a year at 4°C.

(b) Affinity purified antiserum: store aliquoted at –20°C for long-term. Protein concentration should not be less than 1 mg/ml and an 'inert' protein such as BSA is often added to protect against protein denaturation on freezing. Alternatively, glycerol can be added to 50% and the solution stored unfrozen at –20°C.

(c) Hybridoma supernatant: store at 4°C with 10 mM sodium azide.

(d) Ascites fluid: store aliquoted at –20°C for long-term or with 10 mM sodium azide for up to a year at 4°C.

2.3 Antibody specificity

It is important to confirm the specificity of an antibody in any particular assay system used. It is not sufficient to assume that just because an antibody specifically recognizes protein X on Western blots that all the labelling seen on embryos is exclusively protein X. Controls against spurious labelling include:

(a) Using pre-immune sera in the case of polyclonal antibodies.

(b) Using control irrelevant monoclonal antibodies matching the test antibody in immunoglobulin class and subclass (kits are available for isotyping; e.g. ISO-1 Immunotype kit, Sigma).

(c) In *Drosophila*, where available, using embryos homozygous for a null mutation in the gene encoding protein X. Sometimes this does not help, e.g. for a labelling pattern seen at a stage after the lethal phase of the null allele.

(d) Blocking the labelling with excess soluble protein X.

(e) Reducing the labelling by using protein X to absorb the protein X-specific antibodies from a polyclonal antiserum.

(f) Using several monoclonal antibodies against different antigenic sites on the protein and showing the same labelling pattern.

(g) Comparing the pattern of transcript accumulation as seen by RNA *in situ* hybridization and the antibody labelling pattern.

2.4 Antibody binding

Antibodies vary widely in affinity and useful antibodies will have an affinity constant of greater than 10^7 mol^{-1}. A general guideline is that the specific antibody concentration in labelling reactions should be in the range of 1–20 µg/ml. The antibody concentration that is used for immunolabelling is usually determined empirically by examining the results of a dilution series. The chosen concentration may also be affected by other factors such as antibody availability. The optimal concentration of antibody depends on the purpose of the immunolabelling:

(a) Using antibodies to provide markers for cell or tissue types demands only that the specific labelling should be clearly detectable above background. In this context antibodies can be used as trace reagents under non-saturating conditions.

(b) Use of antibodies to gain, as far as possible, a quantitative image of protein distribution requires that antibodies should be used in saturating concentrations. Selecting an antibody dilution that gives a clear labelling pattern may not necessarily give an accurate impression of the protein distribution if there is, for example, a specific pattern of high-expressing cells superimposed on low level widespread expression. In such a case it may be tempting to pick a dilution that reveals only the high level expression and to dismiss the low level labelling as 'background'. In practice, it is often very difficult to discriminate low level-specific labelling from background non-specific labelling. Where possible, a titration of the antibody concentration on wild-type embryos should be compared with a titration on embryos homozygous for a protein null mutation.

2.5 Fixation

Fixation is a delicate balancing act. On the one hand it is necessary to preserve tissue morphology and allow permeabilization of cells but on the other hand fixation can destroy antigenic determinants and cause permeability problems. The most commonly used fixative is formaldehyde which is generally available as a 37% formaldehyde solution containing 10–15% methanol to inhibit the formation of the polymer, paraformaldehyde. To avoid the methanol and formaldehyde oxidation products that accumulate on storage, formaldehyde solution can be made directly by depolymerizing the solid paraformaldehyde (see *Protocol 1*). The extent of fixation can be varied easily by changing the fixation time and this can have a dramatic effect on the antibody labelling. Note that formaldehyde cross-linking is reversible and some antigens, notably soluble cytoplasmic proteins, may be lost during subsequent prolonged incubations.

2.6 Detection

The choice lies between immunofluorescence and immunoenzyme labelling. Both exhibit similar sensitivity. Immunofluorescence gives higher resolution and is easier for double labelling as it permits the simultaneous and independent detection of different antigens. However, the preparations tend to bleach under observation and are not permanent. Using a conventional fluorescence microscope, superficial structures are easy to visualize but labelling of internal structures is affected by out of focus fluorescence which degrades the image.

Confocal fluorescence microscopy (4) solves this problem and produces very high resolution images but it is difficult to screen large numbers of labelled preparations. Immunoenzyme labelling using predominantly horse-radish peroxidase or alkaline phosphatase labelled antibodies requires only a conventional light microscope, provides permanent preparations, and the bright-field illumination allows the labelling to be correlated with the tissue morphology. In many situations immunoenzyme labelling is most suitable for an overview, e.g. of different developmental stages, whereas confocal immunofluorescence is most valuable for looking at details and subcellular localization.

3. Embryo immunolabelling

3.1 Embryo fixation

Penetration of antibodies into the embryo requires the removal of the egg membranes, the chorion and vitelline membrane, and the permeabilization of cell membranes. Current general antibody labelling methods are based on Mitchison and Sedat's development (5) of a procedure for the mass devitellinization of embryos (*Protocol 1*). Patel is an excellent source of methods and wisdom (6).

Protocol 1. Fixation of embryos

Equipment and reagents

- Egg baskets: take a 50 ml screw-cap polypropylene tube and cut off the cap end, around the 40 ml mark, with a razor blade. Cut a large hole in the cap then cover the top of the tube with nylon mesh (150 μm mesh) and secure by screwing the cap on top. This forms a basket which retains eggs but lets yeast through.

- Fixative: 4% formaldehyde in PBS made by depolymerizing and dissolving 4 g para-formaldehyde in 100 ml PBS by heating to 70°C whilst stirring in a fume-hood. Store the 4% formaldehyde:PBS in aliquots at −20°C.

Method

1. Prepare two-phase fixing medium by adding 2 ml 4% formalde-hyde:PBS and 6 ml heptane into a 10 ml glass scintillation vial with snap-on plastic cap and shake well.[a]

2. Collect embryos into egg basket. Wash with tap-water. Dechorionate with undiluted bleach (3% sodium hypochlorite) for 2 min. Rinse well with tap-water.

3. Transfer embryos to fixing medium either by placing the basket in a Petri dish containing water and collecting the embryos off the surface of the water with a paintbrush or spatula, or by unscrewing the egg basket, briefly blotting the nylon mesh on a tissue, and plunging the

---

Protocol 1. *Continued*

mesh into the vial containing fixing medium. The embryos fall off the mesh onto the interface.

4. Fix rolling at room temperature for 20 min.[b]

5. Remove fixative (lower phase) as completely as possible with drawn-out Pasteur pipette, rapidly add 6 ml methanol, and shake vigorously for 20 sec. The devitellinized embryos sink to the bottom of the tube. Aspirate heptane and most of the methanol, then wash embryos three times in methanol. The embryos can be stored at 4°C (or at –20°C) in methanol for weeks, months....

[a] For small numbers of embryos, fixation can be done in microcentrifuge tubes using 300 μl of fixative and 1 ml of heptane.
[b] Fixation time can be varied from 5 min to an hour to select optimal fixation for a particular antibody labelling.

Alternative fixation procedures are discussed in *Table 1*.

Protocol 2. Hand devitellinization of embryos

1. After fixation (*Protocol 1*, step 4), remove fixative and wash embryos in heptane.

2. Transfer to a flat plastic surface, allow the heptane to evaporate, and pick up the embryos on double sided sticky tape. Invert the tape, cover the embryos with a drop of water, and devitellinize by nicking the vitelline membrane with a sharp razor blade.

3. Transfer the embryos to a microcentrifuge tube containing PBS, 0.1% Triton X-100 using a Gilson pipette with a sawn off yellow tip.

4. Proceed with antibody labelling (*Protocol 3* or *4*).

3.2 Labelling of intact embryos

3.2.1 Immunofluorescence labelling

The most commonly used fluorochromes are fluorescein isothiocyanate (FITC), rhodamine isothiocyanate (TRITC), and Texas Red. Other newer fluorochromes which are brighter and more photostable include Cy5 and Cy3 (Molecular Probes, Inc.). Immunofluoresence labelling of whole embryos is described in *Protocol 3*.

3.2.2 Immunoenzyme labelling

The most commonly used enzyme is horse-radish peroxidase (HRP) which produces a brown precipitate from the substrate, diaminobenzidine (DAB). Immunoperoxidase staining is described in *Protocol 4*.

Table 1. Alternative fixation procedures

(a) The original Mitchison and Sedat (5) buffer for fixation was PEM rather than PBS and gives better results with some antigens and may stabilize microtubules PEM:0.1 M Hepes, 2 mM EGTA, 1 mM MgSO$_4$, pH 6.9.
(b) The classical fixative, formalin (37% Formaldehyde stabilized with 10–15% methanol) is usually just as good a source of formaldehyde for the fixation buffer as paraformaldehyde, but some antigens are sensitive to methanol.
(c) EM Buffer Fix. This can give a considerable improvement of the labelling to certain antigens (particularly nuclear antigens) and also we routinely use this for the detection of β-galactosidase. 5× EM Buffer:800 mM KCl, 200 mM NaCl, 20 mM EDTA, 5 mM Spermidine–HCl, 2mM Spermine–HCl, 1% β-mercaptoethanol, 150 mM Pipes pH 7.4. Fix solution:3 ml Heptane, 825 μl H$_2$O, 300 μl 5× EM Buffer, 375 μl 16% Formaldehyde (Methanol free Ultrapure EM grade, Polysciences Inc.). Fix for 30 min rolling at room temperature.
(d) A common modification is to add 1% NP-40 to the fixative which may increase the rate of penetration of the fix (7).
(e) For antigens that are sensitive to methanol, the embryos can be devittelinized by hand (*Protocol 2*).

Protocol 3. Immunofluorescence labelling of whole embryos

Equipment and reagents

- PTX: PBS, 0.1% Triton X-100
- Secondary antibody: fluorescein or Texas Red labelled affinity-purified antiserum raised against immunoglobulin of the same species as the primary antibody. For rabbit antisera primary antibodies use AffiniPure donkey anti-rabbit IgG (H+L) and for mouse monoclonal antibodies use AffiniPure goat anti-mouse IgG (H+L) (Jackson ImmuneResearch Laboratories, Inc.).

- PBTX: PBS, 0.1% BSA, 0.1% Triton X-100
- Primary antibody
- Mounting medium: Citifluor AF1 (Agar Scientific Ltd.), containing an anti-photobleaching reagent. Alternatively, use 70% glycerol plus 0.5 mg/ml phenylene diamine to reduce photobleaching of the fluorochrome.

Method

1. Rehydrate embryos from methanol with two washes in PTX. Use about 20 μl embryos and 1 ml washes in microcentrifuge tubes.[a]

2. Block non-specific protein binding sites and permeabilize cell membranes by incubating for 1–8 h in PBTX rotating at 4°C.

3. Incubate in primary antibody diluted in PBTX rotating at 4°C overnight. Optimal antibody dilution will vary for different antibodies.

4. Rinse three times in PBTX followed by two 30 min washes rolling at 4°C.

5. Incubate in secondary antibody diluted in PBTX for 1–3 h rolling at 4°C. Longer incubations tend to increase background labelling. For secondary antibodies from Jackson Immunoresearch we dilute about 1:400.[b]

6. Rinse three times in PBTX followed by two 30 min washes rolling at 4°C.[b]

Protocol 3. *Continued*

7. Mount in glycerol with anti-fade reagent and then seal the coverslip with nail varnish.

8. Observe with conventional or confocal fluorescence microscope (see *Figure 1*) or store in dark (wrap in foil) at 4 °C.

[a] For early stage embryos, rehydration through a decreasing methanol series (90%, 70%, 50% in PTX) gives better morphology.

[b] The DNA dyes, Hoescht 33258 or diamidino-phenylindole (DAPI; Sigma), provide useful general fluorescent nuclear labels. These can be added with the secondary antibody or just before mounting. Use at 1 ng–1 µg/ml.

Protocol 4. Immunoperoxidase labelling of whole embryos

Reagents

- PBTX (see *Protocol 3*)
- Secondary antibody: HRP labelled affinity-purified antiserum raised against immunoglobulin of the same species as the primary antibody.[a] For rabbit antisera primary antibodies use AffiniPure donkey anti-Rabbit IgG (H+L) and for mouse monoclonal antibodies AffiniPure goat anti-mouse IgG (H+L) from Jackson ImmunoResearch Laboratories, Inc.
- PT: PBS, 0.1% Tween 20

- Peroxidase developing solution: for single labelling immunoperoxidase use the Pierce Immunopure metal-enhanced diaminobenzidine (DAB) substrate kit, or use DAB solution (1 ml 0.3 mg/ml DAB stock solution, 2 µl 30% H_2O_2). DAB stock solution is most conveniently made with DAB tablets (10 mg, Sigma). For Ni intensification see *Protocol 13*. Dissolve one tablet in 33 ml PBS with 16.5 µl Tween 20. Store at –20 °C. NB: DAB is a potential carcinogen. Wear gloves and inactivate waste with bleach.

Method

1. Follow *Protocol 3*, steps 1–5.[b]

2. Rinse three times in PBTX followed by two 30 min washes in PT rolling at 4 °C.

3. Develop in Pierce reagent or DAB solution. This may take from a few seconds to a few minutes. Colour development can be followed under the dissection microscope and stopped by dilution in PT, or by adding 2 µl/ml of 20% sodium azide. If reaction proceeds too rapidly then perform the reaction on ice or dilute the Pierce reagent.

4. Wash embryos three times in PT.

5. Dehydrate embryos through 50%, 70%, 95%, and 100% ethanol, leaving embryos for 1 min in each solution. Remove any last traces of H_2O with 100% isopropanol.

6. Clear in xylene and mount in DPX (Merck)[c] (see *Figure 2a*).

[a] As an alternative to peroxidase-coupled secondary antibodies the Vector ABC method (Vector Laboratories) uses biotinylated antibodies followed by incubation with preformed avidin–peroxidase complex offering the potential of signal amplification.

[b] Sodium azide inhibits the peroxidase reaction and so it is important that the washes following the secondary antibody and reaction buffer are azide-free.

[c] Other embedding media alternatives are methyl salicylate (see ref. 6) or Araldite.

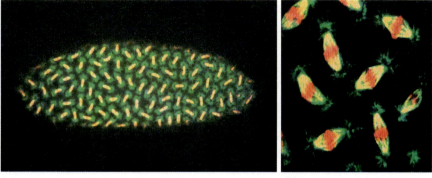

(a)   (b)

Figure 1. Double label immunofluorescence labelling of intact embryos. The green label is anti-tubulin followed by species-specific FITC-conjugated secondary antibody, and the red label is anti-chromatin followed by species-specific TRITC-conjugated secondary antibody. (a) Whole embryo image produced by overlaying a series of confocal sections. (b) Higher power image of a single confocal section. (Images provided by Tom Weaver.)

(a)   (b)

Figure 2. Immunoperoxidase labelling of embryos with anti-connectin antibody. (a) Intact embryo; labelling developed with DAB substrate. (b) Embryo flat preparation; labelling developed with DAB plus Ni. (Images provided by Srikala Raghavan and Lisa Meadows.)

3.2.3 Post-labelling dissection

For many tissues, such as the central nervous system and mesoderm, visualization of the antibody labelling is easier in dissected flattened preparations than in whole embryos (*Protocol 5*).

Protocol 5. Post-labelling dissection of embryos

Equipment

• Dissection needles constructed from fine insect pins (minutia pins 100A1, Watkins and Doncaster) inserted into a short length of glass capillary tubing and secured with Araldite

Method

1. Take embryos from *Protocol 4*, step 4.

Protocol 5. *Continued*

2. Resuspend embryos in 50% glycerol and allow to settle (this may take up to an hour).

3. Resuspend embryos in 70% glycerol and allow to settle overnight at 4°C. The embryos can be stored for months at 4°C at this stage.

4. Transfer the embryos to a depression slide. Select embryo for dissection and transfer to a microscope side together with a minimal amount of glycerol. Dissect embryo open dorsally and flatten out the epidermis on one side. Then carefully remove gut to reveal the central nervous system and mesoderm and flatten out the other side of the epidermis.

5. Place support coverslips on either side of the dissected embryo to mount preparations. Then place a coverslip over the embryos and add 10 μl of 70% glycerol from the side.

6. Remove the support coverslips to flatten the preparation. Excess glycerol can be taken off with the edge of a tissue.

7. Seal with nail varnish (see *Figure 2b*).

3.2.4 Capillary tube mounting of embryos

This method (*Protocol 6*) allows whole embryos or dissected central nervous system preparations to be viewed from a variety of orientations (8).

Protocol 6. Capillary tube mounting of embryos

Equipment

- Capillary tubes: 65 mm long with an internal diameter of 0.2 mm for embryos, central nervous systems of early larvae, or ventral nerve cords of late larvae; or with an internal diameter of 0.24–0.32 mm for imaginal central nervous systems or late larval brain hemispheres. Larger capillaries with an internal diameter of 0.8 mm are used as storage tubes.
- Storage slides are made by fixing storage tubes to a microscope slide with a drop of Araldite (see *Figure 3*)

- Observation slides are prepared by fixing a short length of polyethylene tubing to a slide (see *Figure 3*). Suitable tubing can be made by taking a 10 cm length of 6 mm (outer diameter) tubing, heating over a flame, and pulling slowly to a length of about 50 cm. Cut a 1 cm length from the thinnest part of the pulled tubing.

Method

1. Suck embryos into the capillary. Sometimes it is necessary to push them with a hair fixed to a small holder.

2. Seal the capillary by sticking the tip carrying the preparations into plasticine.

3. Clean the outside of the capillary thoroughly with acetone.

4. Fix a small 'flag' of tape to the other end of the capillary; this will be used to rotate the specimen (see *Figure 3*). Keep capillaries on storage slides.

5. For observation, insert an embryo-loaded capillary into the polythene tube on an inspection slide. Use immersion oil even for low magnification and use the 'flag' to adjust the orientation of the specimen under the microscope.

3.3 Labelling of flat preparations or filets

3.3.1 Labelling of flat embryo preparations

Unfixed embryos can be devitellinized and dissected to produce flat preparations (*Protocol 7*) allowing easier visualization of internal structures (9). These preparations have been particularly useful for the analysis of central nervous system and muscle development. They also allow greater flexibility in the choice of fixation procedure, reduce problems associated with incomplete antibody penetration, and are suitable for antigenic determinants which are sensitive to the methanol used in the mass devitellinization protocol.

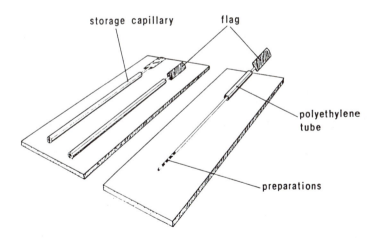

Figure 3. Capillary tube mounting of embryos. For storage, preparation capillaries are inserted into larger capillaries (about 60 mm in length) which have been fixed to glass slides (*left*). For microscopy, a piece of polyethylene tubing is fixed to a slide to hold the preparation capillary (*right*). A piece of tape (flag) attached to the end of the preparation capillary allows it to be turned easily. Figure from Andreas Prokop and Gerhard Technau (8).

Protocol 7.   Labelling of flat preparations of embryos

Equipment and reagents

- Poly-L lysine (PLL) coated coverslips: dip clean coverslips in 1 mg/ml PLL (Sigma P 1542) then air dry. The coverslips should wet evenly; if they don't then flame both sides before dipping. Coverslips can be stored desiccated at room temperature for weeks.
- 4% formaldehyde:PBS (see *Protocol 1*)

- Balanced saline solution (BSS; ref. 10): 55 mM NaCl, 40 mM KCl, 15 mM $MgSO_4$, 5 mM $CaCl_2$, 10 mM Tricine, 20 mM glucose, 50 mM sucrose. Titrate pH to 6.95 with 1 M NaOH, sterile filter, and store at 4°C. Can be made as 10 x stock.
- PTX, PBTX (see *Protocol 3*)

Method

1. Collect, dechorionate, and rinse embryos as in *Protocol 1*, step 2.

2. With a fine paintbrush transfer embryos to a drop of BSS on black sticky tape in a small Petri dish. The black tape is stuck, sticky side up, to the bottom of the Petri dish using double sided Scotch tape. Allow embryos to adhere to tape then cover tape with BSS. Under the stereomicroscope remove the vitelline membrane from an embryo of the desired stage by slitting it open with a pulled glass needle.

3. With a Gilson pipette with a sawn off yellow tip, or with a drawn-out glass capillary mouth pipette, transfer the embryo gently to a drop of BSS on a PLL coated coverslip. The glass capillary of the mouth pipette can be made non-stick by flushing with 10% BSA or saliva. If the embryos are allowed to sediment gently onto the PLL they will often land in the desired dorsal-up orientation. Dissect open the dorsal midline and flatten with the needle or by blowing buffer down onto the embryo from the mouth pipette. Remove the gut.

4. Fix in a small Coplin jar in 4% formaldehyde:PBS for 10–40 min.

5. Rinse in PTX then block in PBTX for 1 h at room temperature.

6. For antibody incubations and washes the solutions are the same as *Protocols 3* and *4* for immunofluorescence or immunoenzyme labelling respectively. The coverslips are placed on Parafilm stretched over a glass plate which retains solutions over the coverslips. Pipette on sufficient antibody solution to cover the preparations. Do not allow to dry out. Incubate in a moist chamber for 1–3 h at room temperature or 4°C overnight.

7. Pipette off the primary antibody and replace with PBTX. Replace the PBTX three times in succession, then twice for 30 min at room temperature.

8. Replace wash with secondary antibody and incubate for 1–3 h.

9. Wash as above.

10. For immunoenzyme labelling develop reaction as *Protocol 4*.

11. Mount the flat preparations by adding mounting medium (for immunofluorescence as in *Protocol 3* or for immunoenzyme labelling as in *Protocol 4*) and overlaying a second coverslip to form a sandwich. The coverslip sandwich can be glued to a microsope slide with nail varnish or simply held in place with plasticine which, by flipping the coverslip, allows the embryo to be viewed from either surface (see *Figure 6*).

3.3.2 Later stages

Cuticle development in late stage embryos (stage 17, ref. 11) presents a barrier to antibody penetration and also the flat preparation of *Protocol 7* does not work once the cuticle has formed. Methods for permeabilization of late embryos and young first instar larvae are given in ref. 6. An alternative approach is to use the flat preparation of *Protocol 7* on stage 16 embryos and then to cover the embryos in *Drosophila* culture medium (Schneider's medium or Shields and Sang M3 medium, Sigma) and allow the embryo to continue development to late embryogenesis before immunolabelling.

Figure 4. Larval flat preparation. A portion of a larval segment is shown. The synaptic boutons on the muscles are visualized using immunoperoxidase labelling using anticysteine string protein and developing with DAB. (Image provided by Srikala Raghavan.)

The flat preparation of larvae (*Protocol 8*) has been used extensively to study the larval body wall muscles (12).

Protocol 8.   Larval flat preparations

Equipment and reagents

- Ringer's solution: 130 mM NaCl, 4.7 mM KCl, 1.9 mM CaCl$_2$, 10 mM Hepes pH 6.9—it is convenient to make 10 x Ringer's solution, filter sterilize, and freeze in 25 ml aliquots
- 4% formaldehyde in PBS (see *Protocol 1*)
- Sylgard dish: cover the bottom of a small Petri dish with Sylgard (Dow Corning Corp.) and allow to set
- Microscissors (Vanna's, Agar Scientific)
- PTX (see *Protocol 3*)

Method

1. Wash larvae in water in a depression slide.
2. Transfer a larva to a Sylgard dish containing Ringers and pin down using insect pins. Make an incision using microscissors either dorsally or ventrally, depending on whether the ventral or dorsal muscles are to be viewed, and cut the larva along the midline. Remove the internal tissues and pin out the larval body walls.
3. Fix for 45 min–1 h in 4% formaldehyde in PBS.
4. Unpin the larva body walls, transfer to an microcentrifuge tube in PTX, and treat as embryos for immunolabelling (see *Figure 4*).

---

4. Immunolabelling of imaginal discs

Analysis of protein distribution in imaginal discs by immunolabelling (*Protocol 9*) has played a major role in our understanding of pattern formation and differentiation mechanisms (13, 14).

Protocol 9.   Fixation and immunolabelling of imaginal discs

Equipment and reagents

- Wire mesh baskets formed by moulding a square of fine wire mesh over the base of a pencil
- 24- or 48-well flat-bottom tissue culture plates
- Glass depression wells or slides or Petri dishes containing Sylgard (Dow Corning Corp.)
- Fine forceps (Dumont, No. 5)
- Sharp tungsten dissection needles: hypodermic needles (use 21G needles on a 1 ml syringe barrel) make good disposable dissection needles
- BSS (see *Protocol 7*)
- Fix: 4% formaldehyde in PBS—made from solid paraformaldehyde as in *Protocol 1* and stored at –20°C[a]

Method

1. Collect larvae and wash in water in a depression slide.

2. Transfer to BSS in a glass depression slide or Sylgard dish. Separate the anterior third of the larvae from the rest by clamping the larva between two pairs of forceps close together then pulling apart. Invert the head end by pushing the mouthparts inwards with one pair of forceps while anchoring the head with the other pair.

3. Carefully dissect away extraneous tissues such as digestive tract, fat body, and salivary glands.

4. When analysing mutant and wild type imaginal discs remove the gut from wild-type but not from the mutant larvae to allow both to be processed together and the subsequent identification of the mutant discs following antibody labelling.

5. Transfer the inverted larval heads to wire mesh baskets in fix in a well of a multiwell tissue culture plate. Fix for 20–30 min at room temperature.

6. Proceed with antibody labelling as in *Protocols 3* or *4*[b] transferring the discs through the block, antibody incubation, and wash solutions in the wells of the multiwell tissue culture plate, keeping the discs in the wire basket.

7. After washing off the secondary antibody:

   (a) For immunofluorescent labelling transfer the larval heads to a glass depression slide or Sylgard dish and dissect out the individual imaginal discs using dissection needles. Mount the discs in Citifluor or 70% glycerol plus anti-photobleaching reagent (see *Protocol 3*).

   (b) For immunoenzyme labelling transfer the larval heads to a depression slide and develop the peroxidase reaction as in *Protocol 4*. Then dissect out the discs as above and dehydrate, clear, and mount in DPX (see *Figure 5*).

[a] An alternative fix is PLP: 2% paraformaldehyde, 10 mM $NaIO_4$, 75 mM lysine, 37 mM phosphate buffer pH 7.4 (15). Also, Brower recommends the addition of 1% NP-40 to the fix (7).
[b] For labelling imaginal discs use PBTX: PBS, 0.1% BSA with 0.3% Triton X-100.

5. Immunolabelling of ovaries

Ovaries present specific fixation and penetration problems as they are large structures and the mature oocytes are surrounded by developing egg membranes. Specific protocols have been devised (*Protocol 10* and refs 16–18).

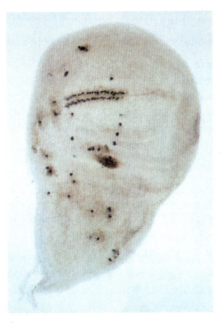

Figure 5. A wing imaginal disc labelled with double immunoperoxidase with two different peroxidase reaction products showing the development of sensory mother cells (labelled with blue/black reaction product representing β-galactosidase expression from the enhancer trap line A101) within areas of Enhancer of split expression (labelled with brown reaction product). First antibody set: rabbit anti-β-galactosidase, followed by HRP-conjugated anti-rabbit Ig, and developed using DAB plus Ni. Second antibody set: mouse monoclonal anti-Enhancer of split, followed by HRP-conjugated anti-mouse Ig, and developed using DAB. (Image provided by Sarah Bray and Jesus de Celis.)

Protocol 10.   Labelling of ovaries

Reagents

- Ringer's solution (see *Protocol 8*)
- Buffer B: 100 mM KH$_2$PO$_4$/K$_2$HPO$_4$ pH 6.8, 450 mM KCl, 150 mM NaCl, 20 mM MgCl$_2$.6H$_2$O
- Fixation buffer: 1 vol. buffer B, 1 vol. 37% formaldehyde, 4 vol. H$_2$O
- PBTX: PBS, 0.5% BSA, 0.3% Triton X-100

Method

1. Dissect out ovaries in cold Ringer's solution. Anaesthetized adult females are held, ventral side up, with a pair of forceps. With another pair of forceps a hole is made in the posterior cuticle and the ovaries gently squeezed out. Dissecting apart the ovarioles aids antibody penetration.

2. Transfer ovaries to a microcentrifuge tube containing cold Ringer's solution on ice.

3. Remove Ringer's solution with drawn-out Pasteur pipette.

4. Add 100 µl of fixation buffer and 600 µl heptane.

5. Shake vigorously to equilibrate phases and then roll for 10 min at room temperature.

6. Remove the fix solution and rinse three times with PBS.

7. Block for 10 min in PBTX at room temperature.

8. Incubate in required concentration of primary antibody diluted in PBTX for 2 h at room temperature followed by overnight rolling at 4°C.

9. Rinse with PBTX three times and then four washes of 15 min.

10. Incubate in secondary antibody at required dilution in PBTX for 2 h at room temperature.

11. Wash as in step 5 then rinse twice with PBS.

12. Develop for immunoenzyme labelling as in *Protocol 4* or proceed to step 13 for immunofluorescence.

13. Add PBS:glycerol (1:1) and leave for about 20 min to equilibrate.

14. Mount and observe.

6. Double and multiple labelling

It is often useful to be able to see the expression of more than one protein in a single preparation. This allows, for example, the direct comparison between expression patterns. With immunofluorescence this can be done by using different fluorochromes (*Protocol 11*). With immunoenzyme labelling this can be done using different enzymes and substrates (*Protocol 12*). Double labelling with immunofluorescence is superior in that the individual labels can be viewed both separately and superimposed.

A number of strategies can be used for distinguishing the labelling due to different primary antibodies:

(a) The simplest strategy uses primary antibodies raised in different species followed by their separate detection with species-specific secondary antibodies.

(b) Primary antibodies can be directly coupled to different fluorochromes or enzymes.

(c) One of the primary antibodies can be biotinylated allowing detection with fluorescent or enzyme-linked avidin or ABC (Vector Laboratories) reagents. For example, using two mouse monoclonal antibodies, the monoclonal A is detected using an anti-mouse Ig secondary followed by

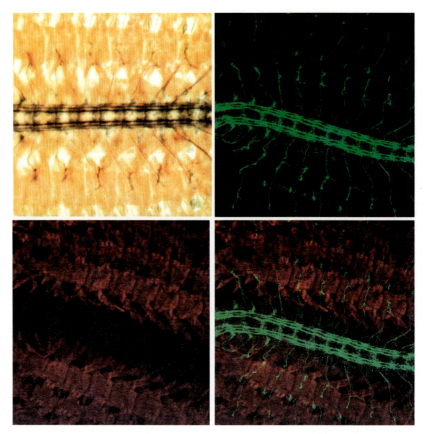

Figure 6. Double labelling on embryo flat preparations. (a) Double immunoenzyme labelling with two different peroxidase reaction products. First antibody set: mouse monoclonal anti-Fasciclin II, followed by HRP-conjugated anti-mouse Ig, and developed using DAB plus Ni to give a blue/black reaction product. Second antibody set: rabbit anti-myosin heavy chain, followed by HRP-conjugated anti-rabbit Ig, and developed using DAB to give a brown reaction product. (b–d) Double immunofluorescence labelling viewed by confocal microscopy. The green label is mouse monoclonal anti-Fasciclin II followed by FITC–anti-mouse Ig. The red label is rabbit anti-myosin heavy chain followed by Texas Red–anti-rabbit Ig. (b) Green channel alone. (c) Red channel alone. (d) Overlap. (Images provided by Srikala Raghavan.)

blocking unoccupied sites with mouse Ig, the biotinylated monoclonal B is then used followed by detection with the avidin reagent.

(d) Class, subclass, or subtype differences between primary monoclonal antibodies may be exploited using appropriate secondary antibodies.

(e) If the antibodies label antigens with distinct subcellular localizations, a single detection method may still allow the simultaneous visualization of the different antigens.

Protocol 11. Double immunofluorescence labelling

1. Prepare, fix, and block preparations for immunolabelling according to appropriate protocols.

2. Incubate with both primary antibodies together, e.g. a mouse monoclonal and a rabbit antiserum.[a]

3. Wash according to appropriate protocol.

4. Incubate with appropriate secondary antibodies[b] using one rhodamine- (or Texas Red)-conjugated antibody and one fluorescein-conjugated antibody.

5. Wash and mount.

6. View using appropriate filter sets (see *Figures 1* and *6*).

[a] For some antibody combinations a lower background labelling can be obtained by performing the antibody incubations sequentially (i.e. primary 1 followed by secondary 1, then primary 2 followed by secondary 2).
[b] The secondary antibodies should only recognize the appropriate primary antibody and should not cross-react with the inappropriate primary antibody or the other secondary antibody. Use secondary antibodies which have been absorbed to remove inappropriate species cross-reactivity, e.g. AffiniPure reagents from Jackson ImmunoResearch Laboratories, Inc. A three colour procedure has been described by Paddock *et al.* (19).

Protocol 12. Double immunoenzyme labelling with peroxidase and alkaline phosphatase labelled antibodies

Reagents

- DAB solution (see *Protocol 4*)
- AP buffer: 5 mM $MgCl_2$, 100 mM NaCl, 100 mM Tris–HCl pH 9.5, 0.1% Tween 20
- BCIP:NBT solution: 1 ml AP buffer, 3.5 μl bromochloroindolyl phosphate (Sigma, 50 mg/ml in 70% dimethyl formamide), 4.5 μl nitroblue tetrazolium (Sigma, 50 mg/ml in 70% dimethyl formamide)—mix just before use

Method

1. Follow *Protocol 11*, steps 1–4 using appropriate peroxidase and alkaline phosphatase labelled secondary antibodies.

2. Develop HRP reaction as brown reaction product using DAB solution.[a]

3. Wash twice for 1 min in PT, followed by twice for 5 min in AP buffer.

4. Develop alkaline phosphatase reaction by adding 300 μl BCIP:NBT. Monitor staining. The reaction is normally optimal after 5–15 min but can be allowed to continue for several hours.

5. Dehydrate, clear, and mount as in *Protocol 4* for embryos, or appropriate protocol for other tissues.

[a] If the first labelling reaction interferes with the second, the labelling reactions can be performed sequentially, and the first antibody set can be stripped off. Wash the embryos five times in PT after developing the first reaction, then incubate in 200 mM glycine–HCl pH 2.2 for 5 min. Wash the embryos five times in PT, then reblock in PBTX, and continue with the second antibody set (20).

Two different peroxidase reaction products can also be used for double immunoenzyme labelling (*Protocol 13*). This procedure, using black and brown peroxidase reaction products can be used with primary antibodies from different species in sequential reactions. It is also useful with primary antibodies from the same species as the darker reaction is performed first, and then it does not matter if the second reaction gives some additional staining over the first reaction product.

Protocol 13. Double immunoenzyme labelling with two different peroxidase reaction products

Reagents

- DAB plus Ni solution: to 1 ml 0.3% DAB in PBS, 0.05% Tween 20 (see *Protocol 4*) add 10 μl of 3 M nickel chloride (stock solution: 3 M NiCl$_2$ in H$_2$O, which can be stored indefinitely at room temperature), and 2 μl of H$_2$O$_2$—mix well and use immediately
- DAB solution (see *Protocol 4*)

Method

1. Follow *Protocol 4*, steps 1–6 for embryos, or appropriate protocol for other tissues.

2. Develop to give black reaction product using DAB plus Ni solution.

3. Wash twice for 1 min in PT followed by twice for 30 min in PBTX.

4. Repeat antibody labelling for second antibody set.

5. Develop to give brown reaction product in DAB solution.

6. Dehydrate, clear, and mount as in *Protocol 4* for embryos, or appropriate protocol for other tissues (see *Figures 5* and *6*).

---

7. Immunolabelling of salivary glands

A particular application of immunolabelling in *Drosophila* is the identification of the chromosomal location of chromatin proteins and gene regulatory proteins on the salivary gland chromosomes (*Protocol 14*). This offers an approach to the study of the role of specific molecules in chromosome architecture and also a way to map the target genes regulated by particular transcription factors (21, 22).

Protocol 14. Labelling of salivary gland chromosomes

Equipment and reagents

- Siliconized coverslips
- Poly-L-lysine coated slides (see *Protocol 7*)
- Pencil with eraser end
- Solution 1: 0.1% Triton X-100 in PBS pH 7.5
- Solution 2: 3.7% formaldehyde (from 37% formaldehyde stock solution prepared by dissolving 1.85 g paraformaldehyde in 5 ml H$_2$O plus 70 μl 1 M KOH by heating to 70°C), 1% Triton X-100 in PBS pH 7.5

- Larval medium (see Chapter 1)
- Solution 3: 3.7% formaldehyde:50% acetic acid
  Note: solutions 2 and 3 should be freshly made and used within 2–3 h
- Blocking solution: 3% BSA, 10% non-fat dry milk, 0.2% NP-40, 0.2% Tween 20–80 in PBS
- Developing solution: 0.5 mg/ml diamino-benzidine tetrachloride, 0.01% H$_2$O$_2$

A. *Preparation of third instar larvae*

1. Let flies lay eggs in bottles with larval medium and a generous helping of live baker's yeast paste.

2. Control the number of eggs layed to give uncrowded bottles (less than 100 larvae/bottle).

3. Grow larvae at 18°C.

4. Use third instar larvae that are still crawling and have not begun pupariation.

B. *Chromosome squashes*

1. Dissect two pairs of glands in solution 1. With one pair of forceps hold the larva by the abdomen, then take hold of the mouth parts with the other pair, and pull. Dissect the glands free from the mouth parts, and dissect away most of the fat body, but don't separate the two glands.

2. Using a tungsten needle with a hook transfer the glands to a drop of solution 2 on a siliconized coverslip.

3. Fix the glands evenly in solution 2 keeping them moving with the tungsten needle for 10–30 sec (time needs to be adjusted for each individual antigen).

4. Fix the glands in a 40 μl drop of solution 3 on a non-siliconized 22 × 22 mm coverslip leaving them for 2–3 min. During this time use tungsten needles to break up the glands and remove any remaining chitinous structures of the pharynx.

5. Take up the coverslip with a poly-L-lysine treated slide. Break up the cells by tapping on the coverslip with a pencil and monitor progess under the stereomicroscope. Hold the coverslip and spread the chromosomes by pressing using the eraser end of the pencil.

6. Remove excess fixative by pressing slides (coverslip down) onto blotting paper.

Protocol 14. *Continued*

7. Examine the extent of spreading under phase-contrast and mark the position of the coverslip with a diamond pencil.

8. Freeze the slide by dipping in liquid N_2 using plastic forceps and flick off the coverslip with a razor blade.

9. Wash slides twice for 15 min in PBS with gentle shaking.

10. Proceed immediately with immunolabelling or store the slides for up to one week in methanol at 4°C.

C. *Immunolabelling*

1. Wash stored slides twice for 15 min in PBS.

2. Block for 1 h in blocking solution at room temperature.

3. Add 40 µl of primary antibody diluted as appropriate in blocking solution. Cover with coverslip and incubate for 1 h at room temperature in a humid chamber.

4. Rinse in PBS then wash with shaking for 15 min in PBS, 300 mM NaCl, 0.2% NP-40, 0.2% Tween 20–80, then for 15 min in the same buffer except at 400 mM NaCl. To remove more background labelling the NaCl concentration can be raised to 500 mM.

5. Rinse in PBS.

6. Add 40 µl diluted secondary antibody in blocking buffer plus 2% normal goat serum. Cover with coverslip, and incubate for 40 min at room temperature in a humid chamber.

7. Rinse in PBS.

8. Wash as in step 4 then rinse in PBS.

9. Add 100 µl developing solution and follow the reaction under phase-contrast. Stop the reaction by dipping slides in PBS then wash for 10 min in PBS.

D. *Cytology*

1. To reveal the background banding pattern to allow the identification of any antibody labelled bands, stain the chromosomes for 10–20 sec in Giemsa solution (diluted 1:130 in 10 mM sodium phosphate buffer pH 6.8).

2. Mount in 99.5% glycerol and immediately examine the slides as the Giemsa stain fades within a few hours (see *Figure 7*). The chromosomes can be washed in PBS and restained. The slides can be stored frozen at −20°C.

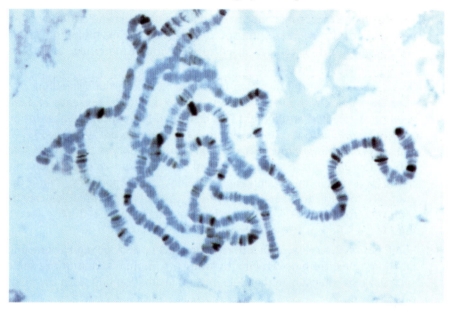

Figure 7. Labelling of salivary gland chromosomes. The chromosomal locations of Polycomb protein are revealed by immunoperoxidase labelling (dark bands) which can be mapped by reference to the blue Giemsa stained chromosomal banding pattern. (Image provided by Renato Paro.)

8. Immunolabelling of sections

Sections are rarely used for immunolabelling in *Drosophila* as optical sectioning in the light microsope for immunoenzyme labelled preparations or confocal microscopy for immunofluorescent labelled material usually allows internal structure to be viewed with sufficient clarity. Sectioning before immunolabelling can be useful however where there are concerns about uniform antibody penetration and methods are given in refs 23 and 24. Sectioning after immunoenzyme labelling can enable high resolution views of internal structures (25).

9. Vexations and variations

9.1 Background labelling problems

Some antibodies give better signal-to-noise labelling than others but steps can be taken to optimize antibody performance:

(a) For an antiserum, affinity purification can dramatically reduce the background (2).

(b) Increasing the detergent concentration in the blocking and antibody bind-ing reactions can decrease non-specific binding. For example, the Triton X-100 concentration can be raised to 1% without inhibiting many anti-gen:antibody binding reactions.

(c) Some secondary antibodies give unacceptable background labelling due to cross-reaction with *Drosophila* antigens or due to the presence of dam-aged or aggregated antibodies. This can often be reduced by absorption on embryos. Dilute the antiserum to 10 x final concentration in PBTX, and add one-fifth volume of fixed, rehydrated embryos. Roll for 3 h at 4 °C. Allow embryos to sediment and take off antibody solution, then spin down for 1 min in microcentrifuge to remove small particles. Dilute the antibody solution to final concentration. Aggregated antibodies can be removed from antibody solutions by centrifugation at 100 000 *g* for 30 min.

(d) Some antibodies cross-react with components in BSA giving background labelling. This can be overcome by using 5% normal serum of the same species as the antibody in the blocking and incubation buffers in the place of BSA.

(e) Background labelling is often associated with poor fixation and increasing the fixation time may help.

9.2 Weak labelling

While weak labelling may reflect a low abundance of antigen, it may also arise from poor recognition of the antigen. There are ways of overcoming some causes of weak labelling:

(a) Signals can often be improved by careful optimization of fixation conditions by varying time of fixation (5 min–overnight), temperature of fixation (4–37 °C), avoiding exposure to methanol, or switching to a different fixative (e.g. glutaraldehyde).

(b) A common problem with monoclonal antibodies is that, while they recog-nize the partially denatured protein on Western blots, they fail to bind to fixed 'native' protein in embryos. Most likely this is due to the antigenic determinants being buried in the native state. Denaturation with urea or protease digestion (26) may help.

(c) If the antibody labels the antigen on Western blots of bacterially expressed protein but fails to label blots of *Drosophila* extracts it is worth considering that post-translational modification, e.g. phosphorylation, may block the antigenic determinant and enzymatic removal of the modi-fication may reveal the determinant.

(d) Weak labelling can also result from dissociation of the antibody from the antigen. This is mainly a problem of primary antibodies and reduction in the length of the steps following primary antibody incubation can help. Stabilization of the bound primary antibody can be achieved by 10 min fixation with 4% formaldehyde:PBS after washing off the primary antibody.

(e) Similarly, as immunofluorescence labelling does not result in a stable reaction product there is a danger that antibodies may dissociate in the time between labelling and observation. A post-labelling fixation can solve this problem.

Acknowledgements

I would like to thank Sean Carroll, Grace Panganiban, and Mark Peifer for advice, and members of my laboratory and Sarah Bray for help with the protocols, and for critical reading of the manuscript. I am greatly indebted to Don Mason who, some time ago, taught me to think about antibodies.

References

1. Bate, M. and Martinez-Arias, A. (ed.) (1993).*The development of Drosophila melanogaster.* Cold Spring Harbor Laboratory Press, NY.
2. Harlow, E. and Lane, D. (1988). *Antibodies: a laboratory manual.* Cold Spring Harbor Laboratory Press, NY.
3. Asai, D. J. (ed.) (1993). *Methods in cell biology*, Vol. 27. Academic Press, London.
4. Matsumoto, B. (ed.) (1993). *Methods in cell biology*, Vol. 38. Academic Press, London.
5. Mitchison, T. J. and Sedat, J. W. (1983). *Dev. Biol.*, **99**, 261.
6. Patel, N. H. (1993). In *Methods in cell biology* (ed. L. S. B. Goldstein and E. A. Fyrberg), Vol. 44, p. 445. Academic Press, London.
7. Brower, D. L. (1987). *Development*, **101**, 83.
8. Prokop, A. and Technau, G. M. (1993). In *Cellular interactions in development* (ed. D. A. Hartley). IRL Press, Oxford, p. 33
9. Broadie, K. and Bate, M. (1993). *J. Neurosci.*, **13**, 144.
10. Chan, L.-N. and Gehring, W. (1971). *Proc. Natl. Acad. Sci. USA*, **68**, 2217.
11. Campos-Ortega, J. A. and Hartenstein, V. (1985). *The embryonic development of Drosophila melanogaster.* Springer–Verlag, NY.
12. Bate, M. (1993). In *The development of Drosophila melanogaster* (ed. M. Bate and A. Martinez-Arias). Cold Spring Harbor Laboratory Press, NY. p. 1013.
13. Cohen, S. M. (1993). In *The development of Drosophila melanogaster* (ed. M. Bate and A. Martinez-Arias). Cold Spring Harbor Laboratory Press, NY. p. 747.
14. Williams, J. A., Paddock, S. W., Vorwerk, K., and Carroll, S. B. (1994). *Nature*, **368**, 299.
15. Tomlinson, A. and Ready, D. F. (1987). *Dev. Biol.*, **120**, 366.
16. Lin, H. and Spradling, A. C. (1993). *Dev. Biol.*, **159**, 140.
17. Verheyen, E. and Cooley, L. (1994). In *Methods in cell biology* (ed. L. S. B. Goldstein and E. A. Fyrberg), Vol. 44, p. 545. Academic Press, London.
18. Orsulic, S. and Peifer, M. (1996). *J. Cell. Biol.*, **134**, 1283.

19. Paddock, S. W., Langeland, J. A., DeVries, P. J., and Carroll, S. B. (1993). *BioTechniques*, **14**, 42.
20. Skeath, J. B. and Carroll, S. B. (1992). *Development*, **114**, 939.
21. Zink, B. and Paro, R. (1989). *Nature*, **337**, 468.
22. Andrew, D. J. and Scott, M. P. (1994). In *Methods in cell biology* (ed. L. S. B. Goldstein and E. A. Fyrberg), Vol. 44, p. 353. Academic Press, London.
23. White, R. A. H. and Wilcox, M. (1985). *EMBO J.*, **4**, 2035.
24. Kelsh, R., Weinzierl, R. O. J., White, R. A. H., and Akam, M. (1994). *Dev. Genet.*, **15**, 19.
25. Leptin, M. and Grunewald, B. (1990). *Development*, **110**, 73.
26. Larsson, L.-I. (1988). *Immunocytochemistry: theory and practice*. CRC Press, Boca Raton, Florida.

8

Population and ecological genetics

J. F. Y. BROOKFIELD

1. Introduction

Drosophila melanogaster is a cosmopolitan species that lives as a human com-
mensal. The family Drosophilidae belongs to the Dipteran Sub-Order Cyclor-
rhapha and is thought to be primitively saprophytic, living on fermenting
substrates. However, immense variety now exists in the food sources exploited
by different species. Of 3020 species in the 61 genera of the family, 1677 are in
the genus *Drosophila*, a result that must partially reflect incomplete sampling of
the others (1). The taxonomic richness of the group potentially allows the comparison of *D. melanogaster* genes with those of other species at many different
levels of evolutionary divergence. The phylogenetic relationships of species of
Drosophila are imperfectly known, but a number of subgenera, species groups,
species subgroups, and species complexes are recognized between the generic
and specific ranks. The *melanogaster* species group, including also *D. simulans*,
D. mauritiana, *D. sechellia*, *D. yakuba*, *D. teissieri*, *D. erecta*, and *D. orena* has a
phylogeny that has been well established through study of the polytene chromosomes and DNA sequence divergence. These species are important in some
tests for selection outlined below. The species group originated in Africa, and
the first bifurcation, separating the ancestor of *D. orena* and *D. erecta* from that
of the others, occurred between 15 and 20 million years ago.

Population genetics measures the levels of genetic variability within populations, and the forces that determine them. The interest in these questions is
twofold. First, genetic diversity gives an indication of population history—if
two populations are similar genetically, this suggests that they have shared
ancestry recently, or that they continue to exchange genes, or both. Secondly,
the neo-Darwinian theory of evolution specifies that evolution proceeds
through changes, by genetic drift or natural selection, in the frequencies of
different genetic types in populations through time. Thus, population genetics
is evolution in microcosm.

Genetic polymorphisms are the traditional focus of population genetics
research. A genetic polymorphism was classically defined (2) as the co-occur-

rence of two different genetic types at such frequencies that the rarest cannot be being maintained merely by recurrent mutation. In large populations, almost all genetic loci show some variation, with harmful alleles being found at low frequencies determined by the balance between mutation and selection. Genetic polymorphism is defined in terms of frequency to exclude such variants, assuming that mutation and selection jointly solely determine allele frequencies. However, the sampling of genotypes from the gametes of the previous generation causes variation in gene frequencies, a process known as genetic drift. High frequencies of deleterious alleles, and even their fixation in populations, can arise if selection is weak and the population size is small. The early years of population genetics were dominated by a debate between Sewall Wright and R. A. Fisher on the importance of drift in evolution. The between-generation variance in gene frequency due to drift varies with the reciprocal of the effective population size (which may be very much less than the census population size), and the question at issue was whether wild populations had sufficiently small effective population sizes for genetic drift to play a major role in changing gene frequencies, in comparison with selection.

The Darwinian principle that fitness is determined by the adaptedness of different genotypes to the environment adds an ecological dimension to studies of selectively important variation. It has stimulated attempts to identify, for genotypes having different fitnesses in the wild, selective agents determining these differences. However, separating the concepts of phenotype and selection, such that genes and environment are seen to create a spectrum of phenotypes, from amongst which the environment selects the fittest, is usually inappropriately complex. It is logically more consistent to view fitness as itself a phenotype, albeit one often environment-dependent.

Empirical population genetics was long constrained by the rarity of genetic polymorphism, with most variability being polygenic, the effects of individual loci being indistinguishable. Indeed, in *Drosophila*, the only polymorphisms easily accessible prior to the 1960s were polymorphic chromosomal inversions, studied memorably in *D. pseudoobscura* by Dobzhansky and co-workers (3). While inversion frequency changes of magnitudes too large to be explicable by drift were indeed found in wild populations, the mechanism for the selection remained mysterious. Furthermore, it was hard to draw inferences from this selective process to those in other species, lacking this kind of variability.

In the 1960s, polymorphisms in abundance were discovered in *Drosophila* and other groups through gel electrophoresis (see Section 2.7). Allelic differences in the charge of polypeptides (known as allozymes) cause differences in electrophoretic mobility. Specific stains for enzyme activity allow the identification of the products of individual gene loci. For the first time, the presence or absence of population genetic variability could be studied in a potentially random sample of loci, whose initial identification had not depended upon the existence of such variation. Lewontin and Hubby (4), applying the technique to five populations of *D. pseudoobscura*, discovered polymorphisms in 30% of loci, and allelic frequencies were such that an individual was heterozygous, on average, at 12% of the sampled loci. These figures were fairly typical of those for invertebrates, with vertebrates showing about half these levels of variability.

The discovery of high levels of allozyme polymorphism, coupled with the observation of high and approximately constant rates of protein sequence evolution, led to the suggestion of Kimura (5), and King and Jukes that the vast majority of protein sequence polymorphism, and amino acid changes in evolution, were selectively neutral. This view was partially stimulated by the issue of genetic load, the number of selective deaths required either for the selective maintenance of multiple polymorphisms, or for multiple adaptive substitutions, which seemed too high for populations to endure. While threshold models of selective epistasis rapidly showed that the genetic load argument for neutrality was not necessarily a serious one, many other predictions of the neutral theory (such as the allozyme frequency spectrum) (6) turned out to be remarkably close to observation.

The neutral theory is sufficiently simple that it yields fairly detailed quantitative predictions about wild populations. One concerns the level of heterozygosity. The inbreeding effect of effective population size (N) reduces heterozygosity by a proportion $1/2N$ each generation. Thus, if h is the proportion of homozygotes, and $(1 - h)$ the proportion of heterozygotes, there will be an increase of $(1 - h)/2N$ in homozygosity through inbreeding. However, in any generation there will be a probability of neutral mutation of approximately $2\mu_N$ in each of the h homozygotes, where μ_N is the neutral mutation rate per gamete. At equilibrium, therefore, this loss of $2h\mu_N$ is balanced by $(1 - h)/2N$, from which rearrangement gives the equilibrium homozygosity as $1/(1 + 4N\mu_N)$, and heterozygosity as $4N\mu_N/(1 + 4N\mu_N)$ (7). An obvious test for neutrality is thus to compare species with different effective population sizes and see if heterozygosity differs as much as expected. In fact, it does not. This, and other observations, led to the development of the nearly neutral theory (8), postulating many mutations with slight, but non-zero, effects on fitness. Such mutations will be effectively neutral if the selection coefficient between them and wild-type, s, is less than $1/N$. Thus, if most mutations have small s values, an increasing proportion of these become effectively neutral (and thus included in μ_N) as the population size diminishes. A change in N between species thus creates a smaller proportional change in $4N\mu_N$ and thus in equilibrium heterozygosity, consistent with observation.

2. Experimental methods

2.1 Maintenance of the genetic variability of wild populations of *D. melanogaster* in the laboratory

A variety of methods exist for sampling *Drosophila* from the wild, usually involving fruit baits, with or without netting (Chapter 1, and discussed in ref. 9).

Three methods are available for the maintenance of genetic variability in the laboratory corresponding to that in the original sample.

2.1.1 Population cages

Here, flies from the wild are placed in a large plastic enclosure. Holes are cut in the lid, and sealed with gauze. This can be removed at intervals to change the food medium supplied in trays in the base of the cages. The population size is typically around 1000 adults, and occasional movements of the population to new cages are necessary.

Neutrality predicts that the loss of heterozygosity in such a cage will be slow. Thus, if N is the effective population size in the cage, the heterozygosity will drop to a proportion $\exp(-t/2N)$ of its initial value after t generations. Such a population can therefore be used, for example, to estimate the additive genetic variance in the original population for a quantitative trait (see Section 2.2.1). However, cage populations lose rare alleles, and, if the experiment aims to assess the total variation in DNA sequence between alleles sampled from the wild, all alleles will be rare. Thus, imagine that n_1 individuals are sampled from the wild, and used to establish a cage population of effective size N. After t generations, n_2 alleles are sampled from the population (where $n_2 << n_1$). The probability that all of the n_2 alleles are derived from different alleles in the original sample is (approximately) only $\exp[-n_2(n_2 - 1)(N + tn_1)/(4Nn_1)]$. For example, if 200 wild-caught individuals found a cage which has a census population size of 1000 (and an effective size of about a third of this), and, after 10 generations, 20 alleles are sampled from the cage, the probability that each of these is derived from a different one of the 400 alleles in the original sample is only about 4%. This calculation also ignores the selected changes that will occur in the population as it adapts to the cage environment, which will further increase the probability of sampling alleles identical by descent.

2.1.2 Chromosome extraction

Chromosome extraction (as shown in Chapter 1) from a sample of flies results in the generation of a collection of lines with complete homozygosity for the extracted chromosomes. Unlike inbreeding (below) the alleles extracted will be an unbiased sample of the variation in the original population, since even recessive lethal chromosomes can be maintained over the balancers. Indeed, typically around 35% of extracted large autosomes from a variety of species (*D. melanogaster*, *D. pseudoobscura*, *D. persimilis*, *D. willistoni*, *D. subobscura*, and *D. prosaltans*) bear one or more recessive lethal.

A problem with chromosome extraction is that it necessarily involves outcrossing. This may mobilize transposable genetic elements through hybrid dysgenesis. The use of females from balancer strains in *PI* genetic backgrounds should prevent (or greatly reduce) movements of the *P* and *I* elements, but there is no way of knowing if other elements are mobilized in the cross, generating additive genetic variance in multiple traits in addition to that present in the original population sampled. Furthermore, unlike cage populations and inbred isofemale lines, chromosome extraction is possible only in those species with appropriate balancer chromosomes.

2.1.3 Inbred isofemale lines

The storage of sperm by wild caught females means that individual females, placed in vials, produce offspring. Transfer of individual virgin females and males to new vials in each subsequent generation ensures sib mating, and this inbreeding will reduce heterozygosity (assuming neutrality) by 25% per generation. After 20 generations, expected heterozygosity will be only 0.3% of its initial level, but comparison between lines will preserve the genetic variability present in the original population.

A disadvantage of inbreeding by sib mating is that within-line variance may be preserved due to selection, in a balanced lethal system, for example. Selection also biases which of the four alleles initially sampled is the one fixed in the inbred line. If one is interested in variation in a gene which has little effect on fitness this is not a problem if, and only if, the alleles under study are in linkage equilibrium (see Section 3.3) with fitness-determining alleles. An advantage of inbreeding is that there is no introduction of other genotypes, and thus no hybrid dysgenesis. However, Nuzhdin *et al.* (10) report that the retrotransposon *copia* increases its copy number greatly in some inbred lines of *D. melanogaster*, creating multiple new insertions and lowering fitness. It may be that in wild *Drosophila* populations catastrophic increases in transposable element numbers are normally prevented by selection, which is greatly reduced following inbreeding.

2.2 Quantitative genetics

2.2.1 Heritability estimation

Almost all traits vary in wild populations of *Drosophila*, partly due to genetics. However, for the majority of traits, the determining loci cannot be identified since their contribution to the phenotype cannot be distinguished from the effects of other loci and of the environment. The methods of study are those of quantitative genetics (for a full account, see ref. 11). The variance in the phenotypic value of a quantitative trait in a population, V_P, is the sum of the variance due to genetic variation, V_G, and that due to environmental differences between individuals, V_E (genotype–environment covariance here being ignored). Of the genetic variance, of most interest is that due to the additive effects of genes, V_A, since only this allows the population to respond to selection on the trait. The heritability of the trait, h^2, is the ratio of the V_A to V_P.

Heritability predicts the covariance in phenotype between relatives, and is of great importance in plant and animal breeding. Its measurement uses one of two general methods. One is to look at correlations between relatives. In *Drosophila*, pairs of males and females from the population of interest are crossed, and, for each pair, the values of the trait in the offspring and parents are measured. The regression slope, b, of the offspring mean on the parental mean estimates the heritability of the trait. Due to the possibility of a maternal effects, it is sometimes better to regress the offspring mean against the paternal value, b now estimating half the heritability. An alternative estimation of heritability comes from the response of the population to artificial selection. The selection differential, S, is the deviation of the mean trait value in flies selected to be parents from the population mean. R, the response to selection, is the deviation of the offspring mean from this population mean. Assuming no natural selection, and no non-genetic causes of parent–offspring correlations, the expected value of the response, R, is just S multiplied by the heritability.

A heritability of a trait is a description of a given population in a given environment, and continued selection will eventually deplete the additive genetic variance in the trait. It is thought to be for this reason that reproductive characters, such as egg number, often show much lower heritabilities than characters not thought to contribute strongly to fitness, such as the number of abdominal bristles. Reproductive characters will inevitably have been strongly selected in the wild, and their additive genetic variance resultingly depleted.

2.2.2 Mapping and identification of quantitative trait loci (QTL)

While techniques of heritability estimation and artificial selection can be applied to all populations, the powerful genetics of *D. melanogaster* allow the mapping of QTLs. The simplest technique involves the following. An inbred line previously selected to have a high mean value of a trait is crossed with another inbred line with a low value of the trait, where the lines also differ at a number of other, unambiguously scorable, loci (which could be defined morphologically or by molecular means). In the F2, there is segregation of the markers and, if individuals bearing the marker allele from the high line show a significantly higher trait value than individuals with the marker allele from the low line, this shows that a QTL is linked to this marker. Generally, for most quantitative traits in *D. melanogaster*, there are multiple QTLs scattered around the chromosomes. One issue that is being resolved is whether there are truly quantitative trait *loci*, whose effects on the trait are always polygenic, or whether quantitative variation is determined by weak alleles at major loci, whose null phenotype includes a profound effect on the trait in question. For bristle number in *D. melanogaster*, the latter appears true, since QTLs affecting bristle number coincide with major bristle loci such as *achaete-scute*, *scabrous*, *Notch*, and *Delta*. Confirmation of the involvement of these loci can be obtained by complementation tests involving the selected lines and major mutations at these loci (12).

2.3 Enzyme electrophoresis

The use of starch and polyacrylamide gel electrophoresis of proteins, coupled with the use of specific stains, allows a large number of loci to be quickly and cheaply scored. Mobility of the proteins through the gel depends upon their mass and charge. Generally, allozymes differ only in charge. The charge on amino acid residues is pH-dependent, and thus alteration in the buffer pH may resolve alleles not previously distinguishable. A major concern is that the detection of the enzyme after electrophoresis is dependent upon continued enzyme activity, and thus cooling of the gel during operation is essential.

Protocol 1.   Preparing, loading, running, and staining a starch gel[a]

Equipment and reagents
- Electrophoresis equipment plus power supply and cooling
- Potato starch (Connaught Laboratories, Toronto, Ontario)
- Buffer: a variety are available (13)
- Stains: see ref. 13 for recipes of 76 specific enzyme stains

A. *Preparation of gel*

1. Prepare gel mould[b] on horizontal surface.

2. Weigh out potato starch to give 10% weight to volume.

3. Add electrophoresis buffer in 1 litre Erlenmeyer flask.

4. Using insulated gloves and eye protection, swirl the flask over a Bunsen burner. After a couple of minutes, there is a marked increase in viscosity of the fluid.

5. Continue heating with shaking until the viscosity reduces.

6. Use aspirator to apply vacuum to the flask. Boiling will recommence.

7. After the boiling has ceased, remove aspirator and pour gel into mould. Apply upper mould (if relevant).[b]

8. Leave gel to set for 2 h.

B. *Loading and running the gel*

1. Place gel in electrophoresis equipment, add buffer to wells but do not apply wicks to gel surfaces.

2. Cut gel parallel to and near to one edge.

3. Homogenize each fly in 100 μl distilled water, and absorb liquid into a 2–4 mm square of Whatman No. 3 filter paper. Insert the samples sequentially into the cut.

4. Add some filter papers loaded with tracking dye (0.01 g amaranth, 0.01 g Brilliant Blue G, 2.5 ml ethanol, 7.5 ml H_2O) among those with experimental samples.

Protocol 1. *Continued*

5. When all samples have been loaded, use a sponge cloth to connect the buffer wells bearing the electrodes to the gel.

6. Apply a cooling pack to the upper surface of the gel.

7. Turn the power supply on, and allow it to warm up, then adjust to an appropriate voltage and current. Do not allow the current to exceed 100 mA, to prevent overheating. Use movement of the tracking dye to determine the length of the run.

C. *Staining*

1. Switch off power.

2. Remove gel from electrophoresis equipment. Slice horizontally into two or multiple slices.

3. Place each gel slice in an individual staining tray.

4. Make up each stain to be used. These consist of enzyme substrates plus cofactors which will create an insoluble coloured precipitate in the presence of the enzyme activity.[c]

5. Add stain to staining tray to cover gel slice.

6. Incubate at 37°C, observing regularly to monitor the development of the colour.

7. Avoid overstaining of the gel. When the results are clear remove stain solution and permanently fix the gel using an appropriate stain-fixing solution.[d]

[a] Summarized from ref. 13.
[b] Many different types of moulds are available. The simplest involve Perspex formers between two glass plates, the lower of which is retained during electrophoresis.
[c] Most of the stains are extremely toxic and/or carcinogenic, and should be used with great care.
[d] A variety of these have been used, including 50% ethanol in water, 50% glycerol in water, and 1:5:5 glacial acetic acid:methanol:water.

2.4 Measurement of fitness

It is fitness variation in the wild that determines the success or otherwise of alleles, and with it the process of evolutionary change. Nevertheless, measurement of fitness in the wild has been notoriously difficult. In some organisms, notably birds, many have tried to measure 'lifetime reproduction success' (LRS), without, it seems, fully appreciating that selective equality of all genotypes nevertheless involves massive environmental variance in LRS. Indeed, even in rapidly evolving populations, the heritability of LRS is vanishingly small. The ease with which large numbers of *Drosophila* can be kept allows laboratory tests for fitness differences between alleles, chromosomes, and genotypes.

2.4.1 Relative fitnesses of alleles

The most straightforward test for fitness differences between alleles is to compete the alleles in population cages. Ideally, isogenic strains should be used, differing only in the allele under study. If not, fitnesses may vary in the course of the experiment, as recombination slowly creates linkage equilibrium (see Section 3.3) between the studied locus and the genetic background.

Assume that fitness of the genotypes is constant, and that we wish to measure the fitness relative to wild-type. If an autosomal recessive allele has frequency q, the frequency of homozygotes will be (from the Hardy–Weinberg formula— see Section 3.1) q^2, and the change in frequency δq, over time δt, resulting from selection of strength s against these homozygotes, will be given by:

$$\delta q = \frac{q^2(1 - s\delta t) + q(1 - q) - q(1 - q^2\, s\delta t)}{1 - q^2\, s\delta t}$$

Integration reveals:

$$st = (q_0 - q_t)/(q_0\, q_t) + \ln[q_0(1 - q_t)] - \ln[q_t(1 - q_0)]$$

where q_0 is the frequency of the recessive allele at the start of the experiment, and q_t is its frequency after t generations (14). The implication is that the mutant allele should be started in the cage at high frequency. For example, if s is 0.3 (a 30% reduction in fitness of the mutant homozygote) a population with 90% mutant alleles would evolve to a frequency of around 50% in ten generations. A population started with 50% mutant alleles would take a further 20 generations to evolve to 16%. The rate of change will be influenced by genetic drift, determined by a variance of $q(1 - q)/2N$, with N being the effective population size, which may be much less than the census population size of the cage. Thus s must be large relative to $1/2N$, and replicates are essential to reduce the variance in estimates of s.

2.4.2 Relative fitnesses of chromosomes

A number of techniques are available to measure the fitness of extracted autosomes from *D. melanogaster*. The simplest involve measurements of fitness over a single generation.

Protocol 2. Knight–Robertson test for fitness of extracted chromosomes[a]

Equipment and reagents
- Set of wild-type extracted homozygous chromosome lines, whose fitness will be compared
- Balanced line for relevant chromosome, e.g. *CyO/PmIV*, isogenic to wild-type lines

Method

1. Isolate 20 virgin males and females of the wild-type line under test, along with 20 virgin males and females of the balancer strain.

Protocol 2. *Continued*

2. Place all 80 flies in bottle.

3. Replicate five to ten times.

4. Count wild-type homozygotes and *CyO/PmIV* flies in F1. Ignore hetero-zygotes between wild-type and balancer chromosomes.

5. Assess fitness as the ratio of +/+ to *CyO/PmIV* flies. Compare amongst chromosomes tested.

[a] From ref.15.

More efficient is the measurement of fitness in cage populations at equilibrium. The balancer chromosome equilibration technique (16) allows this. The population contains only two second chromosomes, the wild-type under test (+) and a balancer (B), such as *CyO*. If the fitness of the extracted chromosome when homozygous is less than that of its heterozygote with the balancer, heterozygote advantage creates a stable equilibrium. If w is the fitness of the wild-type homozygote relative to the heterozygote ($w < 1$), and p is the frequency of the balancer chromosome, the stable equilibrium is $p = (1 - w)/(2 - w)$.

However, this is the frequency in the zygotes, and not the observed frequency in adults. In order to estimate w, the zygotic frequency of the balancer must be known. Tracey and Ayala (17) sampled both adults and eggs from equilibrium cage populations. The eggs were kept under non-competitive conditions, and the ratio of B/+ to +/+ measured in the emerging adults. Under the same conditions the ratio of B/+ to +/+ adults emerging from a cross between B/+ parents was measured. From three ratios, w in the cage population and its components of viability and fertility can be measured in the following way.

Protocol 3. Measurement of fitness components from a balancer equilibration population cage[a]

Equipment and reagents
- Population cage
- Heterozygous strain between balancer (B) and extracted wild-type chromosome

Method

1. Establish a population cage(s) by addition of heterozygous strain.

2. Monitor the frequencies of B/+ and +/+ genotypes for 10–20 generations until a stable equilibrium is established.

3. Count the B/+ and +/+ adults in the cage. Call H the proportion of balancer heterozygotes among these flies.

4. Remove eggs from the cage and grow the offspring under non-competitive conditions. Call *h* the proportion of balancer heterozygotes among the flies emerging.

5. Under the same conditions as step 4, grow the offspring of a cross between B/+ flies. Call *r* the proportion of balancer heterozygotes among the flies emerging. Clearly, *r* > *h*.

6. Calculate the fitness of the wild-type homozygotes relative to that of the balancer heterozygotes, *w*, by $w = (r - h)/[r(1 - h)]$.

7. Calculate the viability of wild-type homozygotes relative to that of the balancer heterozygotes, *V*, by $V = 2h(1 - r)(1 - H)/[rH(1 - h)]$.

8. Calculate the fertility of wild-type homozygotes relative to that of balancer heterozygotes, *F*, by $F = H(r - h)/[2h(1 - r)(1 - H)]$.

9. The viability component, *V*, measures the fitness of +/+ homozygotes between the zygotes and the time when adults are censused, and *F* measures all fitness components which act subsequently.

[a] Modified from ref. 18.

2.4.3 Relative fitnesses of genotypes

In measuring the relative fitnesses of whole genotypes there must be some competition between the genotype under study and a reference genotype, with this competition including both the viability and fertility components of fitness. The problem arises, therefore, that the mating integral to the measurement of fertility will necessarily break up the diploid genotypes whose fitness is to be measured. Two methods have been suggested to overcome this problem. The first is competition between species, in which different genotypes of *D. melanogaster*, for example, are competed in population cages with a standard strain of *D. simulans*. The second involves the use of compound autosomes, where the physical joining of autosomal homologues means that crosses to flies with wild-type chromosome complements will be infertile.

Protocol 4. Measurement of fitness of a genotype using compound autosomes[a]

Equipment and reagents
- Set of homozygous wild-type lines, whose fitness will be compared
- Compound autosome strain, e.g. C45 (C(3L)RM; *ri* , C(3R)RM.*ry²*)

Method

1. Isolate 20 virgin males and females of the line under test, along with 20 males and females of the compound autosome strain.

2. Place all 80 flies in bottle.

Protocol 4. *Continued*

3. Replicate five to ten times.

4. Count wild-type and C(3L)RM; *ri* , C(3R)RM.*ry²* flies in F1.

5. Assess fitness as the ratio of +/+ to C(3L)RM; *ri* , C(3R)RM.*ry²* flies. Compare amongst strains tested.

[a] Adapted from ref. 18.

2.5 Measurement of DNA sequence variation

The measurement of DNA sequence variation in *Drosophila* populations uses what are standard techniques equally applicable to other molecular investigations in this genus, some of which are covered in Chapter 6. The use of PCR and direct sequencing is particularly advantageous in *Drosophila* since homozygous strains can be produced by inbreeding or chromosome extraction.

3. Mathematical and statistical analyses

3.1 Hardy–Weinberg equilibrium and estimation of allele frequencies

Normally, our information about wild populations comes from a collection of genotypes of a small sample of size n that have been randomly sampled from the population of size N. From these data, we generally wish to infer the properties of the very much larger population. A distinction can be drawn between *parameters*, which are the unobservable values of variables describing this larger population, and *statistics*, which are variables derived from the sample and used to estimate these parameters. A first step is to estimate the frequencies of alleles in the population. Suppose that we have seen i different alleles in the n individuals in the sample. If $n_{i,i}$ and $n_{i,j}$ represent respectively the frequencies of i/i homozygotes and i/j heterozygotes in the sample, then, in the absence of dominance, an unbiased estimate, f_i, of the frequency of allele i in the population is given by gene counting as:

$$f_i = n_{i,i} + \frac{1}{2} \sum_{j, i \neq j} n_{i,j}$$

If dominance is present, individuals showing only a single dominant allele could be homozygotes or heterozygotes with recessive alleles. The Hardy–Weinberg principle predicts the frequencies of diallelic genotypes from the allele frequencies in the case of random mating, and no selection or mutation. If p_i is the population frequency of the ith allele, the probability that an individual is an i/i homozygote is p_i^2, and the probability of an i/j heterozygote is $2p_i p_j$. The Hardy–Weinberg principle allows the estimation of allele frequencies as MLEs (maximum likelihood estimates) even in cases of dominance. The principle of the maximum likelihood estimation is that for each value of a parameter there is a likelihood of producing the observed data. The MLE is the value of the parameter that maximizes this likelihood.

Why should there be deviations from Hardy–Weinberg equilibrium? First, there could be heterozygote advantage for viability, which will produce a heterozygote excess. Population geneticists are very interested in heterozygote advantage, since it provides a mechanism (see Section 2.4.2) for a balanced polymorphism. Heterozygote excess is also expected if there is disassortative mating.

One cause of apparent heterozygote deficiency are null alleles. If the frequencies of such alleles are reasonably low, null homozygotes may be absent from a small sample, yet their effect be visible in an elevation in the number of apparent homozygotes for other alleles. Brookfield (19) offers a method for the estimation of null allele frequencies at microsatellite loci from heterozygote deficiency, in the absence of any null homozygotes in the sample, which illustrates maximum likelihood estimation. Assume Hardy–Weinberg proportions, and define the frequency of null alleles as r. If a visible allele has a true frequency of p_i, the expected value of its estimate from gene counting, f_i, is $p_i/(1 - r)$. Thus, when $r = 0$, the expected frequency of heterozygotes is:

$$\sum_{i \neq j} f_i \cdot f_j = H_e$$

The expected frequency of heterozygotes with two visible alleles is $H_e(1 - r)^2$. Call this x. The expected frequency of zero-banded individuals is r^2. Expected frequency of one-banded individuals is $1 - r^2 - H_e(1 - r)^2$. Call this y.

If we see n_1 apparent heterozygotes (i.e. heterozygotes between two visible alleles) and n_2 apparent homozygotes, the likelihood of these data is $(n_1 + n_2)! \, n_1!^{-1} \, n_2!^{-1} x^{n_1} y^{n_2}$. The maximum likelihood estimate of r thus maximizes $n_1 \ln x + n_2 \ln y$, and when (dln(likelihood)/dr) is zero, $r = [n_2 H_e - n_1(1 - H_e)]/[(1 + H_e)(n_1 + n_2)]$. But $n_1/(n_1 + n_2) = H_o$, the observed frequency of heterozygotes. Thus, $r = (H_e - H_o)/(1 + H_e)$.

Chakraborty et al. (20), also discuss a method of estimate the frequency of null alleles but their maximum likelihood estimate of the frequency of null alleles, r, is $(H_e - H_o)/(H_o + H_e)$. The difference between the estimates is that the former assumes that the failure to find any null homozygotes is an empirical result, whereas the latter assumes that any cases of apparent null homozygotes would be ascribed to experimental error.

3.2 Population subdivision

The most likely cause of an overall heterozygote deficiency is population subdivision, with the effect that alleles sampled from the same subpopulation will be correlated, and overall homozygosity increased (known as the Wahlund effect). The statistic F_{ST} is frequently used to quantify population subdivision, which normally arises as a result of isolation by geographical distance. F_{ST} can be measured by the variance in allele frequencies between populations, scaled by the variance expected from the mean allele frequency, but its interpretation may be complex (21). If subpopulations of effective size N exchange genes through migration at a rate m, an expected equilibrium F_{ST} arises in island models when homogenization by migration balances divergence through genetic drift, in which case $F_{ST} = 1/(1 + 4Nm)$. Pairs of subpopulations can also be characterized by pairwise F_{ST} values, which can yield pairwise migration rates Nm.

Another way of characterizing population subdivision is using Nei's genetic identity, I, which, at a given locus is:

$$\sum_i x_i y_i / \left(\sum_i x_i^2 \sum_i y_i^2 \right)^{0.5}$$

where x_i and y_i are the frequencies of the ith allele in the two populations being compared. This can be converted to an identity across loci by replacing $(\Sigma x_i y_i)$, (Σx_i^2), and (Σy_i^2), with their arithmetic means across loci. Nei's standard genetic distance, D, is defined by $D = -\ln(I)$ (22). D has the useful property that if two populations become separated, then, under neutrality, it increases approximately linearly with time in generations, at a rate μ_N, whatever the population size (23). However, the population size must be constant with respect to time in each subpopulation, and, since this is unlikely to happen, rates of change of D are expected to be highly variable, speeding up if population size drops.

D. melanogaster populations throughout the world have been sampled for allozyme variability, with a mean level of polymorphism within populations of 43% and heterozygosity of 10%. F_{ST} values are typically between 8–10%, although high variability is seen between loci, suggesting the influence of selection (24).

3.3 Linkage disequilibrium

Linkage disequilibrium (sometimes called gametic phase disequilibrium) is a correlation within a population in the presence or absence of alleles at different loci. Consider the case in which there are two loci A and B, each with two polymorphic alleles A and a and B and b. Gametes can be of four types, AB, Ab, aB, and ab. Call the frequencies of the four types p_{AB}, p_{Ab}, p_{aB}, and p_{ab}, respectively. The level of linkage disequilibrium is defined as D (not to be confused with Sections 3.2 or 3.7.1) where $D = p_{AB}\, p_{ab} - p_{Ab}\, p_{aB}$. Linkage disequilibrium can arise through selection or through drift, and is particularly likely to do so if the loci are tightly linked. Under neutrality, however, the expected value of D declines exponentially at a rate determined by the recombination rate between the loci, such that $D_t = D_0(1 - c)^t$, where D_t is the linkage disequilibrium t generations after the time it is D_0, and c is the recombination rate between the loci. In *D. melanogaster*, the levels of linkage disequilibrium within populations between allozyme loci is generally low, although polymorphic sites within the same gene typically show high levels of *D*.

3.4 Estimation of evolutionary distances and rates

Differences between homologous DNA sequences from different species represent evolutionary change. There is a non-linearity in the relationship between observed sequence divergence and total evolution in the sequence, which derives from two sources. Multiple mutations at a site cannot be distinguished from single mutations. Thus changes A→G→T cannot be distinguished from A→T. Furthermore, back mutations, of the form A→G→A, will be invisible. Thus the observed genetic dissimilarity between sequences must be converted to a distance measure estimating the total amount of evolution. Many such statistics have been suggested, but most are based on the Cantor and Jukes one-parameter model, in which:

$$K = -3/4\ln(1 - 4/3p)$$

where p is the proportion of bases diverged, and K is the estimated number of substitutions per site. This formula is designed for non-coding sequences. In coding sequences, changes are classified into amino acid changes per amino acid site, K_A, and the number of synonymous changes per synonymous site K_S, estimated by:

$$K_A = -3/4\ln[1 - (4M_A/3N_A)]$$
$$K_S = -3/4\ln[1 - (4M_S/3N_S)]$$

where M_A is the number of differences seen in the N_A amino acid replacement sites and M_S is the number of differences seen in the N_S synonymous sites (25). (Many more sophisticated estimators of these quantities have been devised, allowing, for example, different rates of transitions and transversions.)

For vertebrates, where the fossil record shows the times when bifurcations in phylogeny occurred, changes in K_A and K_S can be plotted against time. This allows an estimation of these evolutionary rates. Rates of change in K_A and K_S are often approximately constant, a result said to indicate a 'molecular clock'. This has been taken as support for the neutral theory, since the rate of evolution under neutrality depends solely on the neutral mutation rate, whereas under selection it depends upon the mutation rate to advantageous alleles, on the sizes of the selective advantages that they convey, and on the effective population size. Generally, K_S is higher and more constant across genes than is K_A. The obvious explanation is that the amino acid sequence is constrained by selection in order to maintain protein function, but the proportion of amino acids subject to this constraint varies amongst genes. DNA sequence conservation potentially allows previously unsuspected functional sequences to be identified as regions conserved in the DNAs of pairs of

Drosophila species sufficiently diverged for synonymous site divergence to have saturated. The converse, that sequences that evolving rapidly are not functional, is not necessarily true, since the function itself could be evolving. Indeed, the finding of a significantly higher K_A than K_S in a coding sequence is evidence for adaptive change. However, even if a gene is evolving its amino acid sequence by selection, one still expects a K_A/K_S ratio below one, since the replacement sites subject to purifying selection would outweigh those changing adaptively.

3.5 Phylogeny estimation

Any collection of DNA molecules that share descent must be connected by a phylogeny, although if recombination has occurred different parts of the sequence will have different phylogenies. From the sequences, the phylogeny can be estimated (26). Such estimation requires an evolutionary model, and, in a general sense, these models share the property that sequences are more likely to be closely related the more similar they are. However, different methods of estimation assume evolutionary models that differ in their details.

Some methods for phylogeny estimation are based on pairwise distances between sequences, such as K_S or K_A. (In phylogeny reconstruction sequences are often referred to as 'operational taxonomic units', or OTUs.) Distance methods are typified by the UPGMA (unweighted pair group method with arithmetic averages).

Protocol 5.   Tree-building with UPGMA[a]

1. Construct table of distances between the OTUs, e.g. *Table 1a*. Choose the two most similar and join into a single branch. Connect the OTUs by a branch of depth equal to half the distance between them.

2. Redraw the distance matrix with one less OTU, replacing two with the branch created in step 1. Calculate the distances between this branch and all others as the average of the two distances prior to joining. See *Table 1b*.

3. Re-examine the distance matrix and join the two with the smallest distance.

4. When recalculating distances between joined branches always average all the pairwise distance between all the OTUs on either side of the join.

5. Repeat steps 2–4 until all the OTUs are connected. *Figure 1a* shows the outcome for these data. See *Tables 1c* and *d*.

[a] Based on the description in ref. 22.

Table 1. Distance statistics used to produce *Figure 1*

Table 1(a) Distance matrix

	B	C	D	E	F
A	0.3	1.2	1.4	1.3	1.5
B		1.3	1.5	1.1	1.4
C			2.3	0.2	1.9
D				2.4	2.5
E					2.0

Join C and E since they have the minimum distance. Replace distances between other OTUs and C and E with their average distance with the C/E pair.

Table 1(b)

	B	C/E	D	F
A	0.3	1.25	1.4	1.5
B		1.2	1.5	1.4
C/E			2.35	1.95
D				2.5

Thus, A and B should be joined next. Replace distances between A and B and the other branches with their arithmetic means.

Table 1(c)

	C/E	D	F
A/B	1.225	1.45	1.45
C/E		2.35	1.95
D			2.5

Thus, A/B must be joined to C/E. Replace the distance between these and D and F with their arithmetic means.

Table 1(d)

	D	F
A/B/C/E	1.9	1.7
D		2.5

Join F to A/B/C/E.

UPGMA assumes constancy of evolutionary rate, which may cause the wrong tree to be generated if there is rate variation. In our example, for example, D becomes an outgroup relative to all the other species as it differs greatly from all of them. However, the distance matrix shows that D is much more similar to sequences A and B, an observation not apparent from the tree. The neighbour-joining method of Saitou and Nei (27) constructs an unrooted tree by joining OTUs whose similarity to each other is greater than their similarities to the other OTUs in the tree. The result, for these data, is a

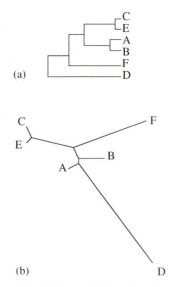

(a)

(b)

Figure 1. The (a) UPGMA and (b) neighbour-joining trees based the distance statistics in *Table 1(a)*.

tree with a different topology, with D joining A and B, as shown in *Figure 1b*. Since branch lengths no longer represent evolutionary time, only tree topology can be taken as an evolutionary hypothesis.

These two methods use an algorithm to convert distance statistics to form a tree. An alternative approach, favoured by many (26), is to examine all possible trees, and assess them on the basis of some statistic, choosing the one that has the highest or lowest value. One such statistic is parsimony, the minimum total number of sequence changes required to create the OTUs, conditional upon a given tree. Such parsimony can be weighted or unweighted, with weighted parsimony giving higher scores to base changes that are rare. Only sites showing changes shared by more than two OTUs affect the total parsimony in a tree-dependent way, and these are referred to as phylogenetically informative. Unfortunately, very many possible trees exist connecting relatively small numbers of OTUs. With only 20, for example, there are about 10^{20} possible trees. Various heuristic methods such as 'stepwise addition', 'star decomposition', and 'branch swapping', allow the computer to search for a most parsimonious tree without necessarily examining all possibilities. Unfortunately, however, often many trees turn out to be equally parsimonious.

The most parsimonious tree may have very variable branch lengths, suggesting variability in rate of evolution, which might be thought to argue against this tree being correct. An answer to such disquiet is to produce a maximum likelihood tree, in which the probability of the data is calculated assuming an evolutionary model (such as one including a molecular clock, for example), and a particular tree. Possible trees can be compared and the one maximizing the probability of the data selected. Maximum likelihood tree estimation is, however, extremely intensive of computer time.

Given an estimated phylogeny, it is natural to wonder what degree of confidence can be attached to the tree being the true tree. If different tree construction methods produce the same tree, this may give confidence. Bootstrapping is a method in which perhaps a thousand randomized data sets are created by sampling the original data with replacement, a new tree being created for each new data set. The support given to individual tree branches is estimated by the percentage of the bootstrapped trees that contain the branch. A very large number of software packages are available for phylogeny estimation (listed in ref. 26).

We have seen (Section 3.2) that populations within species can also yield distance statistics such as D or pairwise F_{ST}. These can be fed into UPGMA or neighbour-joining programs to yield trees of populations. Tree estimation searches for the 'best' tree conditional upon an assumption that the true evolutionary history of the populations is one of population bifurcation without subsequent gene flow. Thus, if one has distance statistics in the form of pairwise F_{ST} values one can use these to estimate gene flow (*Nm*—see Section 3.2), or one can use the same statistics to build a tree. However, since these analytical approaches have contradictory assumptions—one assuming that gene flow between populations balances divergence by drift and one assuming that there is no gene flow at all, the models cannot be simultaneously correct. They can be simultaneously incorrect, however.

The haploidy of mitochondrial DNAs makes them useful for PCR-based approaches to investigating sequence divergence. We should remember, however, that many models of genetic distance assume neutrality, and, with more than 30 genes in the mtDNA molecule, advantageous substitutions may occur relatively frequently. Without recombination, their frequency process determines the fate of base substitutions throughout the molecule. Furthermore, since mitochondrial DNAs are maternally inherited, variants will spread, even if themselves neutral, through any mechanism that differentiates between maternal lines, such as the cytoplasmic incompatibility generated by Ricketttsial endosymbiont *Wolbachia*, seen in many insect species including *D. melanogaster* and *D. simulans*.

3.6 Estimates of DNA sequence diversity

Given a collection of DNA sequences from the same population, one can estimate the proportion of bases that differ between randomly chosen sequences, which is called the gene diversity π. On average, in *D. melanogaster*, the gene diversity is very approximately 0.5%, but differs greatly between genes and chromosomal regions. In *D. simulans*, it is higher. Another statistic that is used to measure the genetic variability is the number of sites in the sequence that are found to vary in a sample of alleles, S. Under neutrality, the expectation of

π is the neutral parameter θ, or $4N\mu_N$ (where the neutral mutation rate is here measured per base), and S is also a simple function of θ. As with K_S and K_A, gene diversity in coding sequences is greater at synonymous than at amino acid replacement sites.

3.7 Tests for natural selection

3.7.1 Tajima's *D* statistic

In 1975, Watterson (28), showed that, for a sample of n neutral alleles, the expected time to common ancestry is a simple function of N, where N is the effective population size. Indeed, the total expected length of all the branches connecting the n sequences is:

$$4N \sum_{i=1}^{n-1} 1/i$$

Here, following Tajima (29), I will call this quantity $4Na_1$. The expected number of variable sites, $\varepsilon(S)$, in the sample is this quantity multiplied by the neutral mutation rate μ_N. Section 3.6 shows that the expected gene diversity, $\varepsilon(\pi)$, is $4N\mu_N$. Thus, neutrality predicts that estimates of $4N\mu_N$ based respectively on the number of segregating sites and on the mean gene diversity should be equal. In other words:

$$S/a_1 = \pi$$

Tajima (29) therefore defines d as $(\pi - S/a_1)$, which has an expected value of zero under neutrality. The variance of d, V_d, is estimated as a complex function of S and n. The statistic D ($= d/\sqrt{V_d}$), has an expected value of 0 and a variance of 1, and can be used to test for selection. An excess of rare variants gives a negative D, while positive values result from unexpectedly large numbers of variants with high frequencies. Deleterious alleles at low frequency created by recurrent mutation will have a larger impact on S than on π, and will create negative values of D, even if all polymorphic sites are neutral. The test is also extremely sensitive to changes in N. An expanding population with only neutral variability will show an excess of low frequency sites, and negative values of D (30).

3.7.2 Hudson, Kreitman, and Aguadé's test (31)

The neutral divergence between alleles sampled from within a species is determined by the neutral mutation rate and the phylogeny of the alleles. The deeper the branching of the phylogeny connecting the alleles, then the greater will be the expected neutral diversity among them. As one scans along a chromosome the phylogeny of alleles will vary, as a result of recombination. Natural selection that has maintained allelic diversity at a given chromosomal site will cause a localized increase in the depth of the phylogeny of the alleles at this site, and an increase in the level of neutral polymorphism. Thus, one way of looking for selection is to work a sliding window of comparison along the chromosome in a sample of alleles, and observe the changes in gene diversity at synonymous sites between different regions.

The main success of the test has been in the interpretation of synonymous site divergence in *Adh* alleles from *D. melanogaster*, where a peak of synonymous divergence coincides with the amino acid replacement distinguishing electrophoretically Fast and Slow alleles. The deepening of the phylogenetic branching in this region is consistent with the selective maintenance of Fast and Slow alleles, as had been previously postulated (either through heterozygous advantage or frequency-dependent selection). That the increase in synonymous diversity results from a phylogenetic cause, rather than a local elevation in neutral mutation rate, is shown by the absence of an elevated sequence divergence compared to *D. simulans* in this region. The comparison here is between regions with different levels of diversity within the same population. Thus, unlike the test above (Section 3.7.1), no difference in mean is expected, given neutrality, simply as a result of changes in population size.

3.7.3 McDonald and Kreitman test

The Hudson, Kreitman, and Aguadé test compares levels of gene diversity between regions of a gene. The McDonald–Kreitman test (32) is a non-parametric test based upon the concept that the numbers of neutral mutations that occur in any part of a phylogeny are dependent only upon the length of the branches in the phylogeny and on the neutral mutation rates. Suppose we sample the coding sequences in collections of alleles from each of a number of closely related species, which differ both in amino acid replacement and synonymous sites. If all variants are neutral, the ratio of replacement changes to synonymous changes should be the same for sites showing polymorphisms in one or more species as the ratio for sites showing only fixed differences between the species.

The test has been applied to a number of *Drosophila* genes and has produced evidence for selection in about half. This has taken the form of an elevation in the numbers of fixed replacement and polymorphic synonymous changes relative to the other two classes, consistent with adaptive changes in amino acid sequence between the species. This deviation from neutral expectation could be alternatively explained by selective effects elevating the number of synonymous polymorphisms. In this context, there is strong, and gene-dependent, codon usage bias in *D. melanogaster* (33), which is probably the result of weak selection (34), and may influence the test.

Protocol 6. The McDonald and Kreitman test for natural selection[a]

1. Align all sequences.

2. Identify variable bases.

Protocol 6. *Continued*

3. Classify all variation into replacement and synonymous changes, and fixed changes between species and polymorphic sites.

4. Construct two-by-two contingency table including all variable sites with axes of polymorphic versus fixed, and replacement versus synonymous.

5. Test for departure from independence of the two axes using a G-test (with Williams correction) with one degree of freedom.

6. A result significant at the 5% level constitutes evidence against the null hypothesis of neutrality.

[a] Adapted from ref. 32.

The test requires that sequences are sufficiently close to rule out multiple substitution at the same site. However, the null hypothesis does not include constancy of population size, nor does it require the distinction between bases to be one of replacement sites versus synonymous sites, providing that the two classes of base are interspersed in the sequence (such that the mean depth of branching of the allelic phylogenies are the same for the two classes). Using a distinction between bases known to bind transcription factors and those not doing so in the *fushi tarazu zebra* enhancer element, Jenkins *et al.* (35) found weak evidence for adaptive evolution in the former.

4. Discussion

Many of the approaches outlined above can be applied to many species other than *Drosophila*, and *Drosophila* is most powerful as an experimental model system in evolutionary genetics when results of its molecular population variability and genetics are integrated. An illustration is an observation of Begun and Aquadro (36). They demonstrated a very strong correlation, across chromosomal regions, between the gene diversity, π, and the rate of recombination. As in the Hudson, McDonald, and Aguadé test described above, the evolutionary divergence between *D. melanogaster* and *D. simulans*, which is no higher in high recombination regions that than in low recombination regions, ruled out a cause of the correlation involving variation in mutation rate. The explanation for the low diversity in low recombination regions thus must reside in a shallower phylogeny of alleles in these regions. There are two types of possible explanation for this. Maynard Smith and Haigh (37) showed that a favourable mutation spreading through the population would result in recent shared descent of all alleles, and thus low gene diversity, over a chromosomal region whose length is defined by the recombination rate, a phenomenon that they called 'hitch-hiking', and which has subsequently been called 'selective sweeps'. The increased length of the region homogenized by a selective sweep in a low recombination region results in an expected gene diversity that is lower in such regions.

An alternative explanation for the recombination—diversity correlation has been suggested by Charlesworth *et al.* (38). They suggest that the selection responsible for this phenomenon is selection against deleterious mutations, which they term 'background selection'. For a long chromosomal region without recombination, the majority of haplotypes in the population will have recessive mutations of low fitness which will rapidly result in their elimination from the population. The level of neutral polymorphism will be determined not by the effective population size of the species, but by the number of haplotypes lacking deleterious mutations. With high recombination, however, linkage equilibrium between the deleterious mutations and linked neutral variability will produce a higher effective population size, and, with it, higher gene diversity. The two models predict different Tajima's D values in the low recombination, low diversity regions, with a strongly negative D expected following a selective sweep, but D expected under background selection to be much nearer to zero. The rate of production of deleterious mutations, many due to insertions of transposable genetic elements, can be estimated in *D. melanogaster*. These rates have been recently shown to be consistent with those required for background selection to explain almost all the reduction seen in π in low recombination regions (39).

References

1. Ashburner, M. (1989). *Drosophila: a laboratory handbook*, p.1076. Cold Spring Harbor Laboratory Press, NY.
2. Ford, E. B. (1964). *Ecological genetics*, p. 84. Broadwater Press, Welwyn Garden City.
3. Dobzhansky, T. (1951). *Genetics and the origin of species*, p.118. Columbia University Press, New York.
4. Lewontin, R. C. and Hubby, J. (1966). *Genetics*, **54**, 595.
5. Kimura, M. (1968). *Nature*, **217**, 624.
6. Kimura, M. (1983). *The neutral theory of molecular evolution*, p.200. Cambridge University Press, Cambridge.
7. Kimura, M. and Crow, J. F. (1964). *Genetics*, **49**, 725.
8. Ohta, T. (1973). *Nature*, **246**, 96.
9. Carson, H. L. and Heed, W. B. (1983). In *The genetics and biology of Drosophila* (ed. M. Ashburner, H. L. Carson, and J. N. Thompson, Jr.), Vol. 3d, p. 1. Academic Press, London.
10. Nuzhdin, S. V., Pasyukova, E. G., and Mackay, T. F. C. (1996). *Proc. Roy. Soc. London B*, **263**, 823.
11. Falconer, D. S. and Mackay, T. F. C. (1996). *Introduction to quantitative genetics*, 4th edn. Longman, London.
12. Long, A. D., Mullaney, S. L., Reid, L. A., Fry, F. J. D., Langley, C. H., and Mackay, T. F. C. (1995). *Genetics*, **139**, 1273.

13. Murphy, R. W., Sites, J. W. Jr., Buth, D. G., and Haufler, C. H. (1996). In *Molecular systematics* (ed. D. M. Hillis, C. Moritz, and B. K. Mable), 2nd edn, p. 51. Sinauer, Sunderland, Mass.
14. Li, C. C. (1955). *Population genetics*, p.256. University of Chicago Press, Chicago.
15. Knight, G. R. and Robertson, A. (1957). *Genetics*, **42**, 524.
16. Sved, J. and Ayala, F. J. (1970). *Genetics*, **66**, 97.
17. Tracey, M. L. and Ayala, F. J. (1974). *Genetics*, **77**, 569.
18. Jungen, H. and Hartl, D. L. (1979). *Evolution*, **33**, 359.
19. Brookfield, J. F. Y. (1996). *Mol. Ecol.*, **5**, 453.
20. Chakraborty, R., De Andrade, M., Daiger, S. P., and Budowle, B. (1992). *Ann. Hum. Genet.*, **56**, 45.
21. Weir, B. S. (1996). *Genetic data analysis II*. Sinauer, Sunderland, Mass.
22. Avise, J. C. (1994). *Molecular markers, natural history and evolution*, p.111. Chapman and Hall, New York.
23. Nei, M. (1987). *Molecular evolutionary genetics,* p.232. Columbia University Press, New York.
24. Singh, R. S. (1989). *Annu. Rev. Genet.*, **23**, 425.
25. Li, W-H. and Graur, D. (1991). *Fundamentals of molecular evolution*, p.53. Sinauer, Sunderland, Mass.
26. Swofford, D. L., Olsen, G. J., Waddell, P. J., and Hillis, D. M. (1996). In *Molecular systematics* (ed. D. M. Hillis, C. Moritz, and B. K. Mable), 2nd edn, p. 407. Sinauer, Sunderland, Mass.
27. Saitou, N. and Nei, M. (1987). *Mol. Biol. Evol.*, **4**, 406.
28. Watterson, G. A. (1975). *Theor. Pop. Biol.*, **7**, 256.
29. Tajima, F. (1989). *Genetics*, **123**, 585.
30. Tajima, F. (1989). *Genetics*, **123**, 597.
31. Hudson, R. R., Kreitman, M., and Aguadé, M. (1987). *Genetics*, **116**, 153.
32. McDonald, J. H. and Kreitman, M. (1991). *Nature*, **351**, 652.
33. Shields, D. C., Sharp, P. M., Higgins, D. G., and Wright, F. (1988). *Mol. Biol. Evol.*, **5**, 704.
34. Akashi, H. (1995). *Genetics*, **139**, 1067.
35. Jenkins, D. L., Ortori, C. A., and Brookfield, J. F. Y. (1995). *Proc. Roy. Soc. London B*, **261**, 203.
36. Begun, D. and Aquadro, C. F. (1992). *Nature*, **356**, 519.
37. Maynard Smith, J. and Haigh, J. (1974). *Genet. Res. Camb.*, **23**, 23.
38. Charlesworth, B., Morgan, M. T., and Charlesworth, D. (1993). *Genetics*, **134**, 1289.
39. Charlesworth, B. (1996). *Genet. Res. Camb.*, **68**, 131.

Behaviour, learning, and memory

JOHN B. CONNOLLY and TIM TULLY

1. Introduction

1.1 Behaviour and behavioural plasticity

Behaviour represents the ultimate in adaptive responses of any animal to its surroundings. How an animal responds to its environment is the consequence of evolutionary changes to a species through its natural history. Adult *Drosophila* demonstrate behavioural responses to:

- light
- odours
- tastes
- mechanical cues
- humidity
- temperature
- gravity
- air flow
- the opposite sex
- under certain circumstances, the same sex!

Some, if not most, behaviours are subject to circadian timing. For example, the locomotor activity of freely-running adult *Drosophila* shows a peak during the relative mid-morning, a sharp decrease during the afternoon, and a final small peak in the early evening.

Larvae are capable of:

- olfaction
- chemosensation
- mechanosensation
- phototaxis
- foraging behaviour for food

Many wild-type populations exhibit a natural polymorphism in the foraging locus producing both 'rovers', who actively search for food, and 'sitters', who tend to be sedentary (1).

The plasticity of behaviour (learning and memory) is perhaps the most sophisticated of all properties of behaviour. The plasticity an animal demonstrates is governed by the previous life history of the individual. Hence, while behaviour itself can be thought of as an innate, 'hard-wired' feature of the nervous system, learning is the modification of behaviour brought about by former experience within the animal's own lifetime. Memory can be thought of as the storage and recall of such past events. Both adult and larvae are capable of learning and can show impressive memory. Adults are apparently capable of forming memories for life, while training of larvae produces memory which persists through metamorphosis.

1.2 Habituation, sensitization, and adaptation

Behavioural plasticity is classically categorized into associative and non-associative learning (2). Non-associative learning occurs in two opposing forms. Habituation is a decrease in a reflex response upon repeated presentation of a stimulus. Sensitization is enhanced responsiveness to many stimuli after presentation of a strong stimulus. Typically, to demonstrate that habituation has occurred and to distinguish sensory or motor fatigue (adaptation) from habituation, an animal must be dishabituated. This is normally achieved by presentation of a strong, novel stimulus to the habituated animal. For example, in *C. elegans* mechanical tapping results in movement away from the stimulus. Such withdrawal movement declines following repeated tapping (3). At face value, this response decrement might be caused by habituation or fatigue. When such habituated worms are electroshocked, however, normal withdrawal to the tap response reappears, indicating that the animals are not simply 'tired' but, in fact, are dishabituated by shock. In addition, the rate of habituation depends on stimulus frequency and stimulus intensity (4).

1.3 Associative learning

Most will be aware of Ivan Pavlov's learning experiments in the 1920s. His dogs demonstrated a strong behavioural response to food by salivating, but relative indifference to a bell-tone. Pavlov taught his dogs to salivate to the bell-tone alone, by presenting food simultaneously with the bell-tone, so that the tone alone became associated with food. Such learning is known as associative learning, where a previously neutral stimulus (called the conditioned stimulus or CS) when subsequently paired with a stimulus which elicits a strong behavioural response (called the unconditional stimulus or US) then becomes associated with that US—producing an altered response to the CS. Associative learning can be further subdivided into Pavlovian (classical) conditioning, the form of learning described above where the animal experiences

CS-US pairings regardless of its responses to them, and operant (instrumental) conditioning, where the animal's responses to the CS determines the occurrence of the US. An example of operant conditioning would be when a rat learns that pressing a foot lever results in a food reward (2).

Despite the complexity of genetic and environmental factors, as well as the myriad of genetic–environmental interactions influencing behaviour and its plasticity, these processes can be measured reliably and routinely in the laboratory. In our experience, analysis of behaviour is no more problematic than DNA ligations and, like subcloning, will work provided protocols are followed consistently and conscientiously.

We describe a reasonable gamut of behavioural responses of *Drosophila* within the confines of this single chapter. We also devote a portion of this chapter to that which we know best: olfactory learning and memory. Along the way, we hope to provide an inkling for how flies must be handled to optimize the measurement of their performance in these behavioural tasks.

2. Maintenance and preparation of stocks

2.1 Genetic background

Population genetic studies have long established that most behavioural responses are highly polygenic (5). As a consequence, genetic background can commonly influence the behavioural performance of *Drosophila* (6–9). Tully and Quinn (9), for instance, reported that the performances of some wild-type strains of *Drosophila* were poorer than those of mutants in the preferred Canton-S genetic background. This becomes a particularly critical issue when screening for single gene mutants defective in a behavioural response. By its very nature, behaviour is highly quantitative and mutants typically show partial, rather than complete, performance decrements. Hence, variability of genetic background during a screen can result in the isolation of 'false positive' mutants via a founder effect, which will under sample the genetic variability in the background (6–8).

A second, but related, issue concerning the polygenic basis of behaviour lies in the fact that the phenotypes of many mutants (including many of the learning and memory mutants) become suppressed within several generations, when maintained under standard free-mating culture conditions (8, 9). This reflects the tendency for mutations to have pleiotropic effects on behavioural and other biological processes. *dunce* and *latheo*, for example, both impaired in olfactory learning, are required during development (10). Such phenotypic suppression is believed to occur through the selection of suppressor alleles at 'modifier' genes, multiple alleles of which already pre-exist in genetic background of wild-type populations.

The wild-type strain used in most behavioural tasks is Canton-S, unless otherwise stated (9). When characterizing single gene mutants in any behavioural task, first equilibrate their genetic backgrounds to that of a wild-type strain by repeatedly backcrossing mutation-bearing flies to the wild-type population so that the 'average effect' of the mutant genotype(s) can be quantified (11). In the case of the EMS-induced learning and memory mutants equilibrate genetic backgrounds by repeatedly outcrossing stocks to an FM7 balancer stock possessing autosomes from an outcrossed (polygenic) Canton-S stock. This ensures that the autosomes from the mutant stock are equilibrated with that of Canton-S. It is not feasible to outcross the X chromosome due to the difficulty of identifying recombinant mutant genotypes.

More recently, engineered *P* elements marked with *white* have allowed the isolation of *P* element insertional mutants, the backgrounds of which can be equilibrated more thoroughly. Backcrossing to a 'Cantonized' *white* stock and selecting for the visible eye colour marker allows free recombination to occur around the *P* element insertion on the mutant chromosome. Backcrossing a mutant stock for five generations (with recombination) appears sufficient to equilibrate its genetic background with that of Canton-S. Use a minimum of 20 pairs for each generation of backcrossing to ensure proper sampling of the heterogeneous genetic background (11).

During screens for mutants, it may be more advantageous to isolate mutants from an isogenic wild-type population (see Chapter 1). In this case, there will be no founder effects, which should reduce considerably the isolation of 'false positives'. Furthermore, since all individuals have the same genetic background within the mutagenized population, there should be no selection for 'modifiers' in the mutant stock. The disadvantage of this approach, however, is that mutations are examined in the context of only a single genetic background and, thus, the average effect cannot be quantified (12). Backcross an isogenic mutant stock to a heterogeneous wild-type population for definitive assessments of that stock in behavioural assays.

2.2 Food, medication, and age

For behavioural studies, flies must be at the peak of condition:

- optimally sized
- well-groomed
- active
- responsive

Performance often is diminished significantly when flies are:

- diminutive
- undernourished
- dehydrated
- overcrowded
- aged

- ill

- wet

The nutritional conditions under which flies are reared will almost certainly affect performance in any behavioural task. The quality and composition of food clearly affects the ability of flies to learn and remember. For the olfactory learning task of Tully and Quinn (9), we have had most success with the 'sugar food' recipe of Cline (13). A molasses-based medium resulted in poorer performances of flies in this task. Poor food quality can reduce performance in visual operant conditioning as well, resulting in performance loss for several generations and requiring several generations to recover fully when switched back to optimal medium (14). Observations such as these will almost certainly extend to any behavioural task. Food quality is also affected by the size of the breeding population. An over-abundance of larvae in the food will inevitably result in smaller, weaker adults showing suboptimal performance in behaviour. This notion is reinforced by studies which show a direct correlation between the size of structures within the brain and the overcrowding conditions of the culture medium (15).

Protocol 1. To rear flies for behaviour experiments

Equipment and reagents

- Bottles
- Vials
- Cellucotton (available from most hospital suppliers)
- Potassium sodium tartarate 4-hydrate
- Calcium chloride dihydrate

- Sucrose
- Glucose
- Yeast
- Agar
- Cornmeal
- Flies

Method

1. Add to 1 litre water while stirring: 8.4 g agar, 8.7 g potassium sodium tartarate 4-hydrate, 0.7 g calcium chloride dihydrate, 31.4 g sucrose, 62.7 g glucose, 31.9 g yeast, 76 g cornmeal.

2. Bring to boil and autoclave for 20 min at 115°C.

3. Allow to cool and when the temperature is about 60°C add 5.1 g propionic acid (mould inhibitor).

4. Dispense in bottles and vials.

5. Use fresh food which has not dehydrated or retracted from the sides of the walls.

6. Add live yeast paste to the surface of the medium 24 h before use.

7. Add 50 flies of each sex to standard half-pint bottles.

8. Remove parents from bottles after five days of breeding to avoid overcrowding.[a]

Protocol 1. *Continued*

9. Add a small volume of water to the bottle at this stage if the food has dried.

10. Add yeast with the water if there are many larvae present to improve the nutritional state of the medium.

11. Add a couple of wads of sterile autoclaved cellucotton to the food when the third instar larvae begin to wander.[b]

[a] The number of flies and how long they should be left to lay eggs will depend on the fecundity of the stock.
[b] This provides extra substrate for pupariation and will increase the number of healthy adults eclosing.

With all behavioural experiments, avoid drugs or chemicals where possible. Some are absolutely necessary, however, such as those that control parasites and infections of food and flies. Do not use mouldy food to culture flies for behavioural experiments. We recommend the addition of Tegosept (see Chapter 1) mould inhibitor to the culture medium (this does not significantly affect performance in olfactory learning).

Outbreaks of mites also will affect the behavioural responses of flies. Mites can be controlled by spraying Tedion (see Chapter 1) on cellucotton prior to use. A useful trick is to add a green food dye to indicate which batches of cellucotton strips have been previously sprayed. Prevention is better than cure; autoclave stocks maintained at 25°C for more than three weeks.

Drosophila and/or food can also become infected with bacteria. We have maintained our flies on a rota of antibiotic pairs since a serious infection of our food and flies by a strain of bacteria drastically affected flies' health and behaviour. These antibiotic pairs are chloramphenicol (USB) and ampicillin (USB) (7 mg/ml of each in 50% ethanol), and deoxycycline (Sigma) and amoxicillin (Sigma) (7 mg/ml and 14 mg/ml, respectively, in distilled water). Add 0.5 ml of fresh antibiotic solution to the surface of food bottles containing 70 ml of food (70 µl to 10 ml food in vials), once it has been cooked, poured, and solidified. After more than five years on this antibiotic regime, we have detected no resistance or adverse effects.

The behaviour of flies is affected by their age. For studies of olfaction and olfactory conditioning, we use one- to four-day-old flies. We have found that the performance of adults drops after eight days. Reduced olfactory acuity and locomotor activity may contribute to this age-dependent reduced performance. Age apparently affects different behaviours to different degrees. It has been reported that one- to two-day-old flies perform poorly in visual operant conditioning, while three- to four-day-old flies perform optimally (14). In courtship experiments, optimal performance is observed when flies are about five days old.

2.3 Preparation of flies

As well as growing flies under optimal nutritional conditions treat flies with care immediately prior to testing. They should be well-groomed, fed, and hydrated. It is preferable to avoid anaesthesia prior to all behavioural experiments. It has its most dramatic effects on conditioned courtship. Anaesthetized flies may never recover normal behaviour (16). When measuring behaviour on populations of mixed genotypes, it should be possible to sort genotypes after behavioural testing is complete. However, anaesthesia is occasionally unavoidable. Flies will dehydrate during exposure to carbon dioxide and other gases, even when gas is bubbled through water. Keep exposure on gas pads to a minimum and allow flies to recover from any effects overnight. With caution, little if any effect of carbon dioxide will be observed.

Protocol 2. Preparation of flies for behavioural experiments

Equipment and reagents

- Food bottles (*Protocol 1*)

- Incubator/room at 25°C and 60% relative humidity for flies which will be heat shocked

Method

1. Prepare feeding bottles of unyeasted normal food (see *Protocol 1*) by removing excess moisture and placing a small wad of paper towels in the middle of the food projecting into the free space of the bottle.[a]

2. Transfer 400–600 flies[b] to freshly prepared bottles the night before the experiment.

3. Do not overcrowd the flies or they will become fatigued and there will be a build up of moisture.[c]

4. Maintain the flies at room temperature and humidity overnight, but maintain flies that are to be heat shocked before testing, at 25°C and 60% relative humidity.

5. Code flies before experiments so that the experimenter is 'blind' with respect to treatment or genotype eliminating any experimenter bias.

6. Transfer the flies to behavioural room conditions about 30 min before testing.[d]

[a] This absorbs remaining free moisture and is a platform for the flies to dry and groom themselves of food debris from the culture bottles.
[b] This is a suitable number for olfactory experiments but the number may vary according to the experiment and the strain of flies used.
[c] If the bottles are inverted the flies should drop on to the stopper but quickly show negative geotaxis. If large numbers of flies remain on the stopper this may indicate deleterious crowding.
[d] This allows the flies to adapt to temperature, humidity, and lighting conditions before testing.

2.4 Preparation of larvae

Protocol 3. Preparation of larvae for behavioural experiments

Equipment and reagents
- Food bottles
- Glass beaker covered with Nytex gauze

15%(w/v) sucrose in distilled water

Method
1. Add 50 pairs of breeding adults to a bottle of standard medium for 24 h at 25 °C.
2. Remove flies.
3. Transfer bottles to 18 °C.
4. Add 15% sucrose solution to the bottle 60 min before training.
5. Swirl and harvest the larvae from the surface of the sucrose.
6. Wash gently with copious amounts of distilled water in a glass beaker covered with Nytex gauze.

---

3. Locomotion

Deficits in locomotor activity may reflect general debilitation, ill-health, muscle system defects, or disruption to the nervous system. Numerous procedures can be employed to measure spontaneous or reactive locomotor responses. Shock reactivity (see *Protocol 12*) is an example of the latter.

3.1 Larval locomotor activity

The locomotor activity of larvae can be measured in a $15 \times 15 \times 2$ cm rectangular Perspex box, with 3% agar covering the base to a depth of 0.5 cm (17). A transparent lid, onto which a grid of 1 cm^2 squares is etched, covers the box. Individual larvae are placed on the surface of the agar and the number of squares entered by the larva is recorded within a 30 second time window over a total of test period of 180 seconds.

3.2 Fly locomotor activity
3.2.1 Open field test

A single adult fly is aspirated gently, via a small entry port, into a $10 \times 10 \times 1$ cm Perspex chamber with a clear lid etched with a grid of 1 cm^2 squares and allowed to acclimatize for 60 seconds. The number of transitions from one square to another is recorded over a 60 second test period (18).

In an alternative version of this assay, a single fly is introduced into a 50 cm cubic acrylglass chamber (19). Each side of the cube is covered with a 5×5 cm^2 grid. The number of times the fly crosses between grid squares is recorded over three sequential blocks of three minutes. Walking activity decreases over the nine minute test period. It has been reported that flies with clipped wings are three times more active than normal flies in this assay while the *no-bridge* brain structural mutant produces defective performance in this test (19).

3.2.2 Flight escape assay

Groups of 20 anaesthetized flies are placed in a Petri dish, and are allowed to recover for one hour. The lid is removed for 30 seconds. The number of flies remaining in the Petri dish is counted (20). The *couch potato* mutant is defective in this paradigm (20).

3.2.3 Dark reactivity

This assay measures locomotor reactivity (escape) of adults in the counter-current apparatus (10) (see *Figure 1*) which consists of a Lucite frame onto which six 100×17 mm polypropylene test-tubes (Falcon, No. 2017) are attached. One of these tubes (the start tube) can slide along the apparatus so that it comes into register with one of five distal tubes. Under dark conditions, approximately 100 flies are tapped into the bottom of the starting tube. The apparatus is laid horizontally and flies are permitted to distribute themselves between the starting and distal tubes over 30 seconds. (The expectation is for $\sim 50\%$ of the flies to run into the distal tube in 30 seconds.) The two tubes then are slid out of alignment, trapping flies in either the start or distal tube. The tubes are slid along again so that the start tube comes into alignment with a second distal tube. Flies are once again permitted to run into the distal tube for 30 seconds. This procedure is repeated five times, separating flies into six fractions. Flies which run into all five distal tubes during the assay score five, while those remaining in the original starting tube receive a score of zero. The dark reactivity score is calculated as the average of scores from all flies tested.

3.2.4 Buridans paradigm (19)

A single fly, with clipped wings, is placed on a circular platform (~ 85 mm diameter) which is surrounded by a water-moat. The platform is brightly illuminated with white light. Two vertical black stripes are located 100 mm from the centre of the platform and facing each other. Wild-type flies will walk continuously between these two inaccessible landmarks—reportedly for up to seven hours without respite. A transition between landmarks is scored when a fly crosses between two lines 66 mm apart and parallel to the diameter of the platform. The walking path of the fly can be recorded on a video scanner, over a 15 minute test period. With such analyses, the time course of walking activity and walking speed can be measured (19).

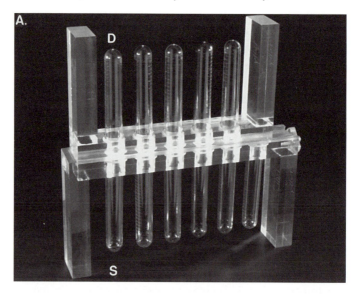

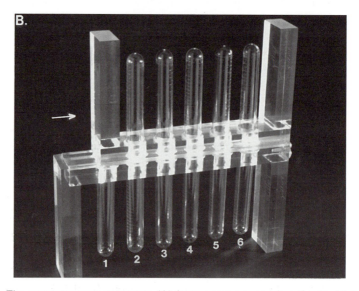

Figure 1. The countercurrent apparatus. (A) Assays commence when the start tube (S) is slid into register with the distal tube (D). Flies are given 30 seconds to move from start to distal tubes. (B) The distal tubes are slid out of register (→) trapping flies remaining in the first start tube (1). Flies in the distal tube are tapped into the second start tube (2) and the distal tubes again slid into register with the start tubes for 30 seconds. The process is repeated five times until the population of flies are fractionated into all six start tubes (1–6).

3.2.5 Circadian locomotor activity

Adult locomotor activity can be monitored continuously to measure their circadian rhythm (21). Newly eclosed flies are entrained by exposure to 12 hours light: 12 hours dark cycle for four days. These flies are placed in glass tubes equipped with infrared light emitters on one side and detectors on the other side. Infrared detectors are linked to a computer, which records breakages of the light beam. Groups of these tubes are loaded onto 'monitor boards' and maintained in incubators at 25°C. Under constant darkness, free-running locomotor activity of individual flies is recorded by 'activity events', where the passage of the fly across the tube breaks the infrared beam. For each fly, the number of activity events is recorded every half-hour over a seven to ten day period.

3.2.6 Geotaxis

The response of flies to gravity can be assayed using a geotaxis maze (22) (see *Figure 2*). A series of 'Y'- and 'T'-tubes (0.32 cm i.d.) interconnected by Tygon tubing present flies with eight, sequential, binary 'up or down' choices. The arm of each 'Y'-tube is 1 cm long. The total size of the maze is 25 cm long and 25 cm high. Openings around the perimeter of the maze are sealed with silicon glue. Flies are drawn phototactically through the maze by a 15 W fluorescent light, placed 30 cm from the exit holes of the maze. Flies are placed in a 12 cm Tygon start tube and given ten minutes to rest. The start tube is then placed at the entrance to the maze (12.5 cm high) and covered. Flies are drawn from the dark start tube into the illuminated maze. At the end of the maze, flies enter one of nine end tubes. These are made from plastic culture tubes (75 × 10 mm) into which the wide end of a pipette tip has been inserted. These pipette tips are cut so that their narrow end forms a 0.32 cm i.d. opening. Thus, once flies enter end tubes, they cannot re-enter the maze. Flies are given three hours to complete the task, but typically do so in less than one hour. Tests produce a fractioned population of flies, based on their geotactic performance. Flies in each end tube receive a score ranging from zero to eight. Flies in the bottom end tube score zero, indicating that all of their choices within the maze were downward, while flies in the top end tube are given a score of eight, reflecting the fact that all of their choices within the maze were upward. For a given genotype, a mean geotaxis score is calculated from the number of flies in each end tube. The *cut*, *thread*, *aristaless*, and *spineless-aristapedia* mutants show defective geotactic performance (22).

4. Mechanosensation

4.1 Larval mechanosensation

Larvae can be assayed for their mechanosensory responses by stroking forward-moving animals across one side of their thoracic segments with

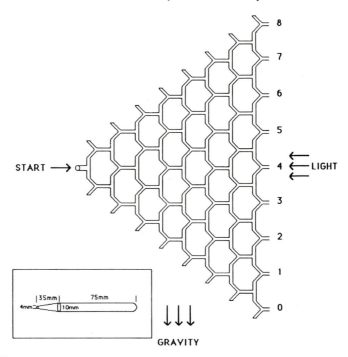

Figure 2. The geotaxis maze. The boxed insertion illustrates the design of one of the end tubes of the maze. See text for further details. Reproduced with permission from ref. 22.

an eyelash (23) (see *Figure 3*). Responses are measured on a scale of zero to four:

- no response is scored as zero
- hesitation as one
- withdrawal of anterior segments and turning as two
- a single reverse contractile wave followed by turning as three
- multiple reverse contractile waves, retreats, and turning as four

uncoordinated (*unc*) and *uncoordinated-like* (*uncl*) have been identified from such screens (23).

4.2 Adult grooming

Adult *Drosophila* respond to mechanical stimulation of the micro- and macrochaetae. Stimulation of bristles on the dorsal abdomen produces a scratch reflex which causes the fly to sweep the stimulated area with one of its legs. Mosaic flies lacking wild-type *unc* in mechanosensory bristles are defective in the scratch reflex. Philis *et al.* (24) have used the scratch reflex to

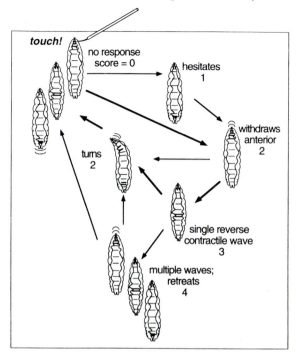

Figure 3. Schematic illustration of larval mechanosensory responses. Larvae are stroked across one side of their thoracic segments with an eyelash. Thicker arrows indicate the most commonly observed responses. More rarely, larvae show several waves of reverse contractions (thinner arrows). Responsiveness is registered on a scale of 0–4 as indicated. Reproduced with permission from ref. 23.

identify mutants defective in grooming behaviour. A population of flies are shaken in an easily seen dust (Reactive Yellow 86, Sigma), sieved, and re-shaken to remove excess dust. Dusted flies then are transferred to a clean chamber and allowed to groom for two hours. Flies are scored for the presence of dust on their notum, head, wings, abdomen, and legs. The mutants *grunt, unsteady, fade out, grooming defective, midline uncordinated,* and *stalled* are defective in grooming behaviour.

4.3 Habituation of the cleaning reflex

Corfas and Dudai (25) have reported that the cleaning reflex undergoes habituation and dishabituation. They have used nozzles to deliver controlled air-puffs to flies mounted on microelectrodes in a micromanipulator and have recorded their responses on a video camera. Capillaries situated about 0.5 cm from the fly release pulses of air flow (63 ml/min) from a pressure pump via three-way air valves (General Valve Corp.) controlled by a pulse generator (Grass S88). Two response parameters are scored.

The magnitude of the response of the foreleg to each stimulus is judged from zero to four:

- no foreleg response is a zero
- stopping of the tarsus before it reached the humeral bristles is one
- the tarsus reaching and stopping at the humeral bristles scores two
- the tarsus reaching and stopping at the antero-noto-pleural bristles scores three
- the tarsus arriving at the tegular bristles scores four

The second parameter measured is the number of stimuli required to eliminate the response (the stimuli to stop responding; SSR).

Repeated presentations of controlled air-puffs directed at the tegulars, antero-noto-pleural, upper humeral, and lower humeral bristles of the notum lead to a waning and eventual disappearance of the response. For wild-type flies, the SSR is about 47. This response decrement can be dishabituated by application of air-puffs to bristles of the dorsocentral notum, producing about an 80% recovery of the response. Increasing the frequency of air-puffs causes the flies to habituate faster, and reducing the flow rate of air-puffs reduces the SSR, indicating the dependence of cleaning reflex habituation on both stimulus frequency and intensity, respectively. Spontaneous recovery of the response (memory decay of habituation) takes about 30 minutes. In *rutabaga* mutants, habituation is normal, but spontaneous recovery occurs within five minutes.

5. Chemosensation

5.1 Larval chemosensation

Larvae are attracted to high concentrations of sugars, such as glucose, sucrose, mannose, and fructose, and to low concentrations of monovalent salts (e.g. KCl), but are repulsed by quinine and high concentrations of monovalent salts (26, 27). Chemosensation in larvae can be measured on Petri dishes containing four quadrants of agarose (see *Figure 10*). Quadrants are cut from Petri dishes containing defined concentrations of chemicals dissolved in 1% agarose. Two quadrants are placed diametrically opposite each other in each Petri plate, along with two quadrants made of 1% agarose only. Plates containing quadrants are then overlaid with 10 ml of 1% agarose to provide a smooth surface for the larvae to crawl on. To ensure diffusion of the chemicals in the agarose, plates should be used two hours after being overlaid. Larvae are placed in the centre of plates and given five minutes to choose between quadrants.

5.2 The proboscis extension reflex

Flies sense non-volatile chemical stimuli, such as sugars and salts, through the tarsi of the legs and the proboscis (28, 29). When a fly is starved, but water-satiated, application of sucrose to the tarsi of its feet elicits the extension of the proboscis (*Protocol 4*).

Protocol 4.   Measurement of the proboscis extension reflex (28)

Equipment and reagents
- Glass bottles
- Insect dissecting needles
- Dental wax

Method

1. Starve 50–100 flies for 21 h at 25°C in a glass bottle containing paper tissue soaked in 10 ml distilled water.

2. Melt a small piece of dental wax onto each of several insect dissecting needles.

3. Anaesthetize flies with CO_2.

4. Soften the wax with heat and under a dissection microscope fix each fly by its dorsal surface to a needle with the wax.

5. Allow the flies to recover in the humid atmosphere ($\sim$ 95% humidity) of a box containing water saturated paper towels at 25°C.

6. Score proboscis extension in the range 1–5 depending on the degree of extension (28).

The response of flies to a weak sucrose solution can be habituated by prior exposure to a strong sucrose solution (29). The mutants *dunce*, *rutabaga*, and *turnip* are defective in habituation of the proboscis extension reflex (*Protocol 5*).

Protocol 5.   Measurement of habituation of the proboscis extension reflex (29)

Equipment and reagents
- 1.3 cm 27 gauge hypodermic needles (0.25 mm o.d.)
- 10 ml plastic syringes
- Dissecting microscope
- Pure sucrose (Sigma)

Method

1. Prepare fresh solutions of 4 mM and 100 mM sucrose in distilled water.

2. Place a mounted fly (*Protocol 4*) under a dissecting microscope (×20 optics) with its head pointing away from the experimenter.

3. Water-satiate the fly by placing a large drop of water ($\sim$ 0.5 mm) in

---

Protocol 5. *Continued*

    contact with both the prothoracic tarsi of the feet and the proboscis for 2 sec, or until the fly stops drinking by retracting its proboscis.[a]

4. To ensure water-satiation of the fly, for 2 sec locally apply a small drop of water from the tip of the needle to the two most distal segments of the prothoracic tarsi of the left foot. If proboscis extension occurs, repeat step 3.

5. Time: 0 min. For 2 sec apply a small drop[b] of 4 mM sucrose to the two most distal segments of the prothoracic tarsus on the left foot, using the inside of the needle tip. Note any proboscis extension.[c]

6. Rinse the left leg by moving a large drop of water over the cuticle of the tibia and tarsus.

7. Repeat step 4, except with right foot.

8. Time: 10 min. For 2 sec apply a small drop of 100 mM sucrose (the habituating stimulus) to the two most distal segments of the prothoracic tarsus on the right foot, using the inside of the needle tip.

9. Repeat step 6, except rinse right leg.

10. Time: 20 min. Repeat step 5 to measure habituation of the response.[d]

11. Place a large drop (~ 0.5 mm) of 100 mM sucrose to the tarsi of both feet as a 'dishabituating stimulus'. Discard flies which fail to extend their probosces.[e]

---

[a] Whichever is longer.
[b] The drop size is adjusted so that the liquid just covers the inside surface of the beveled needle tip.
[c] Proboscis extension of position two or greater (see ref. 28) is measured as a response.
[d] The habituation index (HI) is calculated as the fraction of flies initially responding to the 4 mM sucrose stimulus minus the fraction responding to 4 mM sucrose following application of the habituating stimulus of 100 mM sucrose.
[e] Fatigue cannot be distinguished from habituation unless 'habituated' flies can subsequently be 'dishabituated'.

Sensitization of the proboscis extension reflex has also been reported (29). Here, flies are given a sensitizing stimulus of 1.0 M sucrose, which is applied to the proboscis for two seconds. A subsequent two second application of a water drop onto both prothoracic tarsi elicits enhanced proboscis extension for up to two minutes. *dunce*, *rutabaga*, and *amnesiac* show altered sensitization of this response (see also ref. 30).

Conditioned inhibition of the proboscis extension reflex can occur over a number of trials if the application of sucrose to the tarsi is immediately followed by delivery of electroshock or quinine hydrochloride to the legs (31).

6. Olfaction

Drosophila adults respond to numerous volatile chemicals, such as alcohols, aldehydes, acetates, ketones, and acids (32). Olfactory responses can be measured using the olfactory trap assay, the olfactory jump response, or in the 'T'-maze. The concentration of an odour is critical in determining whether an odour is repulsive or attractive to adult flies. The major olfactory organs in the adult are the antennae and maxillary palps. Mutants defective in olfaction include *pentagon*, which is defective in sensing benzaldehyde, *Scutoid*, which poorly senses ethyl acetate and acetone, and *paralytic*[smellblind], which is impaired in both adult and larval olfactory responses (26).

6.1 Larval olfaction

Larvae can show strong responses to odours (26, 27). Larval olfactory preferences can be measured on Petri dishes containing solidified 1.1% agarose, with a point source of odour on one side and a control point source on the other side. Odour sources can be applied to discs of chromatography paper (3MM Chr, Whatman), produced by a hole puncher. At the end of a five minute test period, larvae on the 'odour' and 'control' sides are counted and PIs calculated as described for 'T'-maze adult assays (see Section 6.5). Larvae find many odours attractive which adults find repulsive, such as ethyl acetate, propionic acid, isoamyl alcohol, and acetone (37).

6.2 The olfactory trap assay

Attractive odours can be measured using the olfactory trap assay (see *Figure 4*) (33).

Protocol 6. Olfactory trap assay

Equipment and reagents
- 1.5 ml microcentrifuge tubes
- Gilson pipette tips (yellow)
- Petri dishes
- Odours in 0.5% agarose[a]

Method

1. Cut the microcentrifuge tube 2.5 mm from its bottom to produce an opening of 4 mm i.d.

2. Cut two 1–200 μl (yellow) pipette tips 9 mm from their narrow ends.

3. Cut one of these 16 mm from its wide end.

4. Insert the small end of this tip into the bottom of the microcentrifuge tube.

5. Place the wide end of the longer tip over the protruding end of the short tip (see *Figure 4*).

Protocol 6. *Continued*

6. Place the odour set in 0.5% agarose in the microcentrifuge tube.

7. Place the trap in 100 × 20 mm Petri dish containing 10 ml 1% agarose.

8. Introduce ten flies into the Petri dish.

9. Score the number of flies in the trap every 10 h for 85 h.

[a] Food, yeast, ethyl acetate, acetone, propionic acid, acetic acid, and ethanol act as attractants.

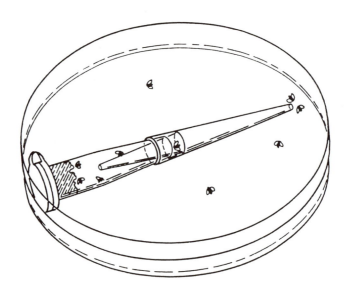

Figure 4. The olfactory trap assay. See *Protocol 6* for details. Reproduced with permission from ref. 33.

6.3 The olfactory jump response

The responses of odours which are repulsive to flies can be measured in the olfactory jump response (*Protocol 7*) (see *Figure 5*) (34).

Protocol 7. Olfactory jump response

Equipment and reagents

- Teflon block
- Glass tubing
- Tygon tubing
- Vacuum pump with flow rate of 1000 ml/min
- Plastic culture tube (100 × 17 mm)
- Nytex gauze
- 1–200 μl pipette tip
- Teflon tubing
- Side-arm flask
- Odour in solution[a]

Method

1. Drill two holes in the Teflon block perpendicular to each other so that air flows through them.

2. Fit a glass tube tightly in one hole and connect this via Tygon tubing to the vacuum pump.

3. Drill a hole in the base of the plastic culture tube half the width of a 1–200 µl pipette tip and tightly fit the tip in the hole.

4. Cover the open end of the culture tube with Nytex gauze and fit it tightly into the Teflon block.

5. Connect the pipette tip via Teflon tubing to the top of a tube, with a side-arm, which contains a solution of the odour.

6. Connect a piece of Teflon tubing from the side-arm and immerse it in the odour solution, such that once the vacuum is applied air is drawn through the side-arm and bubbled through the odour.

7. Aspirate a fly into the culture tube and allow it to walk half-way up the tube.

8. Give a 4 sec pulse of the odour. Flies will jump to avoid a repulsive odour.

9. A response is scored when an animal jumps from the side of the tube to the Nytex gauze at the base of the tube or when it jumps from one wall of the tube to another.

[a] Benzaldehyde yields optimal jump responses, with ethyl acetate and propionic acid also producing significant effects.

6.4 Habituation of the olfactory jump response

Flies show a decrease in the olfactory jump response with repeated presentation of the same odour. This has many of the properties of habituation (4, 10) (see Section 1.2). While this habituation can be examined via manual delivery of odours, a semi-automated version has been developed (General Valve Corp.). A computer controlled three-way solenoid valve switches between streams of 'fresh' air and air bubbled through 5% benzaldehyde (Fluka) in heavy mineral oil. Flies are habituated by repeated exposure to four second pulses of benzaldehyde followed by ten seconds of fresh air. Flies are habituated when they fail to jump in four consecutive trials. The habituation score is calculated from the number of trials required to reach the four non-jumps criterion. After habituation, flies are dishabituated by vortexing in a test-tube for 25 seconds. Dishabituation is measured two minutes after habituation has been reached by exposing the flies to one final four second pulse of odour.

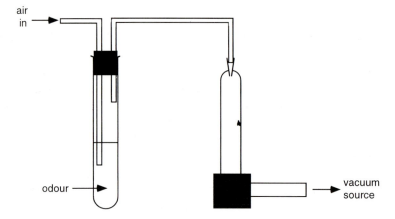

Figure 5. A simple version of the jump response apparatus. In this apparatus, a single fly is aspirated through the hole in the base of an inverted test-tube and allowed to walk about half-way up the tube. Air is drawn continuously through the tube via a vacuum pump. When tubing from the odour-containing vessel is inserted into the hole at the top of the tube, air is drawn ('bubbles') through the odour, causing airborne currents of odour to pass over the fly in the tube. Flies typically jump as an escape response to pulses of odours. Adapted with permission from ref. 34.

6.5 Behaviour in the 'T'-maze

If flies are introduced into the centre of the maze, offering them a simple binary choice between arms, they will walk into either arm without preference. Hence, in the absence of attractive or repulsive cues, flies will distribute themselves 50:50 in either arm by chance alone. If sensory cues are present in either arm of the maze, any alteration in the simple 50:50 distribution is taken as a behavioural response brought about by that cue. Olfaction can be measured by placing an odour source on one end of the arm and a control odour (typically air) on the other arm (*Figure 6*) (10, 35). If the distribution of flies shifts towards the control arm of the maze, this indicates that the odour source is repulsive to the flies. Similarly, responses to light, tastes, or electroshock can be measured in the 'T'-maze. Detailed descriptions of 'T'-maze equipment and preparation and are presented in Section 7.3.

Protocol 8.   Measurement of olfactory acuity[a]

Equipment and reagents

- 'T'-maze or teaching machine (General Valve Corp.)
- Rotor-style vacuum pump[b] (General Electric)
- Odour cups
- Odour blocks (General Valve Corp.)
- Collection and cover tubes
- Flowmeter
- Pure odour sources[c]

Method

1. Prepare flies in glass bottles as described in Section 2.3 the night before testing.

2. Place flies in behaviour room for > 30 min before tests commence.

3. Adjust air flow rate on the lower port of the teaching machine to 1500 ml/min using a flowmeter. Afterward, ensure central compartment is in register with the upper entrance.

4. Gently tap a fly bottle on its side to knock flies to the glass walls and aspirate 100–150 flies.[d]

5. Time: 0 sec. Immediately transfer aspirated flies into the upper entrance, gently tap machine on its side and lower the elevator, so that the central compartment is out of register with the upper entrance, trapping the flies inside for 90 sec.

6. Gently attach an odour block to one arm of the teaching machine.[e]

7. Time: 85 sec. Connect the vacuum source to the lower port of the teaching machine.

8. Time: 90 sec. Lower the elevator swiftly and smoothly so that the central compartment is in register with the choice point of the teaching machine.

9. Time: 210 sec. Raise the elevator so that the central compartment is out of register with the arms of the teaching machine.

10. Remove the odour block from the teaching machine arm and disconnect vacuum source from lower port of the teaching machine.

11. Tap the teaching machine successively on each side while detaching collection tubes from the machine.

12. Transfer flies from each collection tube to a foam-stoppered polypropylene tube via a funnel.[f]

[a] Assays are typically performed under far-red light (> 650 nm).
[b] Any make is adequate provided a pulse-free source of vacuum is generated.
[c] Pure odour sources should be used to minimize variations in quality among batches which will inevitably introduce variability into behavioural responses. (We have observed variations in quality of odours even in analytical grade reagents.) Use heavy mineral oil when testing dilutions of odours. Such odour dilutions should be made fresh daily and mixed thoroughly (> 10 min on a shaker) before testing.
[d] To minimize disruption to flies, always tap bottles and machines on rubber or foam matting.
[e] Care must be taken when moving odour blocks to avoid spillages. Odour blocks should be placed alternately on the left and right arms during tests of a given genotype to avoid the introduction of side biases into tests. Do a 'dry run' between each behavioural test.
[f] Flies are counted under anaesthesia later. From empirical observations, eight PIs (four on each side) are sufficient for an accurate measurement of olfactory acuity to a particular odour at a particular concentration.

Because these assays involve the responses of populations of flies, an index can be calculated from the number of flies in the arm containing the cue compared to the number of flies in the control arm (10).

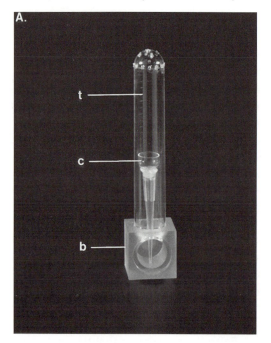

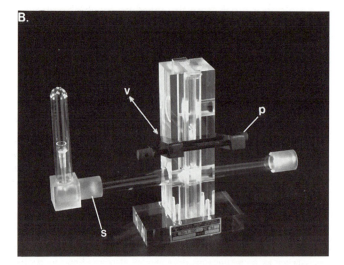

Figure 6. Olfactory acuity in the 'T'-maze. (A) Odour block assembly (see also Section 7). A cover tube (t) fits snugly over an odour cup (c) placed on an odour block (b). (B) 'T'-maze. The odour block assembly is snugly attached to the sleeve (s) of a collection tube. 1500 ml/min vacuum (v.) is drawn through the odour block assembly and collection tubes via a pump attached to lower vacuum port of maze. An airtight seal is maintained with the use of a parallel clamp (p).

In the case of olfactory acuity to a repulsive odour, the 'probability correct' (COR) is calculated as the number of flies avoiding the odour divided by the total number of flies in the 'T'-maze arms:

$$COR_{ODOUR} = [\text{control arm}]/([\text{odour arm}] + [\text{control arm}]).$$

This calculation is modified for assays involving olfactory attractants to:

$$COR_{ODOUR} = [\text{odour arm}]/([\text{odour arm}] + [\text{control arm}]).$$

The performance index (PI) is calculated as:

$$PI_{ODOUR} = \frac{COR_{ODOUR} - 0.5 \times 100}{0.5}$$
$$= [2 \times COR_{ODOUR} - 1] \times 100$$

This results in a quantitative measure of the behavioural response, which is normalized from 0 (no response) to 100 (maximal response). Normally, mutants defective in any particular behaviour show reduced, but not eliminated responses. Hence, assays using the 'T'-maze provide a highly quantifiable, objective, and reliable way of measuring the behavioural phenotype of a mutant.

7. Associative olfactory learning and memory

7.1 Olfactory learning and memory in larvae

Associative olfactory learning and memory in *Drosophila* larvae was first reported by Aceves-Piña and Quinn (36). Electroshock delivered to the larvae through plates containing agarose made conductant with lithium chloride served as the US. Airborne currents of 3- octanol and amyl acetate served as CSs. Training produced acquisition scores of about 20, and 15 minute memory retention scores of above 10 in Canton-S third instar larvae. *dunce, cabbage,* and *turnip* mutants fail to learn this task, while *amnesiac* fails to show any 15 minute memory. Interestingly, heterozygous *turnip* larvae learn as well as Canton-S, but show no memory 15 minutes later. First instar larvae are also capable of olfactory learning.

Tully *et al.* (37) have improved the experimental design of this olfactory-based assay by delivering air flow through baffled air currents. Their training protocol produces acquisition scores of about 30 in Canton-S larvae. In this modified assay, *dunce* and *amnesiac* show no learning. The odours used as CSs (ethyl and isoamyl acetates) are attractive to naive larvae and repulsive to naive adults. Eight days after Canton-S larvae have been trained in this apparatus, their adult olfactory preferences are altered to produce PIs of about 30. *dunce* and *amnesiac* mutants, unable to learn as larvae, fail to show a memory through metamorphosis.

7.2 The teaching machine

Quinn, Harris, and Benzer (38) were the first to report that adult flies are capable of associative learning. Later, Tully and Quinn (9) introduced a substantially modified and improved assay, which is now considered one of the methods of choice for the study of Pavlovian learning and memory in *Drosophila*. In the latter assay, flies are trained and tested in a single apparatus known as the 'teaching machine', which is a modified version of a 'T'-maze (see *Figure 7*). It is constructed of a Lucite frame, which contains a sliding elevator with a small compartment to transfer flies from their training location at the top of the machine to the choice point of a 'T'-maze at the base of the machine where they are tested. Aversive electroshock serves as the US and CSs are provided by olfactory cues (typically 3-octanol, 4-methylcyclohexanol, or benzaldehyde).

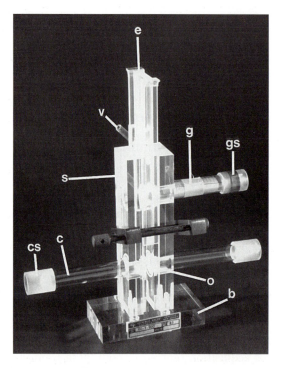

Figure 7. The teaching machine. The machine consists of a sliding elevator piece (e), two side pieces (s), and a base (b). Two vacuum ports (v) at the back of the elevator piece allow vacuum to be drawn through the grid tube (g) during training and collection tubes (c) during testing. Teflon O-rings (o) on the inner face of each side piece form airtight seals with the elevator piece. Odour block assemblies (see *Figure 6A*) attach snugly to the grid tube via a grid tube sleeve (gs) during training. During testing, odour block assemblies are attached to collection tubes via collection tube sleeves (cs).

A population of adults is sequestered in a tube which contains an electrifiable grid (the 'grid tube') on its inner surface. Air currents containing odours can be drawn over the trapped flies using a vacuum source. The flies are trained by delivering electric current to the grid as one of two specific odours (known as the CS^+) is drawn over them. Current to the grid is switched off and a different airborne odour (the CS^-) is delivered to the flies without shock. After a brief rest, flies are transferred to the choice point of two convergent currents of both odours used during training (the CS^+ and CS^-) and allowed to choose. Prior to training, naive flies show an equal preference for both odours, as measured by their 50:50 distribution between both odours in the 'T'-maze. However, immediately after training over 90% of flies avoid the specific odour (CS^+) which was associated with electroshock. If 100% of flies avoid the odour which has been paired with electroshock, this is considered 'perfect learning', while a 50:50 distribution of flies is taken as 'no learning'.

While 'T'-mazes can be constructed at most machine shops, the preparation and maintenance procedures described below refer to the commercially available 'T'-maze from General Valve Corp., Part No. 80–1–900 (see *Figure 7*). This has been designed as a 'teaching machine' to train and test associative olfactory learning and memory in flies, but is also used to measure olfactory acuity (*Protocol 8*) and shock reactivity (*Protocol 12*).

7.3 Teaching machine preparation and maintenance

The teaching machine is manufactured from Lucite and consists of a base, two side pieces, and a central sliding elevator piece (see *Figure 7*). Each side piece has a machined opening three-quarters of the way down to accommodate a collection tube, with one of the side pieces containing a second, upper opening to accommodate a grid tube. Two Teflon (20 mm) 'O'-rings (Scientific Instrument Services, No. T-020) on the inner surface of each side piece allow airtight seals to be formed with the central piece where the collection and grid tubes attach to the machine. In order to avoid air leaks these (20 mm) 'O'-rings should be replaced daily (in heavy use, more often). When replacing these rings, care should be taken to avoid introducing potential air leaking scratches onto the side pieces. Each side piece also possesses an outer (17 mm) 'O'-ring (Scientific Instrument Services, No. T-017), which serves to support the horizontal collection tubes. These rings are changed when collection tubes appear to droop.

To wash side pieces and replace 'O'-rings, each side piece can be detached from the base by unscrewing two screws, visible under the base. The central piece of newly acquired machines contains sharp edges which will rapidly shave the 20 mm 'O'-rings, resulting in air leaks. File down these sharp edges at a 45° angle. The central piece also contains an upper port which is connected to 38 0.5 mm machined holes, designed to draw vacuum through the grid tube, and a lower port connected to the central compartment via 60 0.5 mm machined holes. The central compartment is used to transfer flies from training in the grid tube to testing at the choice point in the 'T'-maze arms of the teaching machine. Holes will occasionally get clogged with fly debris. These can be unclogged using fine electrical wire or by flushing distilled water through the ports. All pieces can be washed with unscented detergent and rinsed copiously in distilled water.

Air leaks between the side and central pieces must be further excluded by using a small parallel clamp, which should be applied as the side pieces are being screwed in place following cleaning. As a rule of thumb, the clamp should be tightened until, when held up to the light and on its side, opaque impressions from all four 20 mm (inner) 'O'-rings can be seen on either side piece of the translucent machine.

7.4 Preparation of collection and cover tubes

Collection tubes form the arms of the 'T'-maze or teaching machine and are used in olfactory and olfactory learning assays (see *Figures 6* and *7*). They are attached to the choice point of the maze so that flies enter the tubes when choosing between odours in each maze arm. Thus, when each experiment terminates flies can be trapped in either of two respectively chosen collection tubes. These tubes can be detached from the machine and the flies transferred to a stoppered test-tube. Once behavioural testing is fully complete, flies can then be anaesthetized and counted. Collection tubes probably represent the single most significant source of variation from materials in olfactory-based assays. Consequently, it is vital to introduce a minimum of variation into their production and maintenance.

Protocol 9.   Manufacture of collection tubes

Equipment
- Polystyrene test-tubes (Falcon, No. 2017)
- Soldering gun (or a Bunsen burner)

Method
1. Melt a standard pattern of one central, four inner concentric, and ten outer concentric holes[a] in the base of the tube using a needle attached to the soldering gun. A less reliable alternative is to use a needle heated in a Bunsen flame[b] (*Figure 8b*).
2. Soak tubes overnight in scent-free detergent.
3. Rinse with copious amounts of distilled water.
4. Air dry the tubes.
5. Break in the tubes by attaching them to the maze and placing a population of flies in the tubes for 30 min.[c,d]
6. Attach 'sleeves' (General Valve Corp.) tightly to the base of the collection tubes (see *Figure 7*).

---

7. Attach odour blocks (General Valve Corp.) (see below) to the tubes via the sleeves.

8. One common source of variation in olfactory-based assays comes when sleeves are defiled by hand-grease. Treat sleeves like RNA preparations!

[a] These holes allow air currents to be drawn through the collection tubes via the vacuum source. Such air currents provide both anemotactic cues (flies walk 'up-wind') and also allow the delivery of airborne odour currents to the maze arms.
[b] The holes should be burned in a well-ventilated area as noxious fumes are produced.
[c] However, performances in olfactory and olfactory learning assays will only gradually increase to optimal levels over four days, when collection tubes have been used in approximately four test runs each day. Alterations in naive olfactory preferences are often observed before tubes are fully 'broken-in'. This delay in optimal performance most likely reflects the gradual breakdown of repulsive odours from the plastic and build-up of attractive fly-based olfactory cues.
[d] The polystyrene collection tubes are brittle. The open end, which fits tightly into the choice point of the machine, tends to deteriorate over time, showing hair-line fractures at first, and inevitably breaking with continued use. Avoid experimental delay by maintaining a continuous supply of 'broken in' stand-by collection tubes.

The odour block assembly consists of an odour block, odour cup, and cover tube (*Figure 6A*). The odour block attaches to the maze arm and also acts as a perch for the odour cup, which contains the odour source. The cover tube fits over the odour cup and snugly onto the odour block. To allow air current to be drawn over the odour cup, cover tubes have the same pattern of holes melted into their base as collection tubes.

7.5 Preparation of odour cups

Odour cups are made from disposable borosilicate glass test-tubes (e.g. Kimble: 10×75 mm large odour cups, and 13×75 mm small odour cups) cut 10 mm from their base, using a standard bench-top glass cutter. Cups are washed thoroughly in distilled water to remove glass dust and baked in a glassware oven until dry. The cups are glued with silicon tile sealant, e.g. General Electric 'Silcone II' Household Glue and Seal (Clear), to the wide end of 1–200 µl (yellow) pipette tips and allowed to dry at room temperature overnight. The next day, the glass from the glued odour cups is immersed in Sigmacote (Sigma) and allowed to stand overnight. The next day, odour cups are baked in a 60°C oven for one hour. Once cool, odour cups are ready to use. Odours should be filled to within 2 mm of the top of the cup using glass syringes fitted with disposable needles: 27 G½ (for octanol) or 18 G1½ (for methylcyclohexanol); (Benton Dickinson, Precision Glide).

For olfactory learning assays, Canton-S flies display equal preferences for 4-methylcyclohexanol and a 50% dilution of 3-octanol. This difference is accommodated by using different sized odour cups 8 mm i.d. (small odour cups) for 3-octanal and 11 mm i.d. (large odour cups) for 4-methylcyclohexanol, such that the surface area for 3-octanol is about half that of 4-methylcyclohexanol.

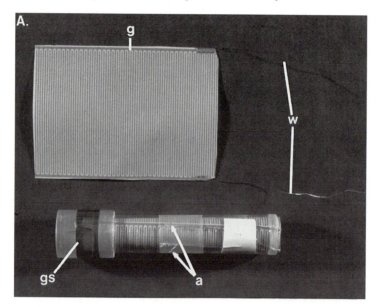

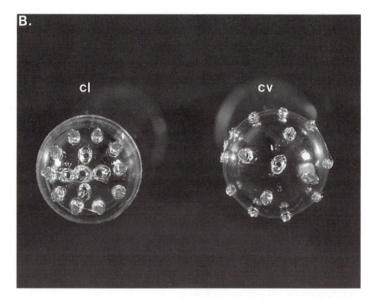

Figure 8. Disposable teaching machine parts. (A) Electroshock grid and tube. Copper wires (w) are soldered to an etched copper grid (g). This is inserted in a grid tube so that exposed copper wires can be connected via alligator clips (a) to an electroshock stimulator. Odour block assemblies are mounted on grid tubes via a sleeve (gs). See Section 7.6 and *Protocol 10* for further details. (B) Pattern of holes burned at the base of cover (cv) and collection (cl) tubes. See Section 7.4 and *Protocol 9* for details.

7.6 Preparation of grid tubes

Electroshock grid tubes (*Protocol 10*) are attached in the upper arm of the teaching machine during training in olfactory learning assays or are attached to either side of the maze arms in shock reactivity assays (*Figure 8A*).

Protocol 10.   Preparation of grid tubes

Equipment
- Two interdigitated copper circuits on epoxy-based paper[a]
- Copper wire strips
- Polystyrene test-tubes (Falcon, No. 2017)
- S48 stimulator (Grass Instruments, USA; Stag Instruments, UK)
- Voltmeter
- Soldering gun

Method

1. Solder 8 cm long copper wire strips to each circuit on either side of the rectangular grid (*Figure 8A*).

2. Wash the grids with unscented detergent and then with copious amounts of distilled water.

3. Avoid further contact of hands and fingers with the clean grids.

4. Roll the copper grids into cylinder, when dry, with the copper circuit facing inwards.

5. Cut the base ~ 1.5 cm from the bottom of a polystyrene test-tube.

6. Melt two small holes 1.5 cm from the top of the tube with a needle mounted in a soldering iron to allow the exit of the wires attached to the grid.

7. Wash the tubes with unscented detergent, rinse with copious amounts of distilled water, and dry.

8. Roll the grids and push them 'wires first' from the bottom to half-way along the top of the tube.

9. Draw the soldered grid wires through the melted holes with fine forceps.

10. Seal the holes and keep the exposed wires in place with tape.

11. Check voltage delivery to the grid by connecting the wires to a S48 stimulator and directly measuring the potential difference between circuits with a voltmeter. Short circuits will illuminate the 'OVERLOAD' indicator on the S48 stimulator.

[a] General Valve Corp. or Teledyne Electromechanisms; although most electronics workshops will etch copper circuits. The distance between the lines of each circuit should be less than the size of a fly (~ 0.5 mm), so that when a potential difference exists between circuits flies unavoidably conduct between the two and are shocked.

7.7 Vacuum source and flowmeters

A vacuum source is attached to the upper and lower ports of the teaching machine, allowing air currents to be drawn through the cover tubes, over the odour cups, into the collection (or grid) tubes, and over the flies. It is important that an uninterrupted air flow is used to minimize disruptions to the flies. Any 'rotor-style' vacuum pump should provide a pulse-free source of vacuum.

Normally, two teaching machines are used in olfactory-based assays, the procedures for each one being staggered by 15 seconds. As a consequence, the vacuum from the pump is bisected into lines to supply each machine via 'Y'-connectors (Kimax, No. 62880–049). As each teaching machine has both an upper port and lower port, each requiring different air flow rates (750 ml/min and 1500 ml/min, respectively), the vacuum lines are split into upper port and lower port lines via 'Y'-connectors. During experiments, a vacuum is connected to the upper or lower ports of each machine via twistcock connectors (Scienceware, No. F19741). A flowmeter connected in an airtight manner to an odour block is used to measure directly the flow rates at the grid and collection tubes. Vacuum flow rates are controlled in each line with adjustable vacuum needle valves (Fisher, No. 14–630–7B).

Protocol 11. Measurement of associative olfactory learning and memory (see *Figure 9*)

Equipment and reagents

- Far-red lighting (> 650 nm)
- Teaching machine (General Valve Corp.)
- Parallel clamp (any machine tool store)
- S48 electroshock stimulator[a] (Grass Instruments)
- Rotor-style vacuum pump[b] (General Electric)
- Needle valves
- Electric grid tubes
- Collection and cover tubes
- Electric grid tube sleeves (General Valve Corp.)
- Collection and cover tube sleeves (General Valve Corp.)
- Odour cups
- Odour blocks (General Valve Corp.)
- Flowmeter
- 3-octanol[c]
- 4-methylcyclohexanol[c]

Method

1. Prepare flies in glass bottles as described in Section 2.3 the night before testing.

2. Place flies in behaviour room for > 30 min before tests commence.

3. Adjust air flow rate on the upper port of the teaching machine to 750 ml/min and on the lower port to 1500 ml/min using a flowmeter.

Connect the 750 ml/min vacuum source to upper vacuum port, ensuring the upper vacuum port of the elevator is in register with the upper grid tube.

4. Connect the cable from the stimulator to the grid tube using alligator clips. Ensure that the stimulator is set to deliver 0.2 pulses per sec of 1.25 sec square wave 60 volt DC shock pulses.

5. Gently tap a fly bottle on its side to knock flies to the glass walls and aspirate 100–150 flies.[d]

6. Time: 0 sec. Immediately transfer aspirated flies into the upper grid tube.[e]

7. Time: 90 sec. Gently attach the odour block containing the CS^+ to grid tube.[f] Immediately electrify the grid by switching on the voltage supply from the stimulator. Ensure the grid does not short circuit by checking the 'OVERLOAD' light remains off on the stimulator.

8. Time: 150 sec. Switch off the voltage supply to the grid tube. Detach CS^+ odour block from the grid tube.

9. Time: 195 sec. Attach the odour block containing the CS^- to the grid tube.

10. Time: 255 sec. Detach the CS^- odour block from the grid tube.

11. Time: 285 sec. Disconnect the alligator clips from the grid tube. Disconnect the 750 ml/min vacuum source from the upper vacuum port. Slide the elevator up so that the central compartment is now in register with the grid tube. Gently tap the teaching machine on its side (cadence of 'dash-dot-dot' works well) and slide the elevator down out of register with the grid tube to trap the flies in the central compartment.[g]

12. Attach the CS^+ and CS^- odour blocks to the arms of the teaching machine.[h]

13. Time: 355 sec. Connect the 1500 ml/min vacuum source to the lower vacuum port.

14. Time: 360 sec. Lower the elevator swiftly and smoothly so that the central compartment is in register with the choice point of the teaching machine.

15. Time: 480 sec. Raise the elevator so that the central compartment is out of register with the arms of the teaching machine.

16. Remove the odour blocks from the teaching machine arms. Disconnect the 1500 ml/min vacuum source from lower vacuum port.

Protocol 6. *Continued*

17. Tap the teaching machine on each side while detaching collection tubes from the arms of the machine. Transfer flies from each collection tube to a foam-stoppered polypropylene tube via a funnel.[i]

[a] Any power pack which, like S48, functions as a 'current clamp' to deliver adjustable voltages to the grids could be used. Set the stimulator to deliver square wave pulses of 60 V DC, with a stimulus duration of 1.25 sec, and an interstimulus interval of 0.2 pulses/sec.
[b] Any make is adequate provided a pulse-free source of vacuum is generated.
[c] Undiluted 3-octanol is used in small odour cups (8 mm i.d.). Undiluted 4-methylcyclohexanol is used in large odour cups (11 mm i.d.). 1% Benzaldehyde in large odour cups can be used as an alternative odour source if a particular genotype shows olfactory acuity defects to 3-octanol or 4-methylcyclohexanol. Flies will also learn to discriminate between 10^{-3} dilutions of ethyl acetate and isoamyl acetate.
[d] To minimize disruption to flies, always tap bottles and machines on rubber or foam matting.
[e] Normally, two populations of flies can be simultaneously trained and tested in two separate machines. The procedural times for the first machine are staggered from the second by 15 sec.
[f] One full PI consists of two half-PIs. The first half-PI is measured with 3-octanol serving as the CS^+ and 4-methylcyclohexanol as the CS^-. The second half-PI is the reciprocal of the first, with 4-methylcyclohexanol serving as the CS^+ and 3-octanol serving as the CS^-. The scores from both half-PIs are then averaged to produce one full PI. Care must be taken when moving odour blocks to avoid spillages.
[g] For measurements of memory, the flies are removed from the machine at this stage and transferred to standard food vials which have been previously equilibrated to the conditions of the behavioural room. They are maintained in food vials within the behavioural room until olfactory memory is to be tested at step 12.
[h] Odour blocks should always be attached to the same testing arms of the machines. For example, 3-octanol is always delivered to the left arm of the machine. This ensures that residual odour does not contaminate a fresh odour current.
[i] Flies are counted under anaesthesia later. From empirical observations, six PIs are required for an accurate measurement of learning or memory.

7.8 Calculating the performance index for olfactory learning

For each reciprocal half-PI assayed, the 'probability correct' [COR] is calculated as the number of flies avoiding the CS^+ divided by the total number of flies in the 'T'-maze arms (10):

$$COR = [CS^+]/([CS^+] + [CS^-]).$$

The two COR values from each half-PI (where either OCT or MCH is the CS^+) are then averaged, and this average is normalized to produce one performance index or:

$$PI = \frac{[(COR_{OCT} + COR_{MCH})/2] - 0.5}{0.5} \times 100$$
$$= [(COR_{OCT} + COR_{MCH}) - 1] \times 100$$

This PI creates an index of scores ranging from 0 (no learning) to 100 (perfect learning). The PI is also a measure of associative learning unaltered by any

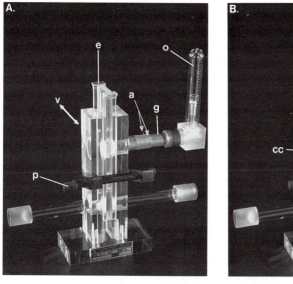
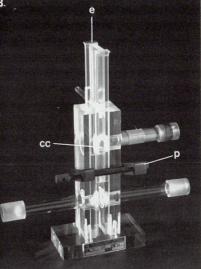
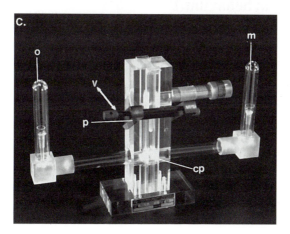

Figure 9. Procedure for associative olfactory learning in the teaching machine. (A) Training. The elevator piece (e) is raised so that vacuum is drawn from the upper vacuum port (v₊) over flies sequestered in the grid tube (g) via the odour block assembly (o). Simultaneously, electroshock can be delivered to the flies in the grid tube via alligator clips connected the copper wires soldered to the grid (a). A parallel clamp (p) maintains an airtight seal. (B) Transfer. After training is complete, the elevator piece (e) is raised so that the central compartment (cc) is in register with the grid tube. Flies are tapped into the central compartment and the elevator piece lowered, trapping the flies therein. Movement of the elevator piece may require adjustment of the parallel clamp (p). (C) Testing. Odour block assemblies containing octanol (o) and methylcyclohexanol (m) are attached to their respective collection tube arms of the maze. Vacuum is applied to the lower port of the elevator piece (v₋). Flies trapped in the central compartment are lowered to the choice point (cp). See *Protocol 11* for further details.

non-associative changes in odour preference which may occur during classical conditioning.

7.9 Spaced and massed training

While the multiple cycles of spaced or massed training can be delivered manually to flies, such training procedures are intensely laborious. A semi-automated computer controlled system has now been developed to conduct extended training protocols. A robotrainer (General Valve Corp., No. 80–5–900) consists of three channels, through which 750 ml/min vacuum can be drawn. One channel delivers room air, the second 3-octanol (diluted 10^{-3} (v/v) in heavy mineral oil), and the third 4-methylcyclohexanol (diluted 10^{-3} (v/v) in heavy mineral oil). Both switching of a relay to deliver electroshock pulses to the grid tube of the robotrainer and switching between 'bubbler' channels are computer controlled (Omnitech Electronics Inc.). Maximal memory scores are produced with a 15 minute rest interval between each of ten cycles of spaced training (39).

7.10 Peripheral behaviours

To establish that impoverished performance of a particular line or treatment of flies is the result of genuine defects in associative processes, flies are also examined for defects in the peripheral behaviours of olfactory acuity (*Protocol 8*) and shock reactivity (*Protocol 12*). Ideally, responses to the entire range of voltages and odour concentrations sensed by flies should be measured (10). Practically, shock and odour cues eliciting minimal and maximal responses are tested. For olfactory acuity controls, responses to training and testing concentrations, and to 100-fold dilutions, of both odours are assayed. In the case of shock reactivity controls, responses to training voltages (60 V) and to 20 V are measured.

Protocol 12.   Measurement of shock reactivity[a]

Equipment
- 'T'-maze or teaching machine (General Valve Corp.)
- S48 electroshock stimulator[b] (Grass Instruments)
- Rotor-style vacuum pump[c] (General Electric)
- Electric grid tubes (× 2)
- Electric grid tube sleeves (General Valve Corp.)
- Collection tubes
- Collection tubes sleeves (General Valve Corp.)
- Flowmeter

Method

1. Prepare flies in glass bottles as described in Section 2.3 the night before testing.

2. Place flies in behaviour room for > 30 min before tests commence.

---

3. Insert grid tubes into either side of the choice point in the teaching machine.

4. Adjust air flow rate on the lower port of the teaching machine to 1500 ml/min using a flowmeter. Afterward, ensure central compartment is in register with the grid tube.

5. Adjust the stimulator to desired voltage, interstimulus interval, and stimulus duration settings.[d]

6. Gently tap a fly bottle on its side to knock flies to the glass walls and aspirate 100–150 flies.[e]

7. Time: 0 sec. Immediately transfer aspirated flies into upper grid tube, gently tap machine on its side and lower the elevator, so that the central compartment is out of register with the grid tube, trapping the flies inside for 90 sec.

8. Connect the cable from the stimulator to one of the lower grid tubes using alligator clips.[f] Electrify the grid by switching on the voltage from the stimulator. Ensure the grid does not short circuit by checking the 'OVERLOAD' light remains off on the stimulator.

9. Time: 85 sec. Connect the vacuum source to the lower port of the teaching machine.

10. Time: 90 sec. Lower the elevator swiftly and smoothly so that the central compartment is in register with the choice point of the teaching machine.

11. Time: 210 sec. Raise the elevator so that the central compartment is out of register with the arms of the teaching machine.

12. Switch off the voltage to the grid and disconnect vacuum source from lower port of the teaching machine.

13. Tap the teaching machine on each respective side while detaching the lower grid tubes from the machine. Transfer flies from each lower grid tube to a foam-stoppered polypropylene tube via a funnel.[g]

---

[a] This assay is also an inherent measure of locomotor reactivity (see Section 3).

[b] Any power pack which, like S48, functions as a 'current clamp' to deliver adjustable voltages to the grids could be used.

[c] Any make is adequate provided a pulse-free source of vacuum is generated.

[d] Typically, the stimulus duration is 1.25 sec, the interstimulus interval 0.2 pulses/sec, the voltage used is square wave DC with optimal avoidance for Canton-S at 60 V, although flies respond to a range from > 0–120 V.

[e] To minimize disruption to flies, always tap bottles and machines on rubber or foam matting.

[f] Grid tubes should be alternately electrified on the left and right arms during tests of a given genotype to avoid the introduction of side biases into tests.

[g] Flies are counted under anaesthesia later. Performance indices are calculated in a similar manner to those for olfactory acuity (Section 6.5). From empirical observations, eight PIs (four on each side) are sufficient for an accurate measurement of shock reactivity to a specific voltage.

7.11 The appearance of memory

The testing of flies immediately after training is taken as a measure of initial learning or acquisition, but in fact probably reflects components of both learning and short-term memory (STM).

Memory is measured by testing flies at increasingly greater time intervals following training. Wild-type flies show a characteristic decay in the odour avoidance responses acquired during training. In the standard conditions of one cycle training in the Tully and Quinn assay, PIs of ~ 85 are observed immediately after training. Three hours later, PIs of about 50 are seen. By 24 hours, however, the preference of flies for either odour is no different from that of naive flies (PIs of ~ 0) (9, 40).

When flies are subjected to ten cycles of massed training, each one delivered immediately after the other, acquisition scores are comparable to those produced by single cycle training but more robust memory is apparent, decaying over a course of three to four days (39). Under spaced training conditions, however, where flies are given ten cycles of the Tully and Quinn assay with 15 minutes rest between each cycle, while acquisition scores are similar to those produced by one cycle training, the memory produced is more refractory to decay. One day (24 hours) after spaced training, PIs of ~ 45 are measured. Seven days (168 hours) following spaced training, PIs of ~ 30 are observed (39).

7.12 Dissecting the phases of memory

Over the course of the last decade, four analyses have been employed to dissect the decay curve in *Drosophila* olfactory memory:

- modified training protocols
- mutagenesis
- pharmacology
- transgenesis

Together they have revealed that memory 'decay' represents the formation of different phases of memory (39–41). Each phase appears to contribute in a largely additive fashion to produce the overall shape of the memory decay curve. Moreover, most phases seem to form sequentially, being dependent of the induction of preceding phases. While both anaesthesia-resistant memory (ARM) and long-term memory (LTM) are dependent on the preceding memory phase, middle-term memory (MTM), they appear to exist independently of each other (39).

Each memory phase has characteristic temporal kinetics. STM appears immediately after acquisition (LRN) and persists for about one hour. Within minutes, MTM is present and persists for about three hours. ARM is evident at about 30 minutes post-training and can last for up to four days after spaced or massed training. Finally, LTM is induced within hours of spaced training and seems to persist indefinitely.

7.13 Mutations impairing olfactory learning and memory

To date, nine mutants defective in olfactory learning have been identified by forward mutagenic screens (42). The first five were generated by EMS: *dunce*, *rutabaga*, *radish*, *cabbage*, and *amnesiac*. *P* element dysgenesis created *latheo*, *linotte*, *nalyot*, and *golovan*. Reverse genetic strategies have led to mutants disrupted in the catalytic and regulatory subunits of cAMP-dependent protein kinase A (PKA), *DC0* and *RI* respectively, and in the *14–3–3* (*leonardo*) gene. These are also defective in olfactory learning and memory (42–45). These mutants have facilitated a genetic dissection of memory formation in *Drosophila*.

As they were the first to be identified, *dunce* and *rutabaga* have been perhaps the most intensively studied of the mutants. Both of their olfactory learning and memory phenotypes are similar (9). Acquisition scores are about 60% of those seen in wild-type flies, but most strikingly, their memory soon after training rapidly disappears, as observed in their memory decay curves which are steeper than that of Canton-S. Hence, while these mutants are clearly defective in acquisition, their memory phenotypes have been used to define STM.

Middle-term memory has been defined by mutations in *amnesiac* and *DC0* , and by reversal learning assays (40, 45, 46) (*Protocol 13*). *amnesiac* was identified in a screen for mutants defective in three hour olfactory memory retention (46). Analysis of its memory decay curve revealed that while acquisition and STM were close to wild-type levels, memory decayed rapidly between 30 and 60 minutes following training. Subtraction of the *amnesiac* memory decay curve from that of Canton-S revealed the presence of a memory phase in wild-type flies which was induced to maximal asymptotic levels 30 minutes post-training, plateaued between 30 and 60 minutes, and gradually decayed away over five hours. More recently, a temperature-sensitive allele of *DC0* revealed that PKA also contributes to MTM (45). When non-specific temperature effects were eliminated from the analysis, the ratio of Canton-S to *DC0* scores over the entire memory decay curve revealed a component of memory which was absent at the restrictive, but present at the permissive, temperature in mutants. This phase of memory was identical to that missing in *amnesiac* flies: MTM.

MTM has also been defined from behavioural studies with reversal learning (*Protocol 13*) (40). Flies are subjected to one training session with one odour as the CS$^+$ and then to a second training session where the other odour, previously the CS$^-$, now serves as the CS$^+$. Groups of flies are subjected to re-training at various time intervals following the original training session. For each group, however, odour preferences are measured immediately after the second training session.

Two hypothetically extreme outcomes on odour preferences might be predicted from this reversal training procedure:

(a) If reversal training eliminated the conditioned odour responses learned during the original training session, then one would expect to observe the same reversal retention scores regardless of the time interval between training and re-training sessions. Hence, plots of reversal retention scores at various time intervals between training and re-training would produce a straight line.

(b) A second outcome would be that conditioned responses from re-training could act additively with those learned in the original training session, so that reversal retention scores would gradually increase with increasing time intervals between training and re-training sessions. In this case, plots of reversal retention scores at increasingly greater time intervals between training and re-training would produce a gradually increasing curve.

Reversal retention observed in Canton-S behaves somewhere in between these two extremes. Of significance, some of memory from the original training procedure is lost from re-training. Moreover, the reversal retention curves of *amnesiac* and Canton-S appear identical. This suggests that the component of memory disrupted in *amnesiac* is also sensitive to reversal learning. This memory component again suggests itself as MTM (40).

Protocol 13. Measurement of MTM by reversal retention

Equipment and reagents
- See Protocol 11

Method
1. Train groups of 100–150 adult flies in one training session of the Pavlovian procedure, as described in *Protocol 11*, steps 1–11.
2. For t = 0 of reversal retention, re-train groups of flies in a second training session (as described in *Protocol 11*, steps 1–11), so that the CS^+ from step 1 is now the CS^-.
3. For reversal retention at t > 0, place flies from step 1 in standard food vials until the second training (as described in *Protocol 11*, steps 1–11), where the CS^+ from step 1 is now the CS^-.
4. Measure conditioned avoidance responses immediately after the second training session in every case (as in *Protocol 11*, steps 12–17).

7.14 Behavioural pharmacology

Anaesthesia-resistant memory (ARM) has been defined by the *radish* mutation and by experiments involving cold shock (40, 47). In all species of animal tested, memory can be disrupted by treatments resulting in unconsciousness, such as hypothermia, anaesthesia, trauma, or electroconvulsive shock. As a poikilotherm, hypothermia or cold shock renders *Drosophila* temporarily unconscious. Administration of cold shock, immediately following training, blocks three hour memory retention (40). As the time interval between training and cold shock increases from zero to two hours, decreasingly severe disruption to three hour memory retention is observed. This suggests memory immediately after training is anaesthesia-sensitive, but that some memory forms within two hours which is anaesthesia-resistant (ARM). While acquisition is lower than wild-type in *radish*, memory decays more rapidly than in Canton-S. *radish* mutants were shown to be completely defective in ARM. Three hour memory retention following cold shock two hours post-training, when ARM is at maximal levels, was zero in *radish* homozygotes and in hemizygotes over a deficiency from the region.

Protocol 14. Measurement of ARM with cold shock

Equipment and reagents
- See *Protocol 11*
- 3.5 × 1.2 cm glass test-tubes

Method

1. Maintain a water-bath containing salted ice-water at 0 °C.

2. Pre-equilibrate glass test-tubes to water temperature > 30 min prior to experiments.

3. Train groups of 100–150 adult flies as described in *Protocol 11*, steps 1–11.

4. Store the trained flies in standard food vials until the desired time of cold shock.

5. Aspirate the flies into the glass tubes and submerge in the iced-water for 2 min at the desired times following training.

6. Transfer the flies back to food vials and maintain them at 25 °C until testing.

7. At 3 h post-training, measure memory in the flies as described in *Protocol 11*, steps 12–17.

The *radish* mutation has also facilitated the distinction between memory induced by massed and memory induced by spaced training (39). Massed training of *radish* mutants results in no memory at 24 hours post-training, while spaced training of *radish* produces 24 hour memory scores of about 30, but normal memory retention two to seven days later.

This suggests that ARM, defective in *radish*, accounts for memory induced by massed training. Like Canton-S, memory established following spaced training in *radish* can be disrupted by feeding of the protein synthesis inhibitor cycloheximide (CXM). Unlike *radish*, CXM fed wild-type flies subjected to spaced training still show some 24 hour memory. This suggests that spaced training induces two memory components in wild-type flies: one decays away over four days and in *radish* is defective (ARM), and the other persists for at least seven days and depends on protein synthesis (long-term memory, LTM) (39).

Protocol 15. Administration of drugs for behavioural experiments[a]

Equipment and reagents

- 3.5 × 1.2 cm glass test-tubes
- Pure glucose (Sigma)

- Chromatography paper (1MM and 3MM Chr, Whatman)

A. *Chronic exposure*

1. Make a solution of 5% glucose (w/v) in distilled water.[b] Dissolve the drug at the desired concentration in this glucose solution.

2. Soak two 1.0 × 2.5 cm strips of Whatmann 1MM chromatography paper in 250 μl of the drug solution.

3. Transfer drug soaked strips of paper to glass vials (feeding vials).

4. 12–15 h prior to behavioural assays, aspirate 100–150 flies into these feeding vials, and maintain at 25°C.

5. 30 min prior to behavioural assays, transfer flies to standard food vials which have been pre-equilibrated to the conditions of the behavioural room: 25°C and > 50% relative humidity.[c]

B. *Acute exposure*

1. Starve (~ 1000) flies for ~12–15 h in a 1 litre flask containing Whatmann 1MM chromatography paper soaked in 2.5 ml distilled water.

2. Dissolve drug at desired concentration in 5% (w/v) glucose in distilled water.[b]

3. Apply 250 μl drug solution to 1.0 × 2.5 cm strips of Whatmann 1MM chromatography paper.

4. Insert one drug soaked strip per glass vial.

5. Transfer ~ 100 starved flies to each glass vial for 2 h.

6. Transfer flies to glass vials containing 1.0 × 2.5 cm dry strips of Whatmann 3MM chromatography paper.

---

7. Maintain flies for 30 min at 25°C before commencing behavioural assays.[c]

---

[a] These protocols have been optimized for olfactory learning and memory, olfactory acuity, and shock reactivity assays.
[b] When the drug to be administered is non-soluble in water, solutions of ethanol or methanol can be used as a solvent.
[c] This allows flies to groom themselves of debris accumulated during feeding.

7.15 Transgenes affecting learning and memory

Inducible expression of transgenes from heat shock promoters have been invaluable reagents in the study of behaviour in adult flies (48–51). The distinction between a mutation acutely affecting behaviour and a mutation chronically affecting behaviour through maldevelopment has muddied many behavioural studies. Discrete temporal expression of dominant loss-of-function, dominant gain-of-function, or rescuing transgenes in adults allows the acute roles of gene products to be monitored in the absence of expression through development.

Drain *et al.* (48) reported that expression of mutant forms of regulatory and catalytic subunits of PKA impinged on associative learning. Recently, the adult learning and memory phenotype of *linotte* has been fully rescued by induced expression of the *lio* wild-type open reading frame in the *lio* mutant background (49). The molecular underpinnings of LTM are beginning to be understood from transgenic studies in *Drosophila*. In the first of these, Yin *et al.* (50) reported that heat shock induced expression of a blocking isoform of CREB (cAMP-responsive element binding protein) prevented flies from forming LTM (formed normally after ten spaced training sessions). In further experiments, Yin *et al.* (51) demonstrated that induced expression of an activator isoform of CREB resulted in the appearance of precocious LTM, formed after only one training session. These experiments, along with the observation that spaced, but not massed, training induced LTM, suggested a model for LTM induction. In this model, both activating and blocking activities of CREB compete within 'memory cells', but during the rest interval of spaced training the activating activity increases and/or the repressing activity decreases allowing overall CREB function to switch to activator form and induce the gene expression required for LTM (51).

Protocol 16. Administration of heat shock prior to behavioural assays[a]

Equipment and reagents
- 37°C water-bath
- Chromatography paper (Whatman 3MM Chr)
- 15 × 85 mm glass vials

Protocol 16. *Continued*

Method

1. Store flies in bottles overnight at 25°C and 50% relative humidity,[b] as described in *Protocol 2*.

2. Place a 10 × 20 mm strip of chromatography paper in each glass vial.[c]

3. Ensure that the water-bath has equilibrated to 37°C at least 30 min before heat shocking flies. Pre-heat the foam-stoppered glass vials in the water-bath.

4. Aspirate 100–150 flies into each pre-heated glass vial, push foam-stoppers half-way down each glass vial, and submerge each vial in the water of the water-bath, ensuring that the trapped flies are completely submerged.

5. Maintain flies at 37°C for 30 min.

6. Transfer flies to a standard glass food vial via a funnel. Maintain vials on their side in behavioural room conditions (25°C and 50% relative humidity). Allow the flies to recover from heat shock, before commencing behavioural assays.[d]

---

[a] These conditions have been established for assays of olfactory learning and memory, olfactory acuity, and shock reactivity. However, heat shock appears to affect different behaviours to different degrees. 1 h recovery is sufficient for olfactory learning performance, but flies must be given a minimum of 3 h recovery for LTM assays. Hence, heat shock administration and recovery conditions may need to be optimized for other behavioural paradigms.
[b] Flies must be maintained under these controlled environmental conditions the night prior to behavioural assays, to ensure consistent induction of transgene expression from heat shock promoters.
[c] The paper strip serves to absorb excess moisture produced in the glass vials during heat shock.
[d] Heat shock has 'non-specific' effects on many behaviours, which dissipate over the recovery period.

In a more recent transgenic study, Connolly *et al.* (52) used the *P*–GAL4 enhancer trap system (see Chapter 4) to target expression of a constitutively activated form of Gα_s to the mushroom bodies (MB) of the brain. This eliminated associative olfactory learning in flies. Previously, chemical ablation of the MBs through development was shown to abolish olfactory learning (53), and mutants defective in cAMP signalling were shown to be impaired in association (54). These studies suggest that cAMP signalling in mushroom body neurones mediates associative olfactory learning.

8. Vision

Flies are highly sensitive to visual stimuli, and respond to a range of wavelengths from ultraviolet ($\sim$ 350 nm) to far-red ($\sim$ 650 nm). Perhaps the most celebrated of the *Drosophila* mutants, *sevenless*, was isolated in a screen for mutants unable to sense ultraviolet light. As well as simple phototactic responses, flies can also recognize specific visual patterns which can be used to produce operant conditioning.

8.1 Larval phototaxis

Larvae are negatively phototactic (*Figure 10*) (26). Their response to light can be assayed on Petri dishes (*Protocol 17*).

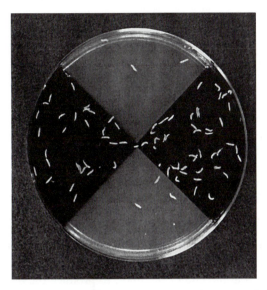

Figure 10. Larval phototaxis assay. Larvae are placed at the centre of a Petri dish containing diametrically opposed dark and translucent quadrants of 1% agarose. Dishes are placed on a light box, and larvae are given 5 min to choose between illuminated and dark quadrants. A response index (RI) is calculated by subtracting the number of larvae on translucent quadrants from the number on dark quadrants divided by the total number of animals. The distribution of larvae shown here represents an RI of 0.85. Reproduced with permission from ref. 26.

Protocol 17. Larval photoaxis

Equipment and reagents

- Petri dishes
- Light box

- Agarose
- Blue, green, and red food dyes

Method

1. Prepare two Petri dishes of 1% agarose, one containing 1 ml/100 ml of each of the blue, green, and red food dyes.
2. Cut the agarose on each dish into quadrants and move the quadrants

---

Protocol 17. *Continued*

to make two dishes each with two opposed transparent quadrants and two opposed dark quadrants.

3. Pour 12 ml of 1% agarose on top to produce a smooth surface for larvae to crawl on.
4. Place the dishes on a light box in a dark room.
5. Place larvae in the centre of the dish and allow them 5 min to choose between transparent and dark quadrants.

8.2 Slow phototaxis in adults

In adult flies, response to light can be measured using slow phototaxis (19, 55). In a simple version of this assay, the apparatus consists of two clear, funnel-shaped containers connected by a small Plexiglas tunnel (19). An infrared light gate in this tunnel detects passages between containers. At the start of the assay, 10–20 flies are placed in one container and the second container is illuminated using heat filtered light. Every 90 seconds for two hours, alternative containers are illuminated and the passage of flies recorded.

A more sophisticated measurement of slow phototaxis can be established using a maze similar to the one described for geotaxis in Section 3.7. In this case, the maze is blackened and laid flat in the horizontal plane. A light source is placed above the maze. Each 'T'-tube consists of one translucent arm and one blackened arm, allowing flies to choose between 'light' and 'dark' at each choice point of the maze. As with geotaxis, a population of flies is fractionated into a series of tubes at the end of the maze to produce the mean phototactic score for that population (55).

8.3 Fast phototaxis in adults

Fast phototaxis measures both response to light and locomotor activity and is measured in a countercurrent apparatus (*Figure 1*) (56) (*Protocol 18*). This consists of a Lucite frame, onto which six 100 × 17 mm polypropylene test-tubes (Falcon) can attach. One of these tubes (the starting tube) can slide along the apparatus so that in comes into register with one of five distal tubes.

Protocol 18. Measurement of fast phototaxis

Equipment and reagents
- Countercurrent apparatus
- Falcon tubes
- Flies

Method

1. Transfer ~ 150 flies into the starting tube and allow them to rest for 5 min.

308

2. Tap the flies to the bottom of the start tube.

3. Lay the apparatus horizontally with the distal tubes directly in front of a 15 W fluorescent light.

4. Allow the flies 15 sec to distribute between the starting tube and the distal tube.

5. Slide the starting tube along to the second distal tube and allow the flies to distribute between the two tubes.

6. Repeat for all five distal tubes to produce a fractionated population of flies in the five distal tubes.

7. Calculate the phototaxis score as: $\Sigma i \times N_i / \Sigma N_i$, where N is the number of flies in the ith tube.

8.4 The optomotor response

The optomotor response (*Protocol 19*) measures the ability of flies to co-ordinate movement in response to visual stimuli (57). Wild-type flies have a tendency to turn in the direction of moving stripes in their visual field.

Protocol 19.   Optomotor response

Equipment and reagents

- Light source (Sylvania, FC12T10 CW RS)
- Food vials
- Empty vials
- Watch glass
- Rotating drum
- Flies

Method

1. Collect newly eclosed flies and store individually in standard food vials for three to six days.

2. Starve flies in empty vials for 3–4 h before testing commences.

3. Place single flies under a 25 mm watch glass located on a platform[a] in the centre of rotating drum of alternating black and white stripes. Each pair of black and white stripes subtends 19° of arc.

4. Rotate the drum at 12 r.p.m.

5. Illuminate the drum with white fluorescent light (Sylvania, FC12T10 CW RS).

6. Score optomotor behaviour by measuring the number of times a fly crosses into a quadrant in the same direction of drum rotation versus the number of times it enters a quadrant in the opposite direction of drum rotation.

Protocol 19. *Continued*

7. Test flies for 5 min with clockwise drum rotation and 5 min with anticlockwise drum rotation, with a 3–5 min rest in darkness between tests.

8. Express the mean optomotor score a the ratio of quadrant crossings in the direction of drum rotation over total number of quadrant crossings.

a The platform on which the watch glass rests is divided into four quadrants.

disconnected mutants and *Ace* mutant mosaics show abnormal optomotor response.

8.5 Plasticity of the visual response

Drosophila displays light-induced modification of phototactic behaviour. Pre-exposure of wild-type flies to standard blue light (in the transmission range of 440–490 nm) for two to four hours results in altered phototactic preferences in 'T'-maze assays. This effect is reversed by subsequent exposure to yellow light (530 nm) for two hours (58).

Folkers reported that *dunce*, *turnip*, and *amnesiac* were defective in visual learning (59). In this assay, flies are subjected to two different wavelengths of light (406 nm and 507 nm), one of which is paired with aversive mechanical agitation. During testing, flies are allowed to choose between both wavelengths of light. Conditioning shifts a 50:50 distribution of naive flies to about 60:40, with such plastic effects lasting for up to two hours.

8.6 The landing response and its habituation

Flies will react to certain visual cues during flight in preparation for landing (60). This landing response is characterized by a forward thrust of the forelegs from a backward held position to an extended horizontal position at about the level of the antennae. If an individual fly is held in position by a wire loop attached to the dorsal side of its thorax by wax such that its position can be adjusted via a micromanipulator the fly will beat its wings, as if in stationary flight. Place the fly, eyes horizontally, in front of a viewing screen, which covers –42° to +42° vertically and –39° to +39° horizontally in the visual field of the fly, and the landing response can be induced by vertical movement of a grey stripe (7.5° in the vertical direction) through the visual field of the fly over an otherwise evenly illuminated viewing screen (Leitz pradovit projector, 250 W halogen bulb, 1 W/m^2). The stripe moves along the screen at 170°/sec with respect to the visual field of the fly. Almost complete habituation to the response is observed following delivery of 40 stimuli every 2 sec. Dishabituation of the response has been observed by a short pulse of mechanical vibration to the mount of the fly (60).

8.7 The flight simulator

Among the most intricate analyses of *Drosophila* behaviour are those in the flight simulator (61) (*Figure 11*). A single fly is glued to a small silver wire hook, which is attached to a torque-meter monitored by a computer. Flies are placed at the centre of a translucent, cylindrical arena, which is uniformly illuminated from behind. Yaw torque of the stationary 'flying' fly is continuously converted to DC voltage by the torque-meter. This signal allows the angular image deviations observed by the fly (which would have occurred in free flight from identical torque movements of the fly) to be calculated by computer. The computer signals this calculated angular position to a motor, which controls movement of the arena comprising the fly's panorama. Thus, flight movements of the stationary fly result in the corresponding 'virtual' movement of its artificial panorama.

8.8 Visual operant conditioning

Operant conditioning in the flight simulator is achieved with an infrared laser beam controlled by the same computer which manipulates the arena (61). The laser is applied to the abdomen of the fly, punishing it with heat, whenever one of two patterns on the arena is in the frontal quadrant of the fly's visual

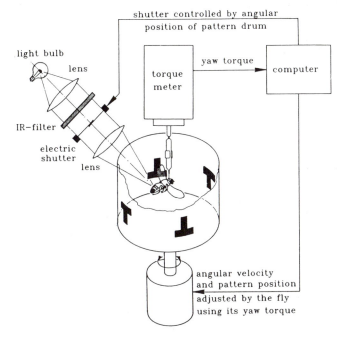

Figure 11. The flight simulator and operant conditioning. See Section 8.7 for details. Reproduced with permission from ref. 61.

system. This conditions the flies to avoid heat by flying towards specific land-marks. Training normally occurs over a four minute period, with testing taking place in the next two minutes. This learning is considered to be operant conditioning, since flies must be in control of the simulator during training for learning to occur. An avoidance performance index is calculated from $PI_A = (t_1 - t_2)/(t_1 + t_2)$, where t_1 and t_2 represent the total amount of the two minute test period spent in non-heat associated and heat associated quadrants, respectively. (Likewise, flies can be operantly conditioned to adopt flight paths favouring attractive odour plumes of fermenting banana.)

9. Courtship

9.1 Courtship

Courtship arguably represents the most complex and elaborate of the behavioural repertoire in *Drosophila* (see ref. 62 for a review). A courting male performs a highly stereotypical and hierarchical sequence of activities. He orientates himself towards a female, extends and vibrates one or both of his wings producing 'love-song', then engages in genital licking, and finally mounts and copulates with a receptive female (*Figure 12*). Each stage in the males courtship depends on performance of the preceding steps. The role in courtship of the female is not a passive one. A receptive female will remain still, and wave her abdomen at the courting male. A female rejecting courtship advances will actively avoid the male, kick him, and extrude her ovipositor until he 'gets the message'.

Mutations affecting male courtship include:

(a) *fruitless*, homosexual courtship.

(b) *don giovani*, males of which are not conditioned by courtship of fertilized females (see Section 9.2) because they fail to produce to appreciate cues from females.

(c) *inactive* and *couch potato* 'stand on the side-lines' rather than actively court.

(d) *period, cacophony, dissonance*, and *croaker* abnormal courtship love-song.

Females with the *spinster* mutation show defects in courtship receptivity (62).

9.2 Courtship conditioning

Despite the stereotypical nature of male courtship, this behaviour shows considerable plasticity. A male previously exposed to a fertilized female, who rejects his courtship advances, subsequently shows reduced levels of courtship to all females, including highly receptive virgins (63). Virgin females produce aphrodisiac pheromones, while fertilized females produce both aphrodisiac

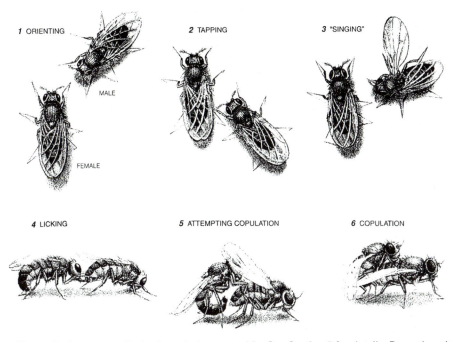

1 ORIENTING
2 TAPPING
3 "SINGING"

MALE

FEMALE

4 LICKING
5 ATTEMPTING COPULATION
6 COPULATION

Figure 12. Sequence of behaviour during courtship. See Section 9 for details. Reproduced with permission from ref. 74.

and anti-aphrodisiac pheromones. During his courtship of a fertilized female, the male learns to associate the aphrodisiac pheromone of the female with the anti-aphrodisiac pheromone (64). Such an exposed male engages in significantly less courtship when later presented with a virgin female who produces the aphrodisiac pheromone. This plasticity was shown to be associative in nature when extracts from fertilized females, by themselves, failed to inhibit subsequent male courtship. Courtship inhibition required the simultaneous presence of both a non-receptive female (the US, if you will) and the aphrodisiac pheromone contained in fertilized female extract (the CS^+) (64). Such conditioning of courtship can last for several hours (63). Furthermore, Ackermann and Siegel (65) showed that males presented with virgins in the presence of quinine sulfate (as a negative reinforcer) showed conditioned courtship suppression when subsequently exposed to virgins.

Interestingly, some of the mutants identified as defective in olfactory learning and memory also fail to show conditioning in courtship. *dunce*, *rutabaga*, *amnesiac*, and *Shaker* continue to court females after exposure to fertilized females. More recently, transgenic expression of an inhibitory peptide to cam kinase II (66) or to protein kinase C (67) was found to impair courtship conditioning.

313

Protocol 20. Measurement of courtship conditioning[a]

Equipment
- 9 × 12 cm 12-well porcelain plate (5 mm deep) (Fisher, No. 13–745)
- Microscope slides
- Dissecting microscope

Method

1. Collect flies under light anaesthesia within 8 h of eclosion.

2. Store females to be used as 'immature virgins' in groups of ten in standard food vials until 14–20 h post-eclosion.

3. Store females to be used as 'fertilized females' in groups of ten in standard food vials for four days. One day prior to conditioning experiments, these females are paired with four-day-old virgin males, observed to mate, and transferred back to groups of ten in food vials.

4. Store males individually in standard food vials for five days post-eclosion before using in conditioning experiments.

5. Aspirate single fertilized females into three wells of the porcelain plate.[b] Slide a microscope slide over the wells of the plate to trap the flies inside, each one forming an 'observation chamber' which can be viewed under a dissecting microscope.

6. Time: 0 min. For a given genotype, aspirate a single male into each well containing a fertilized female and a single male into each of the three empty wells.

7. Time: 60 min. Aspirate into the remaining six fresh wells virgin females which have been subjected to prolonged exposure to anaesthesia (> 20 min).[c] Aspirate the males from their previous wells into the wells containing the immobilized virgin females. Measure continuously the amount of time each male spends in courtship, including orientation, wing vibration, genital licking, and attempted copulation.

8. Time: 70 min. Terminate the measurement of courtship for each male.[d]

---

[a] A single 12-well plate is used to produce $N = 3$ each genotype. Typically, data is collected from $> N = 20$.

[b] Alternatively, a mating wheel can be used (68).

[c] Ether or CO_2 can be used as anaesthetics.

[d] For males not pre-exposed to a fertilized female, the ratio of time spent in courtship of virgins over the total test period of 10 min is the initial courtship index (CI_1). For males pre-exposed to fertilized females, the ratio of time spent in courtship of virgins over the total test period of 10 min is the experienced courtship index (CI_2). The courtship conditioning index (λ_c) is the normalized ratio of experienced courtship indices over initial courtship indices for a given group:

$$\lambda_c = (1 - CI_2/CI_1) \times 100$$

(But see also refs. 7 and 69. This index is biased towards initial courtship levels.)

9.3 Inhibition of courtship to immature males

Gailey *et al.* (70) reported that naive virgin males respond to sexually immature males (< six hours post-eclosion) but that this response diminishes significantly, so that the exposed males subsequently avoid further courtship of immature males for up to four hours. This courtship inhibition effect is specific to immature males since the response of exposed males to virgin females remained intact. Such plasticity seems to be non-associative learning, probably habituation, as exposure to the chemosensory cues from immature males alone produce the effect. Measurement of courtship inhibition is essentially analogous to that of courtship conditioning (see *Protocol 20*), with immature males replacing both 'fertilized females' and virgin females used in courtship conditioning assays. Courtship inhibition indices are calculated in an identical manner to courtship conditioning indices. *dunce*, *rutabaga*, *amnesiac*, and *turnip* are all impaired in inhibition of courtship to immature males.

9.4 Acoustic priming

Female receptivity is another aspect of courtship behaviour subject to plasticity. Females which have been previously stimulated by male love-song mate more quickly when introduced to males than do females unexposed to love-song (71). This effect, known as acoustic priming, lasts for about five minutes in wild-type flies and is believed to be a form of sensitization. The *dunce*, *rutabaga*, and *amnesiac* mutants, as well as transgenic flies expressing an inhibitory peptide to cam kinase, show altered responses to acoustic priming.

10. Statistics

Data handling is one of the most important tools used in behavioural studies. With unjustified or misguided statistical analyses, significant effects will be overlooked or insignificant effects pursued. We cannot do justice to a discussion of statistics within the remit of this chapter. The reader is referred to an accessible and comprehensive treatment of the subject in ref. 72. For specific applications to olfactory learning and memory see ref. 73. The JMP 3.1 statistical software package from SAS Institute is also recommended.

Acknowledgements

We thank Josh Dubnau, Ralph Greenspan, Chris Jones, Mike Regulski, and Brian Svedberg for the benefit of their experience.

References

1. de Belle, J. S. and Sokolowski, M. B. (1987). *Heredity*, **59**, 73.
2. Mackintosh, N. J. (1983). *Conditioning and associative learning*. Oxford University Press, NY.

3. Rankin, C. H., Beck, C. O., and Chiba, C. M. (1990). *Behav. Brain Res.*, **37**, 89.
4. Thompson, R. F. and Spencer, W. A. (1966). *Psychol. Rev.*, **173**, 16.
5. Hirsch, J. (ed.) (1967). *Behaviour—genetic analysis*. McGraw Hill Book Company, NY.
6. Dura, J.-M., Preat, T., and Tully, T. (1993). *J. Neurogenet.*, **9**, 1.
7. Gailey, D. A., Villella, A., and Tully, T. (1991). *J. Comp. Physiol. A*, **169**, 685.
8. de Belle, J. S. and Heisenberg, M. (1996). *Proc. Natl. Acad. Sci. USA*, **93**, 9875.
9. Tully, T. and Quinn, W. G. (1985). *J. Comp. Physiol A*, **157**, 263.
10. Boynton, S. and Tully, T. (1992). *Genetics*, **131**, 655.
11. Falconer, D. S. (1981). *Introduction to quantitative genetics*, 2nd edn. Longman Press, NY.
12. Dubnau, J. and Tully, T. (1998). *Annu. Rev. Neurosci.*, (in press).
13. Cline, T. W. (1978). *Genetics*, **90**, 683.
14. Guo, A., Li, L., Shou zen, X., Chun hua, F., Wolf, R., and Heisenberg, M. (1996). *Learn. Mem.*, 3, 49.
15. Heisenberg, M., Heusipp, M., and Wanke, C. (1995). *J. Neurosci.*, **15**, 1951.
16. Gilbert, D. (1981). *Drosophila Inf. Serv.*, **56**, 45.
17. Godoy-Herrara, R., Burnet, B., Connolly, K., and Gogarty, J. (1984). *Heredity*, **52**, 63.
18. Hayward, D., Delaney, S., Cambell, H., Ghysen, A., Benzer, S., Kasprzak, A., *et al.* (1993). *Proc. Natl. Acad. Sci. USA*, **90**, 2979.
19. Struss, R., Hanesch, U., Kinkelin, M., Wolf, R., and Heisenberg, M. (1992). *J. Neurogenet.*, **8**, 125.
20. Bellen, H., Vaessin, H., Bier, E., Kolodkin, A., D'Evelyn, D., Kooyer, S., *et al.* (1992). *Genetics*, **131**, 365.
21. Hamblen-Coyle, M., Konopka, R., Zwiebel, L., Colot, H., Dowse, H., Rosbash, M., *et al.* (1989). *J. Neurogenet.*, **5**, 229.
22. McMillan, P. A. and McGuire, T. R. (1992). *Behav. Genet.*, **22**, 557.
23. Kernan, M., Cowan, D., and Zuker, C. (1994). *Neuron*, **12**, 1195.
24. Phillis, R. W., Bramlage, A. T., Wotus, C., Whittaker, A., Gramates, L. S., Seppala, D., *et al.* (1993). *Genetics*, **133**, 581.
25. Corfas, G. and Dudai, Y. (1989). *J. Neurosci.*, **9**, 56.
26. Lilly, M. and Carlson, J. (1989). *Genetics*, **124**, 293.
27. Monte, P., Woodard, C., Ayer, R., Lilly, M., Sun, H., and Carlson, J. (1988). *Behav. Genet.*, **19**, 267.
28. Dethier, V., Solomon, R., and Turner, L. (1965). *J. Comp. Physiol. Psychol.*, **60**, 303.
29. Duerr, J. S. and Quinn, W. Q. (1982). *Proc. Natl. Acad. Sci. USA*, **79**, 3646.
30. Vargo, M. and Hirsch, J. (1982). *J. Comp. Physiol. Psychol.*, **96**, 452.
31. Médioni, J. and Vaysse, G. (1975). *C. R. Soc. Biol.*, **172**, 961
32. Carlson, J. R. (1996). *Trends Genet.*, **12**, 175.
33. Woodward, C., Huang, T., Sun, H., Helfand, S. L., and Carlson, J. (1989). *Genetics*, **123**, 315.
34. McKenna, M., Monte, P., Helfand, S. L., Woodard, C., and Carlson, J. (1989). *Proc. Natl. Acad. Sci. USA*, **86**, 8118.
35. Helfand, S. L. and Carlson, J. (1989). *Proc. Natl. Acad. Sci. USA*, **86**, 2908.
36. Aceves-Piña, E. O. and Quinn, W. Q. (1979). *Science*, **206**, 93.
37. Tully, T., Cambiazo, V., and Kruse, L. (1994). *J. Neurosci.*, **14**, 68.

38. Quinn, W. G., Harris, W. A., and Benzer, S. (1974). *Proc. Natl. Acad. Sci. USA*, **71**, 708.
39. Tully, T., Preat, T., Boynton, S. C., and Del Vecchio, M. (1994). *Cell*, **79**, 35.
40. Tully, T., Boynton, S., Brandes, C., Dura, J. M., Mihalek, R., Preat, T., *et al.* (1990). *Cold Spring Harbor Symp. Quan. Biol.*, **55**, 203.
41. DeZazzo, J. and Tully, T. (1995). *Trends in Neurosci.*, **18**, 212.
42. Tully, T. (1996). *Proc. Natl. Acad. Sci. USA*, **93**, 13460.
43. Skoulakis, E. M. C., Kalderon, D., and Davis, R. L. (1993). *Neuron*, **11**, 197.
44. Skoulakis, E. M. C. and Davis, R. L. (1996). *Neuron*, **17**, 931.
45. Li, W., Tully, T., and Kalderon, D. (1996). *Learn. Mem.*, **2**, 320.
46. Quinn, W. G., Sziber, P. P., and Booker, R. (1979). *Nature*, **277**, 212.
47. Folkers, E., Drain, P., and Quinn, W. G. (1993). *Proc. Natl. Acad. Sci. USA*, **90**, 8123.
48. Drain, P., Folkers, E., and Quinn, W. G. (1991). *Neuron*, **6**, 71.
49. Bolwig, G., Del Vecchio, M., Hannon, G., and Tully, T. (1995). *Neuron*, **15**, 829.
50. Yin, J. C. P., Wallach, J. S., Del Vecchio, M., Wilder, E. L., Zhou, H., Quinn, W. G., *et al.* (1994). *Cell*, **79**, 49.
51. Yin, J. C. P., Del Vecchio, M., Zhou, H., and Tully, T. (1995). *Cell*, **81**, 107.
52. Connolly, J. B., Roberts, I. J. H., Kaiser, K., Forte, M., Tully, T., and O'Kane, C. J. (1996). *Science*, **274**, 2104.
53. de Belle, J. S. and Heisenberg, M. (1994). *Science*, **263**, 692.
54. Davis, R. L. (1993). *Neuron*, **11**, 1.
55. Halder, N. M. (1964). *Genetics*, **50**, 1269.
56. Benzer, S. (1967). *Proc. Natl. Acad. Sci. USA*, **58**, 1112.
57. Greenspan, R., Finn, J., and Hall, J. (1980). *J. Comp. Neurol.*, **189**, 741.
58. Willmund, R. and Fischbach, K. F. (1977). *J. Comp. Physiol. A*, **118**, 261.
59. Folkers, E. and Spatz, H.-Ch. (1981). *J. Insect Physiol.*, **27**, 615.
60. Wittekind, W. and Spatz, H.-Ch. (1988). In *Modulation of synaptic transmission and plasticity in nervous systems* (ed. G. Hertting and H.-Ch. Spatz), p. 351. Springer–Verlag, Berlin.
61. Wolf, R. and Heisenberg, M. (1991). *J. Comp. Physiol. A*, **169**, 699.
62. Hall, J. C. (1994). *Science*, **264**, 1702.
63. Siegel, R. W. and Hall, J. C. (1979). *Proc. Natl. Acad. Sci. USA*, **76**, 3430.
64. Tompkins, L., Siegel, R. W., Gailey, D. A., and Hall, J. C. (1983). *Behav. Genet.*, **13**, 565.
65. Ackerman, S. L. and Siegel, R. W. (1986). *J. Neurogenet.*, **3**, 111.
66. Griffith, L. C., Verselis, L. M., Aitken, K. M., Kyriacou, C. P., Danho, W., and Greenspan, R. J. (1993). *Neuron*, **10**, 501.
67. Kane, N. S., Robichon, A., Dickinson, J. A., and Greenspan, R. J. (1997). *Neuron*, **18**, 307.
68. Hotta, Y. and Benzer, S. (1976). *Proc. Natl. Acad. Sci. USA*, **73**, 4154.
69. van Swinderen, B. and Hall, J. C. (1995). *Learn. Mem.*, **2**, 49.
70. Gailey, D. A., Jackson, F. R., and Siegel, R. W. (1982). *Genetics*, **102**, 771.
71. Kyriacou, C. P. and Hall, J. C. (1984). *Nature*, **308**, 62.
72. Sokal, R. R. and Rohlf, F. J. (1995). *Biometry: the principles and practice of statistics in biological reseach*. W. H. Freeman and Company, NY.
73. Tully, T. and Gold, D. (1993). *J. Neurogenet.*, **9**, 55.
74. Greenspan, R. J. (1995). *Sci. Am.*, **272**, 72.

10

Cell culture

LUCY CHERBAS and PETER CHERBAS

1. Introduction

In recent years cell lines have emerged as important tools for *Drosophila* research, tools that are being applied in such diverse projects as promoter analysis, studies of signal transduction mechanisms and hormone action, somatic cell genetics, and even as substrates for the development *in vitro* of the malarial parasite *Plasmodium* (1). In this chapter we introduce the *Drosophila* cell lines and provide those detailed protocols needed to maintain, transfect, and clone, and otherwise manipulate *Drosophila* cells. Our direct experience has been almost exclusively with the embryonic cell lines (e.g. Kc and S2 cells) that are in widespread use, and our laboratory protocols reproduced here were developed for those lines. Recently imaginal disc and CNS-derived lines have become available, and promise to become important research tools. Lacking direct experience with those lines we have, for the most part, elected to direct the reader to the appropriate literature.

The reader should also be aware of other reviews touching on the material in this chapter. Sang reviewed thoroughly the origins of the embryonic cell lines (2). Ashburner published a number of protocols, as well as an extensive list of cell lines available at the time of publication (3, 4). We have recently reviewed transfection procedures for *Drosophila* cells in greater detail than could be included here (5).

2. Cell lines

2.1 Embryonic lines
2.1.1 Establishment

Most of the *Drosophila* cell lines in common use today, including Schneider's line 2 (S2) (6) and Kc (7), were established from primary cultures prepared from minced or enzymatically disaggregated late embryos. We refer the interested reader to protocols published elsewhere; in particular, Schneider and Blumenthal (8), who summarized the procedures of several laboratories, and the detailed protocol of Bernhard *et al.* (9). Sang published a magisterial review of the biology of these primary cultures and their evolution into permanent lines (2).

2.1.2 Histological origins of the embryonic lines

The embryonic cell lines consist of those immortalized cells that emerged from the highly diverse collection of cells in the originating primary cultures. While the initial isolates may have represented a variety of cell types, selection in culture and subsequent clonings have yielded pure cultures whose histological origins can only be inferred from catalogues of the cells' gene expression. By this criterion, Kc cells are very like larval haemocytes and some haematopoietic (lymph gland) cells (10). This relationship was suggested by the observation that the genes *Eip71CD* and *Eip55*—primary responders to ecdysone in Kc cells—are similarly regulated by ecdysone only in the haemocytes and a portion of the lymph gland of metamorphosing third instar larvae. Moreover, the regulatory elements that mediate induction of *Eip71CD* are tissue-specific and the same elements are used in Kc cells as in the haemocytes/lymph gland (11). Thus, with respect to these genes, Kc cells behave like pre-metamorphic haemocytes/lymph gland. Similarly: a number of extracellular matrix proteins, such as tiggrin and the collagen gene product DCg1, synthesized primarily by the haemocytes and fat body in larvae, are expressed in Kc cells (12–16); β3 tubulin is expressed in Kc cells and in a few larval cell types, including haemocytes (17). The interested reader may wish to consult ref. 10 for additional examples and discussion.

All of the embryonic lines we have tested have qualitatively similar ecdysone responses, although the quantitative differences are substantial. Like Kc cells, cells of the S2 and S3 lines secrete extracellular matrix proteins (15) and a qualitative survey of gene expression in various embryonic cell lines (18) (see Appendix 1) suggests that they are quite similar. However, significant differences have been noted. For example, the cecropins—defensive peptides that play important roles in the immune response—are inducible in larval fat body and in a subset of larval haemocytes. They are also inducible in S2 cells (as well as in the tumorous haemocyte line mbn-2) (19), but are not inducible in Kc cells. Similarly, a scavenger receptor-mediated endocytotic mechanism characteristic of embryonic macrophages has been demonstrated in S2, but not in Kc, cells (20). In addition, embryonic lines can differ in ways that probably have nothing to do with their origins but represent either genetic accidents or epigenetic instability. In the first category we note that Kc cells (but not S2) express a truncated version of *Notch* made from a rearranged gene (S. Artavanis-Tsakonas, personal communication; note that the names of the two lines were inadvertently interchanged in ref. 21). In the second category we note that individual lines differ dramatically in their susceptibilities to calcium phosphate transfection (22) and that the morphological properties of cells—for example, their adherence to surfaces—are notoriously subject to (often inadvertent) selection.

In summary, cells of the various embryonic lines are strikingly similar in many respects; we consider it likely that all are derived from haemocytes, haematopoietic cells, macrophages, or some undescribed mesodermal precursor of all these lymphoid cells. However, these lines are not identical and an investigator should be prepared to test a variety of lines in order to find one with optimum properties for the experiments planned.

2.2 Lines from defined cell types

A number of cell lines have been derived from specific, defined tissues. The first of these were made from tumorous blood cells of the mutants *l(2)mbn* and *l(3)mbn* (23). These lines maintain some, though not all of the properties of the parental cells. More recently, permanent lines have been developed from isolated imaginal discs (24, 25) and from the central nervous systems of wild-type larvae (26). Milner has published a detailed protocol for the preparation of imaginal disc lines (27). As in the case of the tumorous blood cell lines, the disc lines retain some but not all of the properties of their parental primary cultures (24, 28, 29). Although characterization of the imaginal disc and neural lines is still in its early stages, it is clear that their existence has vastly expanded the variety of cell properties available for developmental and molecular studies (see *Appendix 1*).

The establishment of permanent lines from isolated non-tumorous tissues appears to be facilitated by the use of medium conditioned either by incubation with a primary embryonic culture (24, 26) or by the addition of fly extracts (25). This approach has not been explored extensively, but it may make possible the establishment of lines from other cell types.

3. General guidelines

For those experienced with mammalian cell culture, working with *Drosophila* cells poses few challenges and will appear very much simpler: *Drosophila* cells are generally maintained at 25 °C but room temperature is adequate and no special gas phase is required. Moreover the cells need not be released from the substratum with enzymes and, because investigator safety is less of an issue, hoods need not be so restraining. However, *Drosophila* cell culture is still cell culture and the investigator is rewarded for good, reproducible technique by experiments that are not compromised by contaminated samples or by lines that refuse to grow. In that respect, the more stringent requirements of mammalian cell culture are an excellent introduction to work with *Drosophila* cells. For those who are new to cell culture, the following is a brief summary of some of the more pertinent considerations.

3.1 Sterility

Standards of sterility are much higher for cell culture than for bacteriology. While it is not absolutely essential, we strongly recommend the use of a laminar flow hood for all manipulations of cells and media. Biosafety hoods, designed to protect both the experimenter and the experiment, work well. However, they are not required for invertebrate cell culture and it is currently acceptable —and both cheaper and more convenient—to use a simple, pharmaceutical hood in which sterile air sweeps toward the investigator and protects the experiment. Obviously, any experiment that involves the introduction of human pathogens should be carried out only under appropriate conditions in a biosafety hood.

Swab the work surface inside the hood with 70% ethanol before and after each use. The hood should contain only essential items, since the more objects present in the hood, the greater the air turbulence and the less the protection from contamination. The function of the hood depends on the integrity of its HEPA filter and filters occasionally develop leaks, with disastrous consequences. While replacement filters are extremely expensive, leaks can usually be repaired. We have our hood professionally certified at least once each year.

Sterilize all glassware by autoclaving or (preferably) by dry heat in a sterilizing oven. Several hours heating in a 250°C oven is distinctly more effective than 25 minutes in a standard autoclave, but the oven can be used only for materials capable of withstanding this extreme heat, generally only glass and metal. If possible, set the sterilizing oven on a timer, and allow the glassware to cool in the sterile environment of the oven. Use disposable sterile plasticware where possible.

Special precautions must be taken in handling spinner flasks, for several reasons. Because they are used for large volumes of culture, the presence of a single bacterium or fungal spore is a much more expensive accident than with 10 ml cultures in Petri plates. In addition, since a culture is normally stirred in a spinner flask for several days, surface residues that would be inconsequential on a piece of glassware that contacts cells only transiently can be problematic. To avoid these problems, fill the spinner flasks with water prior to autoclaving, and then autoclave each spinner flask twice, separated by a day. Discard the water, which is intended to leach most extractable chemical residues from the walls of the flask, immediately before the flask is used.

3.2 Cleanliness

Standards for cleanliness—in particular those for cleaning glassware—should be very high. Use a separate room for tissue culture and use reserved glassware which is not intermingled with the general laboratory supplies. Wash glassware with a minimal-residue detergent designed for tissue culture; we use 7X (ICN Biomedicals). After washing rinse thoroughly using pure water.

The frequency of contamination in cell cultures is influenced by the level of airborne spores and bacteria in the room. This level can be minimized by taking a few precautions to avoid the establishment of bacterial and fungal cultures in the tissue culture room.

(a) If culture medium is poured down a sink in the tissue culture room, flush thoroughly (*c.* 10 minutes) with a stream of tap-water to avoid establishing rich bacterial cultures in the U-trap under the sink.

(b) Seal all rubbish, particularly if it has come in contact with culture medium, and remove from the room as soon as possible; waste baskets are often sites of bacterial and mould cultures.

(c) Use heavy plastic liners in the waste baskets in the culture room; seal the liners and remove at least once a day. If a liner develops a leak, clean the waste basket thoroughly.

(d) Clean up any medium spills immediately.

(e) Seal and discard contaminated cultures (if they are in disposable plates) or autoclave (if they are in non-disposable vessels, such as spinner flasks).

(f) Fill bins containing used glassware awaiting washing with a dilute bleach solution, to minimize the growth of bacteria.

(g) It is also useful occasionally to swab down all the laboratory surfaces with a disinfectant (e.g. Lysol).

(h) Finally, minimize general air turbulence in the room by restricting the use of the room as much as possible.

While most of these strictures can be violated for a single short-term experiment, for long-term work with cells, uninterrupted by serious contamination or cell growth problems, it is simplest to maintain a well-organized environment for cell culture.

3.3 Water quality

Water quality is an important practical consideration. Media prepared with insufficiently pure water seldom work and, what is more troublesome, minor water problems may reveal themselves only slowly as chronic and elusive decreases in growth rate and/or aberrations of cell morphology. Given the time that can be lost to diagnosing such ills, it turns out to be more efficient simply to eliminate water as a potential source of problems. Prepare pure water using a Milli-Q system (Millipore Corp.) and you will experience no problems using water filtered through this or similar systems when they are kept in proper operating order. Glass stills can yield water of at least equivalent quality, but they require careful and frequent maintenance; colleagues have reported problems with their use for *Drosophila* cell culture and one such report has been published (30).

4. Maintenance of cell lines

4.1 Media

A variety of media have been used for embryonic cell lines; in many cases, lines established in one medium can be adapted to—or even transferred, without adaptation, to—another of the standard media. Recipes for a number of these media are given in refs 3 and 8. (Note: the recipe for medium M3 in ref. 3 contains a critical typographical error. The concentration of tissue culture yeastolate in M3 is 1 g/litre, not 10 as published.) Most workers find it convenient to use commercially prepared media for maintenance of the standard lines. Sigma Chemicals sells D22 (31), M3 (32), and Schneider's medium (6); Schneider's medium is also available from BioFluids and Gibco. This list of sources is not exhaustive; we have not searched for all potential vendors nor tested samples of all commercial media. Medium is sold either as a liquid, complete except for serum, or as a powder. Medium prepared in the laboratory is similar in cost (counting materials only) to the commercial powder. Dissolve the commercial powder in pure water, adjusted for pH, and sterilize by filtration through a nitrocellulose filter (0.2 μm pore size, e.g. Millipore type HA). Cloning in soft agar (see below) requires 1.25 × concentrated medium, which must be prepared from powder.

Recently, a variety of commercial media have been developed to support the growth of lepidopteran cell lines and some of these do not require serum (or exhibit reduced serum requirements). Unfortunately, their compositions are proprietary. We have found that one of these, CCM-3 (HyClone) unsupplemented by serum, is an excellent medium for the growth of Kc cells, and is far less expensive than any other combination of medium plus serum. Other lines adapt to CCM-3 with variable success; in many cases (e.g. S2, S3) it is necessary to supplement the medium with 1% fetal calf serum, and to adapt the cells to the new medium over a period of several weeks (L. C., unpublished data). We have had no success in adapting any of the imaginal disc lines to growth in CCM-3 (L. C., unpublished data). We are unaware of any systematic tests of the various lepidopteran media for *Drosophila* cells, and it is quite possible that others among the commercially available serum-free media may also be useful.

4.2 Serum

In most cases, media must be supplemented with fetal calf serum (Flow Labs, Gibco, HyClone, Sigma), at concentrations of 5–12%, depending on the line. Following the procedures established for mammalian cell culture, we test a sample of each new lot of serum for several weeks on Kc and S2 cells before purchasing a larger batch of that particular lot. However, in contrast to the reports of mammalian cell culturists, we have never (in 25 years) found a lot of serum that failed to support normal growth. Hence, it appears that *Drosophila* lines—at least Kc and S2—are quite insensitive to the variations in lots of fetal calf serum. Nonetheless, since fetal calf serum is a major financial investment, it seems prudent to test a sample, and then to purchase a 6–12 month supply of the tested lot. Laboratories that use large quantities of serum will find it advantageous to check with a number of dealers before ordering;

prices and quantity discounts vary widely. Heat treat serum before use as described in *Protocol 1*.

Protocol 1. Heat treatment of fetal calf serum

Reagents
- Fetal calf serum, stored frozen ($\leq 20\,°C$)

Method
1. Store serum frozen ($-20\,°C$ to $-80\,°C$) until ready to use.[a]
2. Thaw a bottle of serum, and bring it to room temperature.
3. Treat the bottle for 30 min at $65\,°C$, swirling occasionally. Keep the bottle in a sealed plastic bag during the heat treatment, to facilitate recovery of the serum if the bottle breaks.
4. If there is any sign of a crack in the bottle, resterilize the serum by filtration through a $0.2\ \mu m$ filter. Use a pre-filter to prevent filter clogging. If in doubt about sterility, test a 5 ml aliquot by incubating it overnight at $37\,°C$ in a sterile flask and checking for bacterial growth under the compound microscope.
5. Store the heat-treated serum at $4\,°C$ (or, for long-term storage, $-20\,°C$).

[a] The serum is stable for a longer time at the lower temperatures, but the bottles (both glass and plastic) are more likely to break, requiring that the medium be resterilized by filtration.

With the exception of CCM3 (see above), we have not found a suitable alternative to fetal calf serum for the long-term culture of Kc and S2 cells. However, newborn calf serum (heat treated according to *Protocol 1*) can be substituted for a period of several weeks without obvious deleterious effects. Since newborn calf serum is about tenfold less expensive than fetal calf serum, we use it for scaling up large cultures for biochemical experiments.

4.3 Other additions to media

Individual lines may require additions to the standard media. Supplement M3 with a mixture of bactopeptone (2.5 g/litre final concentration) and yeast extract (2 g/litre final concentration, including the 1 g/litre already present in M3), bringing it closer to the composition of Schneiderís medium. The addition of this mixture (BPYE) is essential for the growth of some lines (e.g. Kc, S3) in M3 medium, and not required for others (e.g. S2). Insulin was included in the medium used for the establishment of the *shi* embryonic line EH34A3 (33), but we have found it easy to adapt EH34A3 cells to medium lacking the insulin supplement. The imaginal disc and nervous system lines are all grown in media containing supplements of insulin and, in the case of the imaginal disc lines established in Milnerís laboratory, fly extracts (24, 25).

In general, with a new cell line, use caution in changing the culture conditions, although in many cases it may be possible to adapt cells to a simpler or more convenient medium.

4.4 Saline solutions

Use Robb's saline (*Table 1*) if it is necessary transfer the cells to a saline solution for up to a few hours without loss of viability. Phosphate-buffered saline (*Table 1*) is simpler though less gentle to the cells; it is useful for washing cells prior to extraction.

4.5 Culture procedures

4.5.1 Culture vessels

Although some lines may be cultured in bacto Petri dishes, some require, and nearly all do better on, tissue-grade plastic surfaces. It is convenient to use Petri dishes rather than T flasks for most purposes, because of the lower cost, because the cells will be transferred by vigorous pipetting and the Petri dishes are more accessible. The culture suspension should in general be about 2–3 mm deep, to permit good gas exchange. Use the volumes stated in *Table 2* for standard plastic microwell plates, Petri dishes, and flasks. Store the plates in sealed plastic boxes (e.g. food storage boxes) inside the incubator to control evaporation. Multiwell plates containing < 1 ml/well are particularly prone to desiccation, and should also be sealed with Parafilm.

None of the embryonic *Drosophila* cell lines are surface-dependent. Even those cells that tend to grow attached to surfaces have always proven adapt-

---

Table 1. Saline solutions

Robb's saline[a]	**Composition (g/litre in brackets)**
Solution A:	26 mM NaCl (3.04 g), 20 mM KCl (2.98 g), 5 mM glucose (1.80 g), 50 mM sucrose (34.2 g), 0.6 mM $MgSO_4.7H_2O$ (0.28 g), 0.6 mM $MgCl_2.6H_2O$ (0.24 g), 0.5 mM $CaCl_2.2H_2O$ (0.15 g)
Solution B:	1 mM $Na_2HPO_4.2H_2O$ (0.36 g), 0.18 mM KH_2PO_4 (0.05 g), adjusted to pH 6.75 with 1 M HCl

Autoclave solutions A and B separately. Allow to cool to room temperature. Then mix the two solutions in a 1:1 ratio.

Phosphate-buffered saline[b]	**Composition (g/litre in brackets)**
10 × PBS:	1.3 M NaCl (76 g), 70 mM $Na_2HPO_4.2H_2O$ (12.5 g), 30 mM $NaH_2PO_4.2H_2O$ (4.8 g), pH 6.7–6.8

Store at room temperature as a 10 × PBS. Dilute to 1 × before use.

[a] From ref. 34.
[b] Modified from ref. 3.

Table 2. Culture volumes

Vessel	Volume
96-well plate	0.1 ml/well
48-well plate	0.3 ml/well
24-well plate	0.5 ml/well
12-well plate	1 ml/well
6-well plate	2 ml/well
35 ml Petri dish	1 ml/dish
60 ml Petri dish	4 ml/dish
100 ml Petri dish	10 ml/dish
150 ml Petri dish	25 ml/dish
25 cm^2 flask	5 ml/flask
75 cm^2 flask	15 ml/flask

able to growth in suspension. Though some laboratories have used rotating bottles with success, we use spinner flasks for large scale preparations for bio-chemical work. Kc, S2, and S3 cells all do well in spinner flasks. To provide sufficient aeration, fill each flask to about 25% of its capacity. Rotate the magnetic stirrer just fast enough to keep the cells in suspension, usually about one revolution/second. Growth of the cell population is generally somewhat slower in spinner flasks than in plates, and a significant amount of cell debris is always detectable in the microscope, presumably from mechanically damaged cells. Because of these problems, because growth in spinners usually requires antibiotics, and because of the cost, we do not maintain cell lines in spinner flasks.

4.5.2 Culture conditions

Incubate plates at 25 °C in air; room temperature is acceptable so long as it is relatively stable and it may be preferable if there is risk of the refrigeration cycle of incubators failing. We routinely remove our cultures (in their food storage boxes) to the bench-top when the building experiences a power failure. Kc cells do not tolerate temperatures above 29 °C. Most lines (including Kc and S2) can be maintained at 18 °C for some weeks if one wishes to slow the growth rate (e.g. during a vacation). Cell growth is negligible at 18 °C, but some lines do not tolerate this temperature well.

Cell lines vary in their sensitivity to low and high cell densities. Cells should always be maintained within the range of exponential growth. For Kc, S2, and S3 cells, this range is from about 5×10^5/ml to about 10^7/ml. These numbers (and typical doubling times of approx. 24 h) dictate routine transfers every three to four days. Cell density is easily measured using a haemocytometer.

Cells of most *Drosophila* lines adhere to the substrate much more loosely than mammalian lines. Dislodge cells from the culture vessel by blowing medium at the surface using a sterile cotton-plugged Pasteur pipette. Determine the cell titre in the suspension using the haemocytometer and prepare diluted cultures.

4.5.3 Freezing and thawing cell lines

Unlike flies, cell lines can be stored indefinitely as frozen stocks. As soon as possible after acquiring a new line prepare a frozen stock for storage using *Protocol 2*.

Protocol 2.   Freezing viable cells

Equipment and reagents

- Sterile cryovials: ampoules capable of holding approx. 1 ml, and designed for storage in liquid nitrogen
- Dewar flask, with tight-fitting lid
- Freezer (–70 °C to –80 °C)

- Liquid nitrogen freezer
- Healthy cells in mid-exponential growth
- Freezing medium: 10% DMSO in the normal growth medium used for the cells

Method

1. Pellet the cells. Resuspend them in freezing medium at 2×10^7/ml.

2. Dispense the cell suspension into sterile cryovials at 0.5 ml/vial.

3. Place the vials of cells into a Dewar flask at room temperature. Tightly stopper the flask, preferably taping the stopper to the flask so that it does not fall off as the flask cools.

4. Place the flask of vials into a –80 °C freezer for at least two days. The aim is to cool the cell suspension very slowly. After 48 h, the vials will be at the temperature of the freezer.

5. Transfer the vials to liquid nitrogen.[a]

6. A few hours after the cells are put in liquid nitrogen, thaw one ampoule to test for viability.[b]

7. To thaw cells, simply bring the vial to room temperature and immediately transfer its contents to 5 ml of medium in a 60 mm Petri dish or a 25 cm^2 T flask.

[a] Cells stored in liquid nitrogen for ten years show no apparent loss of viability. Alternatively cells may be stored at –80 °C; we have had no experience with this technique, but we would assume that *Drosophila* cells, like mammalian cell lines, lose viability at the rate of a few per cent per year, and hence occasionally have to be thawed and refrozen.
[b] There is considerable variation in the quality of successive freezings of a single cell line; if viability is low, it is probably worth repeating the freezing.

The success of *Protocol 2* varies considerably among lines; in some cases, only a few per cent of cells remain viable, and it may take a week or more for a freshly thawed vial to develop into a healthy culture. S2 cells typically survive this procedure very well; Kc cells give more variable results.

4.5.4 Cloning cells

Cloning *Drosophila* cells is difficult simply because none of the available culture media supports growth of these cells at low densities: growth is slow when the starting cell concentration is 10^5/ml and essentially no growth occurs when it is 10^4/ml. To overcome this problem, one must supplement the medium. Some workers have accomplished this by conditioning the medium (S. Lindquist protocol in ref. 3). We have had more success using feeder cells. Feeder cells are cells that have been rendered reproductively dead by massive irradiation; morphologically—and more importantly, metabolically—they remain normal for extended periods. The feeder cells condition the medium, without contributing their own progeny to the resulting clones. In its traditional application to mammalian cells (35, 36) the feeders were chosen to be morphological or genetically distinguishable from the cells to be cloned. For *Drosophila* use cells similar to those to be cloned and rely on parallel control experiments to confirm that the feeders are reproductively dead.

Two different cloning strategies are available. Clones may be grown in soft agar for cell lines that are not markedly surface-adherent (see *Protocol 3*). This works well for S2 but not for Kc cells. Soft agar cloning is easy and inexpensive. A more general procedure is to plate the cells at limiting density in 96-well plates (*Protocol 4*).

Protocol 3.   Cloning in soft agar

Equipment and reagents

- Bacteriological grade 100 mm Petri plates
- X-ray source: a Cs source works well
- Cells to be cloned
- 1.25 × medium: for S2 cells, this is 1.25 × M3 medium containing 15.6% fetal calf serum (see *Protocol 1*)—the medium may be supplemented with selective agents for cloning stably transformed cells
- Culture of cells suitable for use as feeder cells: the cells should carry resistance to any selective agents to be used in the cloning medium
- 1.5% Noble agar in water, melted in a boiling water-bath or microwave oven, and brought to 45–50 °C before use

Method

1. Prepare feeder cells by irradiation (24 kR, Cs source). Irradiate the cells in 5 ml Robb's saline (*Table 1*).[a] If possible, the feeder cells should carry resistance to any selective agents that will be used in the cloning medium to maximize their effectiveness.

2. Suspend the feeder cells at $1–1.2 \times 10^6$ cells/ml in 1.25 × medium.

3. Make serial dilutions of the cells to be cloned in the feeder cell suspension, to give a final concentration of about 10–50 viable cells/ml.[b]

4. Place 8 ml of the cell suspension on one side of a Petri plate, then 2 ml of molten agar on the other. Gently swirl to mix. If the agar starts to

329

Protocol 3. *Continued*

set, stop swirling to avoid generating an irregular surface in the agar. The final mixture, containing 0.3% agar, is very soft and must be handled with care. Repeat the procedure with each sample to be cloned. Include a control plate containing only feeder cells. Allow each plate to set for at least 10 min in the sterile hood before moving it; then stack the plates very gently in a sealed plastic box and store at 25°C.

5. Usually, clones are visible after about two weeks as small white spheres. They can easily be distinguished from bubbles or from bacterial colonies by examination of the plate in an inverted microscope. There should be no clones in the plate containing feeder cells only. When the clones have reached a diameter of about 1 mm, pick isolated clones as plugs of agar, using a Pasteur pipette or a 50 µl capillary pipette, and disperse the plug as well as possible in 0.1 ml medium in a well of a 96-well plate.

6. As the clones grow, they can be transferred to progressively larger wells (see above). We usually prefer to dilute only two- to threefold in the early transfers, since it is very difficult to estimate the cell number, and it is important to avoid over-dilution of the cells. Once the clone consists of an even layer of healthy cells, the cells can be handled in the usual way.

[a] Irradiation in medium works equally well (S. Lindquist, in ref. 3).
[b] The cloning efficiency for S2 cells in soft agar is around 50%.

Protocol 4. Cloning in 96-well plates

Equipment and reagents

- Sterile 96-well plates (e.g. Corning, Costar, or Falcon)
- Multiple pipettor (8- or 12-well design; optional but very helpful)
- Culture of cells suitable for use as feeder cells: the cells should carry resistance to any selective agents to be used in the cloning medium

- X-ray source: a Cs source works well
- Cells to be cloned
- Medium: for Kc cells, this is M3 medium supplemented with BPYE and 5% fetal calf serum (see *Protocol 1*), or CCM-3 medium—the medium may be supplemented with selective agents for cloning stably transformed cells

Method

1. Irradiate feeder cells as described in *Protocol 3*. Suspend in medium at 10^6/ml.

2. Make serial dilutions of the viable cells in the feeder cell suspension, to give a final concentration of approx. 1 viable cell/ml.[a]

3. Dispense into 96-well microtitre plates at 0.1 ml/well, using a multiple

pipettor if available. Seal the plates with Parafilm, place in a sealed plastic box, and store at 25°C. Include one plate with feeder cells only.

4. Clones should be clearly visible in an inverted microscope within about two weeks. There should be no clones in the plate containing feeder cells only. Begin transferring clones to progressively larger wells after the diameter of the clone reaches about one-third the diameter of the well.

[a] The cloning efficiency of Kc cells in 96-well plates is about 90%.

5. Transfection

We have recently published an extensive discussion of techniques and strategies for transfection of *Drosophila* cells (5). Here, we provide a few specific protocols, and some suggestions on the design of transfection experiments. We refer the reader to the earlier review for discussions of promoters, reporters, and selection systems.

5.1 Procedures

Transfection is used either for transient expression (the measurement of gene expression in a mixed population of cells, within a few days of the transfection) or stable transformation (the selection of cells carrying the transfected DNA to establish a stable population or clone of transformed cells). The transfection protocol is essentially the same in either case; the experimental strategies are quite different for the two.

Three transfection procedures are now in common use for *Drosophila* cell lines; calcium phosphate:DNA co-precipitation, electroporation, and lipofection.. Very little comparative data is available to evaluate the merits of the three procedures. In principle, one would wish to know for each procedure:

• how much DNA is taken up and expressed by the cells
• how that DNA is distributed among the cells of the population

We find that a given reporter construct gives about the same level of reporter expression in Kc167 cells when introduced by calcium phosphate:DNA co-precipitation or by electroporation. Søndergaard (37) found lipofection to give 2.5–170-fold higher expression of a given reporter in S2 cells than calcium phosphate. Estimates of the distribution of DNA among cells are more difficult to obtain and to interpret. When Kc167 cells were transfected by calcium phosphate:DNA co-precipitation with a series of plasmids carrying a *lacZ* reporter gene transcribed from a series of different promoters, the proportion of cells that stained for lacZ ranged from 0.1% to about 20%, depending on the strength of the promoter (A. J. Andres, personal communication). Hence, following calcium phosphate transfection, there is a broad distribution of levels of uptake among the cells of the population, with the number of staining cells simply reflecting the sensitivity of the assay. We know of no comparable data for electroporation or lipofection.

5.1.1 Calcium phosphate

Calcium phosphate:DNA co-precipitation (*Protocol 5*) is easy and very inexpensive, and gives high transfection efficiencies in some cell lines. Its principal disadvantages are:

(a) The precipitate is somewhat toxic to the cells and may affect their physiology; for example, mock transformation with a calcium phosphate precipitate containing no DNA causes a substantial decrease in the induction (by ecdysone) of the gene *Eip71CD* (22).

(b) Some cell lines simply cannot be transfected by this procedure. This includes most Kc cells and S3 cells; the only Kc clone we have succeeded in transfecting with calcium phosphate is Kc167.

(c) The amount of DNA used per plate of cells is fixed; one can vary the amount of a plasmid of interest only by including carrier DNA, and there is an upper limit to the amount of DNA that can be transfected.

Protocol 5. Calcium phosphate transfection[a]

Equipment and reagents

- Cotton-plugged Pasteur pipettes
- Compressed air supply
- Exponentially growing *Drosophila* cells at about 3×10^6/ml: we use 10 ml of cell culture in a 100 mm Petri plate for transient expression, or 3 ml of cell culture in a 60 mm Petri plate for stable transformation[b]
- 2 × BBS: 280 mM NaCl, 1.5 mM Na_2PO_4, 50 mM BES pH 6.95, filter sterilized and stored at –20°C
- Sterile pure water

- 2 M $CaCl_2$, filter sterilized and stored at –20°C
- Plasmid DNA, dissolved in TE buffer (10 mM Tris–HCl, 1 mM EDTA pH 8.0) and stored at –20°C. We generally use DNA purified on a CsCl gradient, but DNA prepared on Qiagen columns also works well. CsCl purified DNA generally requires no further sterilization; if it is necessary, plasmid DNA can be sterilized by overnight ethanol precipitation.

Method[c]

1. Prepare the DNA:calcium phosphate co-precipitate as follows. In a 15 ml sterile plastic tube, mix 62 μl 2 M $CaCl_2$, 20 μg DNA, and water to 0.5 ml. Agitate the solution by bubbling air through a sterile cotton-plugged Pasteur pipette whose opening is at the bottom of the tube; adjust the rate of air flow so that at least two or three bubbles are in transit at all times. While continuing the aeration, add 0.5 ml of 2 × BBS dropwise. Remove the pipette, cap the tube, and let it sit for 30 min. The precipitate should be visible only as a faint blue coloration (Tyndall effect); if the precipitate settles visibly during the 30 min, it is too coarse and will not work well.

2. Add 1 ml of the precipitate dropwise to a plate of cells; mix gently by tipping the plate back and forth.

3. Incubate the cells overnight (12–16 h) at 25 °C. At this point the precipitate should be easily visible in an inverted microscope, looking not unlike bacterial contamination.

4. Wash the cells twice with fresh medium, trying to disturb them as little as possible. If the cells adhere well to the surface of the plate, one can simply pipette the medium off the plate and replace it with fresh medium. If a substantial proportion of the cells are in suspension, they should be collected by centrifugation and added back to the plate. One can omit serum from the medium used for washing, in order to save money, but the cells should be in their normal growth medium at the end of the washing procedure.

5. For transient expression, cells are usually harvested 12–24 h after washing; the timing depends on the nature of the experiment. For stable transformation, the selective agent is added two days after washing.

[a] Based on the technique of Wigler *et al.* (38), as modified in ref. 39.
[b] When one adds 1 ml of precipitate to 3 ml of cells, for stable transformation, the cells are much sicker than transiently transfected cells, in which 1 ml of precipitate is added to 10 ml of cells. The former cells must be handled particularly carefully.
[c] The procedure is given for 1 ml of precipitate; it can be scaled up as desired.

5.1.2 Electroporation

Electroporation (*Protocol 6*) is a general procedure which appears to be applicable in principle to any cell line. It permits the use of a wide range of DNA concentrations, and is relatively non-toxic to the cells. We have used up to 0.3 mg DNA/plate, with the uptake apparently linear up to that point (40). Once the equipment is available, the procedure is very simple and inexpensive. The principal disadvantages are:

(a) The equipment is expensive.

(b) The parameters for electroporation must be optimized for each cell line. To our knowledge, the procedure has so far been optimized only for Kc cells (*Protocol 6*); the same procedure can be used for S2 cells (X. Hu and L. Cherbas, unpublished data) and MLDmD-8 cells (L. Cherbas, unpublished observations), but in the case of S2 cells the conditions described for Kc cells in Protocol 6 should be modified by increasing the voltage to 715 volts.

5.1.3 Lipofection

Lipofection has been used for transfection of *Drosophila* cells in several laboratories. Lipofection permits the use of a wide range of DNA concentrations, and appears to be generally applicable. A number of commercially available reagents have been used for this purpose. Søndergaard (37) evaluated several commercial reagents in an S2 cell transient expression assay, and found a 70-fold range of expression for a given reporter plasmid; Lipofectin (Gibco/Life Technologies) was the most effective. The principal disadvantage of this procedure is the high cost of commercial reagents. A less expensive procedure involves the preparation of liposomes in the laboratory from readily available components (41); this procedure has been used successfully for transfecting several imaginal disc lines (V. Panin and K. Irvine, personal communication), but no data are available to compare the transfection efficiencies of commercial and non-commercial reagents.

Protocol 6. Electroporation of Kc cells[a]

Equipment and reagents

- Hoefer PG200 Progenitor power supply with PG250 electroporation chamber and PG220C cuvette electrode (3.5 mm gap)[b]
- Clinical centrifuge
- Kc cells, plated at 10^6/ml in M3 + BPYE + 5% FCS 48 h before electroporation
- Plasmid DNA, as described in *Protocol 5*
- M3 medium supplemented with BPYE, without serum
- Sterile pure water
- 70% ethanol

Method

1. If non-disposable electrodes are to be used, sterilize them by immersion for at least 20 min in 70% ethanol. Rinse the electrodes in water, then in medium (without serum), before they come in contact with the cell suspension.

2. Remove the cells from the plates, centrifuge to pellet the cells, and wash twice in approximately half the original volume of M3 + BPYE (no serum). Finally, resuspend the cells in M3 + BPYE (no serum), 0.8 ml/electroporation. Dispense this suspension into electroporation cuvettes at 0.8 ml/cuvette. Add the DNA, cap the cuvettes, mix by inversion, and allow to sit for 5 min.[c]

3. Insert the electrodes, and shock cells the cells at 440 V/cm, 1200 μF, 1 sec (i.e. complete decay). Allow the cells to recover for 10 min.

4. Transfer the cells into the original volume of complete medium (M3 + BPYE + 5% FCS), i.e. 10 ml per starting plate, and dispense 10 ml/plate.

5. Harvest or add selective agents as described in *Protocol 5* for calcium phosphate transfection.

[a] Method of S. Bodayla, M. Vaskova, and P. Cherbas, unpublished, cited in ref. 5.
[b] The procedure should be readily adaptable to other commercially available electroporation apparatuses by making appropriate adjustments for the dimensions of the chamber.
[c] The concentration of cells during electroporation does not seem to be important; we have electroporated the contents of one to eight plates successfully in a single cuvette.

5.2 Strategies

Transfection of *Drosophila* cells is now a widely used technique for the analysis of promoter function, gene interactions, and protein function. The following examples give a sense of the variety of applications:

(a) The structure of ecdysone response elements was determined in part by assaying elements inserted into a reporter construct, using transient expression in the ecdysone-responsive cell line Kc (22).

(b) The interaction of Notch and Delta proteins was demonstrated by assaying the aggregation of cells from separate populations of S2 cells transiently transfected with plasmids expressing each protein (21). Portions of the Notch protein essential for this interaction were mapped by assaying the aggregation of a clone of stably transformed Delta-expressing S2 cells with populations of S2 cells transiently expressing deleted versions of Notch (42).

(c) The stabilization of Armadillo protein by the product of *wg* was demonstrated by expressing *wg* in the imaginal disc line cl-8, by co-culture of S2 cells transformed to express *wg* and untransformed cl-8 cells, and by culture of untransformed cl-8 cells in medium conditioned by the *wg*-expressing S2 line (43).

(d) The spreading behaviour of S2 cells transformed to express PS2 integrins was used to identify tiggrin as an integrin ligand (16).

(e) *P* element transposition was used to obtain insertion of a single copy of an exogenous sequence into Kc cells (44).

(f) The tendency of exogenous DNA to recombine near a homologous chromosomal site has been used for targeted mutagenesis in Kc cells (45).

In designing experiments of this nature, one should bear in mind that the function of exogenous promoters generally reflects the function of the endogenous copy of the same promoter. Thus, analysis of the *Eip71CD* promoter in Kc cells has permitted the identification of elements required for the basal expression and ecdysone induction of *Eip71CD* transcription in Kc cells and in haemocytes, but the elements involved in the very different pattern of expression of the same gene in epidermis cannot be detected by expression of test constructs in Kc cells (10). It is therefore necessary to begin by testing individual cell lines to determine their expression patterns for genes of interest. Appendix 1 is a compilation of expression data for non-housekeeping genes, from a variety of published and unpublished sources, which we offer as a starting point in designing such experiments. The table is not complete; undoubtedly we have inadvertently omitted many data. We encourage workers in the field to communicate to us additions to this table, so that a revised version can be published elsewhere. As indicated in Appendix 1, there are very few qualitative differences in the expression patterns of the different embryonic lines, although substantial quantitative differences exist (not shown in the table). For example, in the case of *Eip71CD*, both the levels of basal expression and the amounts of induction by ecdysone differ among the embryonic lines by more than an order of magnitude (46). These quantitative variations may be important in choosing a cell line for use in a particular experiment. The data from imaginal disc lines and especially from nervous system lines are much more fragmentary than those from embryonic lines, but already it is clear that imaginal disc and CNS lines are qualitatively very different from embryonic lines.

Acknowledgements

We are grateful to our many colleagues who have shared unpublished data with us, and to James Henderson for reading the manuscript.

References

1. Warburg, A. and Miller, L. H. (1992). *Science*, **255**, 448.
2. Sang, J. H. (1981). *Adv. Cell Culture*, **1**, 125.
3. Ashburner, M. (1989). *Drosophila: a laboratory manual.* Cold Spring Harbor Laboratory Press, NY.
4. Ashburner, M. (1989). *Drosophila: a laboratory handbook.* Cold Spring Harbor Laboratory Press, NY.
5. Cherbas, L., Moss, R., and Cherbas, P. (1994). *Methods Cell Biol.*, **44**, 161.
6. Schneider, I. (1972). *J. Embryol. Exp. Morphol.*, **27**, 353.
7. Echalier, G. and Ohanessian, A. (1970). *In Vitro*, **6**, 162.
8. Schneider, I. and Blumenthal, A. B. (1978). In *The genetics and biology of Drosophila* (ed. M. Ashburner and T. R. F. Wright), Vol. 2a, p. 265. Academic Press, London.
9. Bernhard, H. P., Lienhard, S., and Reganass, U. (1980). In *Invertebrate systems in vitro* (ed. E. Kurstak, K. Maramorosch, and A. D¸bendorfer), p. 13. Elsevier/ North Holland, Amsterdam.
10. Andres, A. J. and Cherbas, P. (1992). *Development*, **116**, 865.
11. Andres, A. J. and Cherbas, P. (1994). *Dev. Genet.*, **15**, 320.
12. Knibiehler, B., Marre, C., Cecchini, J.-P., and Le Parco, Y. (1987). *Rouxís Arch. Dev. Biol.*, **196**, 243.
13. Blumberg, G., Mackrell, A. J., and Fessler, J. H. (1988). *J. Biol. Chem.*, **263**, 18328.
14. Lundstrum, G. P., Bachinger, H.-P., Fessler, L. I., Duncan, K. G., Nelson, R. E., and Fessler, J. H. (1988). *J. Biol. Chem.*, **263**, 18318.
15. Fessler, J. H., Nelson, R. E., and Fessler, L. I. (1994). *Methods Cell Biol.*, **44**, 303.
16. Fogerty, F. J., Fessler, L. I., Bunch, T. A., Yaron, Y., Parker, C. G., Nelson, R. E., *et al.* (1994). *Development*, **120**, 1747.
17. Sobrier, M. L., Chapel, S., Couderc, J. L., Micard, D., Lecher, P., Somme-Martin, G., *et al.* (1989). *Exp. Cell Res.*, **184**, 241.
18. Cherbas, L. and Cherbas, P. (1980). *Adv. Cell Culture*, **1**, 91.

19. Samakovlis, C., Kimbrell, D. A., Kylsten, P., Engstrom, A., and Hultmark, D. (1990). *EMBO J.*, **9**, 2969.
20. Abrams, J. M., Lux, A., Steller, H., and Krieger, M. (1992). *Proc. Natl. Acad. Sci. USA*, **89**, 10375.
21. Fehon, R., Kooh, P. J., Rebay, I., Regan, C. L., Xu, T., Muskavitch, M. A. T., *et al.* (1990). *Cell*, **61**, 523.
22. Cherbas, L., Lee, K., and Cherbas, P. (1991). *Genes Dev.*, **5**, 120.
23. Gateff, E., Gissmann, L., Shrestha, R., Plus, N., Pfister, H., Schröder, J., *et al.* (1980). In *Invertebrate systems in vitro* (ed. E. Kurstak, K. Maramotosch, and A. Dübendorfer), p. 517. Elsevier/North Holland Biomedical Press, Amsterdam.
24. Ui, K., Ueda, R., and Miyake, T. (1987). *In Vitro Cell. Dev. Biol.*, **23**, 707.
25. Currie, D. A., Milner, M. J., and Evans, C. W. (1988). *Development*, **102**, 805.
26. Ui, K., Nishihara, S., Sakuma, M., Togashi, S., Ueda, R., Miyata, Y., *et al.* (1994). *In Vitro Cell. Dev. Biol.*, **30A**, 209.
27. Milner, M. J. (1996). In *Cell and tissue culture: laboratory procedures* (ed. G. A. Doyle, J. B. Griffiths, and S. G. Newell), p. 24A:1.1. Wiley, New York.
28. Ui-Tei, K., Nishihara, S., Sakuma, M., Matsuda, K., Miyake, T., and Miyata, Y. (1994). *Neurosci. Lett.*, **174**, 85.
29. Ui-Tei, K., Sakuma, M., Watanabe, Y., Miyake, T., and Miyata, Y. (1995). *Neurosci. Lett.*, **195**, 187.
30. Cullen, C. F. and Milner, M. J. (1991). *Tissue Cell*, **23**, 29.
31. Echalier, G. (1976). In *Invertebrate cell culture* (ed. E. Kurstak and K. Maramorosch), p. 131. Academic Press, NY.
32. Shields, G. and Sang, J. H. (1977). *Drosophila Inf. Serv.*, **52**, 161.
33. Woods, D. F. and Poodry, C. A. (1983). *Dev. Biol.*, **96**, 23.
34. Robb, J. A. (1969). *J. Cell Biol.*, **41**, 876.
35. Puck, T. T. and Marcus, P. I. (1955). *Proc. Natl. Acad. Sci. USA*, **41**, 432.
36. Puck, T. T. and Marcus, P. I. (1956). *J. Exp. Med.*, **103**, 653.
37. Søndergaard, L. (1996). *In Vitro Cell. Dev. Biol. - Animal*, **32**, 386.
38. Wigler, M., Pellicer, A., Silverstein, S., Axel, R., Urlaub, G., and Chasin, L. (1979). *Proc. Natl. Acad. Sci. USA*, **76**, 1373.
39. Chen, C. and Okayama, H. (1987). *Mol. Cell. Biol.*, **7**, 2745.
40. Swevers, L., Cherbas, L., Cherbas, P., and Iatrou, K. (1996). *Insect Biochem. Mol. Biol.*, **26**, 217.
41. Campbell, M. J. (1995). *BioTechniques*, **18**, 1027.
42. Rebay, I., Fleming, R. J., Fehon, R. G., Cherbas, L., Cherbas, P., and Artavanis-Tsakonas, S. (1991). *Cell*, **67**, 687.
43. van Leeuwen, F., Samos, C. Y., and Nusse, R. (1994). *Nature*, **368**, 342.
44. Segal, D., Cherbas, L., and Cherbas, P. (1996). *Somatic Cell Mol. Genet.*, **22**, 159.
45. Cherbas, L. and Cherbas, P. (1997). *Genetics*, **145**, 349.
46. Lee, K. (1990). Ph. D. Thesis, Indiana University.
47. Debec, A. (1978). *Nature*, **274**, 255.
48. Peel, D. J. and Milner, M. J. (1992). *Roux's Arch. Dev. Biol.*, **201**, 120.
49. Richard-Molard, C. (1975). *Arch. Virol.*, **47**, 139.
50. Mosna, G. and Dolfini, S. (1972). *Drosophila Inf. Serv.*, **48**, 144.
51. Best-Belpomme, M., Courgeon, A.-M., and Echalier, G. (1980). In *Progress in ecdysone research* (ed. J. Hoffmann), p. 379. Elsevier, Amsterdam.
52. Best-Belpomme, M. and Courgeon, A.-M. (1977). *FEBS Lett.*, **82**, 345.

53. Cherbas, P., Cherbas, L., and Williams, C. M. (1977). *Science*, **197**, 275.
54. Berger, E., Ringler, R., Alahiotis, S., and Frank, M. (1978). *Dev. Biol.*, **62**, 498.
55. Bhanot, P., Brink, M., Samos, C. H., Hsieh, J.-C., Wang, J., Macke, J. P., *et al.* (1996). *Nature*, **382**, 225.
56. Theopold, U., Pinter, M., Daffre, S., Tryselius, Y., Friedrich, P., Nassel, D. R., *et al.* (1995). *Mol. Cell. Biol.*, **15**, 824.
57. Samakovlis, C., Asling, B., Boman, H. G., Gateff, E., and Hultmark, D. (1992). *Biochem. Biophys. Res. Commun.*, **188**, 1169.
58. Tryselius, Y. and Hultmark, D. (1997). Insect Molec. Biol. **6**, 173.
59. Petersen, U.-M., Bjorklund, G., Ip, Y. T., and Engstrom, Y. (1995). *EMBO J.*, **14**, 3146.
60. Koelle, M. R., Talbot, W. S., Segraves, W. A., Bender, M. T., Cherbas, P., and Hogness, D. S. (1991). *Cell*, **67**, 59.
61. Savakis, C., Demetri, G., and Cherbas, P. (1980). *Cell*, **22**, 665.
62. Savakis, C., Koehler, M. M. D., and Cherbas, P. (1984). *EMBO J.*, **3**, 235.
63. Theopold, U., Dal Zotto, L., and Hultmark, D. (1995). *Gene*, **156**, 247.
64. Best-Belpomme, M., Courgeon, A.-M., and Rambach, A. (1978). *Proc. Natl. Acad. Sci. USA*, **75**, 6102.
65. Theopold, U., Samokovlis, C., Erdjument-Bromage, H., Dillon, N., Axelsson, B., Schmidt, O., *et al.* (1996). *J. Biol. Chem.*, **271**, 12708.
66. Theopold, U., Ekengren, S., and Hultmark, D. (1996). *Proc. Natl. Acad. Sci. USA*, **93**, 1195.
67. Ireland, R. C., Berger, E., Sirotkin, K., Yund, M. A., Osterbur, D., and Fristrom, J. (1982). *Dev. Biol.*, **93**, 498.
68. Andres, A. J. (1990). Ph. D. Thesis, Indiana University.
69. Dushay, M. S., Asling, B., and Hultmark, D. (1996). *Proc. Natl. Acad. Sci. USA*, **93**, 10343.
70. Hirano, S., Ui, K., Miyake, T., Uemura, T., and Takeichi, M. (1991). *Development*, **113**, 1007.
71. Bunch, T. A. and Brower, D. L. (1992). *Development*, **116**, 239.
72. Rothberg, J. M., Jacobs, J. R., Goodman, C. S., and Artavanis-Tsakonas, S. (1990). *Genes Dev.*, **4**, 2169.
73. Rosetto, M., Engstrom, Y., Baldari, C. T., Telford, J. L., and Hultmark, D. (1995). *Biochem. Biophys. Res. Commun.*, **209**, 111.

Appendix 1. Gene expression in *Drosophila* cell lines. The data in this table is entirely non-quantitative; '+' means that the gene product was detectable in the assay used, and '−' means it was not detected. References and the nature of the assay are indicated below (letters in parentheses). Where known, the response to ecdysone (E) or lipopolysaccharide (L) is indicated as ↑, ↓, or =.

Line origin	1182-6 embryo (a)	Ca embryo (b)	cl-8 wing i.d. (c)	D embryo (a)	D1 embryo (d)	Eadh[N1] embryo (e)	EH34A3 embryo (f)	Emal embryo (e)	G embryo (g)	G2 embryo (h)	GM3 embryo (i)
AbdA											
AChE	−E↑ (p)	+E= (q)									
Antp											
arm			+(u)								
AttA											
BR-C											
CalpA											
CecA1											
CecA2								− (z)	− (z)		
CecB											
CecC											
ci			+ (aa)		− (aa)		− (aa)				
CP1											
DCg-1			+ (v)								
Dfz2											
Dif			− (ff)	− (o)							
Dl										− (o)	
dl											
dpp			− (aa)		− (aa)		− (aa)				
E63-1											
E74A											
E74B											
E75A											
EcR											
Eip55											
Eip71CD				+E↑ (jj)		+E↑ (kk)	+E↑ (kk)			+E↑ (kk)	+E↑ (kk)
ELAV				+(mm)	+(mm)		+(mm)				+(mm)
en			− (aa)		− (aa)		− (aa)				

Appendix 1. *Continued*

Line origin	1182-6 embryo (a)	Ca embryo (b)	cl-8 wing i.d. (c)	D embryo (a)	D1 embryo (d)	Eadh[N1] embryo (e)	EH34A3 embryo (f)	Emal embryo (e)	G embryo (g)	G2 embryo (h)	GM3 embryo (i)
FKBP39											
fng			– (ff)								
fu			+ (aa)		+ (aa)		+ (aa)				
Gal		– E ↑ (oo)									
hemomucin											
hh			– (aa)		– (aa)		– (aa)				
HLH106											
hsp22											
hsp23											
hsp26											
hsp27											
IMP-E1											
lab											
Lys*											
mad											
N			+ (ff)		+ (o)				– (o)		
oaf											
pb											
PS1			+ (c)								
PS2			+ (c)								
ptc			+ (aa)		– (aa)		– (aa)				
Rel											
Scr											
Ser			– (ff)								
slit											
smo											
tiggrin											
Tl											
Ubx											
wg			– (aa, ff)		– (aa)		– (aa)				
β3 tubulin											

Line	Kc	mbn-2	MLDm D-3	MLDm D-6	MLDm D-8	MLDm D-9	MLDm D-12	MLDm D-14	MLDm D-33
origin	embryo (b)	b.c. tumor (j)	mixed i.d. (k)	mixed i.d. (k)	wing i.d. (k)	wing i.d. (k)	wing i.d. (l)	wing i.d. (l)	i.d. (l)
AbdA	- (o)		- (o)	- (o)	- (o)		- (o)		- (o)
AChE	-E ↑ (r)								
Antp	- (o)		- (o)	- (o)	- (o)		- (o)		- (o)
arm									
AttA		-L ↑ (w)							
BR-C	+ E ↑ (x)								
CalpA		+L= (y)							
CecA1	- (z)	-L ↑ (z)							
CecA2	- (z)	-L ↑ (z)							
CecB		-L ↑ (z)							
CecC		-L ↑ (z)							
ci	- (bb)								
CP1		+L= (cc)							
DCg-1	+E ↓ (x,dd)								
Dfz2									
Dif		+L ↑ (ee)							
Dl	- (o)		- (o)	- (o)	- (o)		- (o)		- (o)
dl		+ (ee)							
dpp	- (aa)		+ (hh)	+ (hh)	+ (hh)	- (hh)	+ (hh)	+ (hh)	+ (hh)
E63-1	- (x)								
E74A	- E ↑ (x)								
E74B	+E ↑ (x)								
E75A	+E ↑ (x)								
EcR	+E ↑ (x)								
Eip55	+E ↑ (II)								
Eip71CD	+E ↑ (II)				+E= (t)				
ELAV	+(mm)	+(mm)			+(mm)		+(mm)		+(mm)
en	- (aa)								
FKBP39		+L= (nn)							

Appendix 1. *Continued*

Line	Kc	mbn-2	MLDm D-3	MLDm D-6	MLDm D-8	MLDm D-9	MLDm D-12	MLDm D-14	MLDm D-33
origin	embryo (b)	b.c. tumor (j)	mixed i.d. (k)	mixed i.d. (k)	wing i.d. (k)	wing i.d. (k)	wing i.d. (l)	wing i.d. (l)	i.d. (l)
fng									
fu	+ (aa)		+(aa)						
Gal	-E ↑ (oo)								
hemomucin			+L= (pp)						
hh	- (aa,bb)								
HLH106	+ (qq)	+L= (qq)							
hsp22									
hsp23									
hsp26									
hsp27	+E ↑ (x)								
IMP-E1	- (ss)								
lab	- (o)		+/- (o)	- (o)	- (o)		- (o)		- (o)
Lys*		- (tt)							
mad					+ (uu)		+ (uu)		+ (uu)
N	+ (o)		+ (o)	+ (o)	+ (o)		+ (o)		+ (ff)
oaf			+ (hh)	+ (hh)	+ (hh)	+ (hh)	+ (hh)	+ (hh)	+ (hh)
pb	- (o)		- (o)	- (o)	+/- (o)		- (o)		- (o)
PS1	- (dd)								
PS2	- (dd)								
ptc	- (aa)								
Rel		+L ↑ (vv)							
Scr	- (o)		- (o)	- (o)	- (o)		- (o)		- (o)
Ser									
slit									
smo	+ (bb)								
tiggrin	+ (aaa)								

Line	Kc	mbn-2	MLDm D-3	MLDm D-6	MLDm D-8	MLDm D-9	MLDm D-12	MLDm D-14	MLDm D-33
origin	embryo (b)	b.c. tumor (j)	mixed i.d. (k)	mixed i.d. (k)	wing i.d. (k)	wing i.d. (k)	wing i.d. (l)	wing i.d. (l)	i.d. (l)
Tl		+L= (bbb)							
Ubx	- (o)		- (o)	- (o)	- (o)		- (o)		+/- (o)
wg	- (aa)								
β3 tubulin	+E↑ (x,ccc)								

Line	MLDmBG-1	S1	S2	S3
origin	CNS (m)	embryo (n)	embryo (n)	embryo (n)
AbdA			- (o)	
AChE		- (s)	- (s); -E↑ (t)	+ E↑ (t,s)
Antp			- (o)	
arm			+ (v)	
AttA				
BR-C				
CalpA				
CecA1			-;-L↑ (z)	
CecA2				
CecB			-;-L↑ (z)	
CecC				
ci		- (aa)	- (aa, bb)	
CP1				
DCg-1				
Dfz2			- (v)	
Dif			+L↑ (ee)	
Dl			- (o, gg, ff)	+ (o)
dl		- (aa)	- (aa)	
dpp				
E63-1				
E74A				

Appendix 1. *Continued*

Line origin	MLDmBG-1 CNS (m)	S1 embryo (n)	S2 embryo (n)	S3 embryo (n)
E74B				
E75A				
EcR			+ (ii)	
Eip55				
Eip71CD		+E ↑ (kk)	+E ↑ (kk)	+E ↑ (ll, kk)
ELAV		+(mm)	+(mm)	+(mm)
en		- (aa)	- (aa)	
FKBP39				
fng			- (ff)	
fu		+ (aa)	+ (aa)	+ (aa)
Gal				
hemomucin				
hh		- (aa)	- (aa,bb)	
HLH106			+ (qq)	
hsp22				+E ↑ (rr)
hsp23				+E ↑ (rr)
hsp26				+E ↑ (rr)
hsp27				+E ↑ (rr)
IMP-E1				
lab			- (o)	
Lys*				
mad				
N			- (o, gg)	- (o)
oaf				
pb			- (o)	
PS1	+ (ww)			
PS2	+ (ww)	- (xx)		

Line origin	MLDmBG-1 CNS (m)	S1 embryo (n)	S2 embryo (n)	S3 embryo (n)
ptc		- (aa)	+ (aa,bb)	
Rel			- (o)	
Scr			- (ff, yy)	
Ser			+(zz)	
slit			+(bb)	
smo				
tiggrin				
Tl				
Ubx			- (o)	
wg		- (aa)	- (aa, ff)	
β3 tubulin				

a.   Ref. 47.
b.   Ref. 7.
c.   Ref. 48.
d.   A. Dübendorfer and J. Sang, personal communication
e.   Ref. 9.
f.   Ref. 33.
g.   Ref. 49.
h.   W. Gehring, personal communication
i.   Ref. 50.
j.   Ref. 23.
k.   Ref. 24.
l.   T. Miyake, personal communication
m.   Ref. 26.
n.   Ref. 6.
o.   M. Vaskova and M. A. T. Muskavitch, personal communication. Assays by immunostaining and in some cases by Westerns. Note: Kc cells have an aberrant N protein, derived from a chromosomal rearrangement (S. Artavanis-Tsakonas, personal communication).
p.   Ref. 51. Assay by enzyme assay.
q.   Ref. 52. Assay by enzyme assay.
r.   Ref. 53. Assay by enzyme assay.
s.   Ref. 54. Assay by enzyme assay.
t.   LC, unpublished. Assay by enzyme assay.

Appendix 1. *Continued*

u.   Ref. 43. Assay by Westerns.
v.   Ref. 55. Assay by Northerns, Westerns.
w.   B. Asling and D. Hultmark, personal communication.
x.   A. Andres, personal communication. Assay by Northerns.
y.   Ref. 56. Assay by Northerns, immunostaining.
z.   Ref. 57. Assay by RNase protection. Results from two different S2 sublines were different.
aa.   D. Casso, personal communication. Assay by Northerns.
bb.   J. Hooper, personal communication. Assay by Northerns.
cc.   Ref. 58.
dd.   Ref. 15.
ee.   Ref. 59. Assay by immunostaining, Northerns.
ff.   V. Panin and K. Irvine, personal communication. Assay by immunostaining, Westerns, and Northerns.
gg.   Ref. 21. Assay by Westerns.
hh.   R. Blackman, personal communication. Assays by Northerns. *dpp* transcripts differ in size among the imaginal disc lines; in some cases the size difference has been shown to result from differential promoter usage.
ii.   Ref. 60. Assay by Westerns.
jj.   K. Lee and L. Cherbas, unpublished. Assay by Northerns.
kk.   Ref. 46. Assay by Northerns.
ll.   Refs. 61, 62. Assay by protein synthesis, Northerns.
mm.   K. White personal communication. Assays by Westerns. In all cases, the level of expression is much lower than seen in brain.
nn.   Ref. 63. Assay by Northerns.
oo.   Ref. 64. Assay by enzyme assay.
pp.   Ref. 65. Assay by binding to FITC-*Helix pomatia* lectin.
qq.   Ref. 66. Assay by Northerns.
rr.   Ref. 67. Assays by Northern.
ss.   Ref. 68. Assay by Northerns.
tt.   Daffre and D. Hultmark, personal communication.
uu.   L. Attisano, J. Wrana, and S. Neufeld, personal communication. Assay by immunoprecipitation
vv.   Ref. 69. M. Dushay and D. Hultmark, personal communication.
ww.   Ref. 70. Assays by immunoprecipitation; assay incapable of distinguishing between PS1 and PS2.
xx.   Ref. 71. Assays by immunostaining.
yy.   Ref. 42. Assays by immunostaining.
zz.   Ref. 72. Assay by immunofluorescence.
aaa. Ref. 16. Assay by immunoprecipitation.
bbb. Ref. 73. Assays by Northerns.
ccc. Ref. 17. Assays by Northerns.

11

Preparation of nucleic acids

T. JOWETT

1. Introduction

This chapter describes protocols for:

(a) Purifying DNA for making libraries, pulsed-field gel electrophoresis (PFGE), Southern blot analysis, amplification.
(b) Purifying DNA and total RNA with and without organic extractions.
(c) Separately extracting both RNA and DNA from the same sample.

  The methods described for preparing mRNA directly from whole animals utilize affinity chromatography in combination with spin columns or magnetic bead technology. These methods are suitable for work on a microscale to provide material for Northern blots or RT-PCR. Emphasis is placed on those methods which are easiest and quickest to perform as well as produce high purity of the final products.

1.1 General precautions

Equipment and reagents which are used in several of the protocols are listed in the *Table 1*.

(a) Autoclave all solutions before use where possible.
(b) Plasticware such as disposable pipette tips and microcentrifuge tubes may be autoclaved before use. For DNA work this is usually not necessary, if the containers are kept closed when not in use.
(c) Wrap glassware in aluminium foil and bake at 200–250°C for 4 h before use.
(d) Siliconize glass centrifuge tubes with dimethyldichlorosilane solution. This produces tighter-packed pellets but prevents them from sticking to the glass.
(e) A refrigerated bench-top centrifuge which will take a selection of rotors including those for microcentrifuge tubes is a valuable addition to the laboratory and is essential for some of the protocols where the g force and temperature are important. Choose one with a rotor which will hold the various spin columns that are available.

Table 1. General equipment and reagents

Equipment

1.5 ml and 0.75 ml microcentrifuge tubes.

12 ml screw-capped round-bottomed polypropylene tubes.

15 ml and 30 ml glass centrifuge tubes (Corex), sterilize by baking at 150°C for 4 h.

50 ml screw-capped centrifuge tubes.

Disposable polyethylene Pasteur pipettes.

Glass homogenizer with PTFE pestle, PTFE pestle for 1.5 ml centrifuge tubes, disposable pestle for Eppendorf tubes No. 0030 120.973, Treff, VWR.

Protective clothing, insulated gloves, rubber gloves, and face mask.

Mortar and pestle.

Refrigerated microcentrifuge which will hold 2 ml tubes and spin columns: Eppendorf, Heraeus, etc.

Tissue homogenizer, e.g. Ultra Turrax®, IKA; Polytron® Kinematika AG; Tissuemizer® Tekmar Inc.; Tissue-Tearor® BIOSPEC Products; Omni Homogenisers, OMNI Int. Inc.

Water aspirator fitted with a holder for disposable 200 μl pipette tip.

Standard reagents and buffers—all reagents should be analytical grade or of higher purity

2-mercaptoethanol: usually supplied as a 14.3 M solution, store at 4°C in a dark bottle, use in a fume-cupboard.

Chloroform: harmful avoid skin contact and inhalation of vapour.

DEPC treated water: to deionized water in a Duran bottle add diethylpyrocarbonate to 0.1% (v/v), shake vigorously, and leave overnight. Autoclave to destroy the DEPC. DEPC reacts with Tris base so, buffers should be made with DEPC treated, autoclaved water and reautoclaved.

DNase-free RNase: dissolve RNase A (Pharmacia, No. 27–0323–01) in deionized water at 10 mg/ml. In a screw-cap vial, heat the solution to 100°C for 30 min, and then allow to cool slowly to room temperature. Store in aliquots at –20°C. Some commercial sources do not require heat treatment.

0.5 M EDTA solution: add 186.1 g of disodium ethylenediaminetetraacetate.$2H_2O$ to 800 ml of water. Stir vigorously and adjust the pH to 8.0 with ~ 20 g of NaOH pellets, adjust volume to 1 litre, autoclave.

Ethanol: 96–100%, and 70%, 75%, 80% solutions.

Glycogen: 20 mg/ml in DEPC treated water, filter sterilize, store at –20°C.

Guanidine isothiocyanate: harmful by inhalation and skin contact, make up solutions in fume-cupboard.

Isopropanol.

Phenol equilibrated with 0.5 M Tris–HCl pH 8.0 and then with 10 mM Tris–HCl, 1 mM EDTA pH 8.0, add hydroxyquinoline to 0.1%, harmful avoid skin contact.

Phenol equilibrated with water containing 0.1% hydroxyquinoline, harmful avoid skin contact.

5 M potassium acetate: to 60 ml of 5 M potassium acetate, add 11.5 ml of glacial acetic acid and 28.5 ml of water. The resulting solution is 3 M with respect to potassium and 5 M in acetate.

8 M potassium acetate.

Proteinase K solution: 10 mg/ml in 10 mM Tris–HCl pH 7.8, 50% glycerol, kept at –20°C.

SDS: for a 10% solution dissolve 100 g in 900 ml water by heating to 68°C, adjust pH to 7.2 with a few drops concentrated HCl, and make volume 1 litre with water.

2 M sodium acetate pH 4.0.

Sodium acetate stock solutions of 2.5 M and 3.0 M, bring the pH to between 4.6 and 5.2 with glacial acetic acid.

Spermidine trihydrochloride (Sigma S 2501): make a 0.5 M stock solution in deionized water, store at –20°C.

Spermine tetrahydrochloride (Sigma S 2876): make a 0.3 M stock solution in deionized water, store at –20°C.

TE buffer: 10 mM Tris–HCl pH 8.0, 1 mM EDTA, sterilize by autoclaving.

1.2 Safety considerations

Several reagents used in the following protocols are particularly hazardous. Protective clothing and gloves should be worn at all times. Used solutions should be disposed of appropriately.

(a) 2-mercaptoethanol is harmful by skin contact and inhalation and should only be used in the fume-cupboard.

(b) Chloroform is very toxic by inhalation, skin contact, and ingestion. It is irritating to the skin and eyes possibly causing conjunctivitis and burning, and is carcinogenic in laboratory animals.

(c) Guanidine isothiocyanate is harmful by ingestion, skin contact, and inhalation and is irritating to the skin and eyes. It liberates hydrogen cyanide gas in contact with acid. Solutions should be made and used in a fume-cupboard.

(d) Liquid nitrogen causes burns. Microcentrifuge tubes which have been dropped into liquid nitrogen may explode if warmed up too quickly. Wear cold-resistant gloves and a face mask.

(e) Phenol and phenolic solutions cause burns. It is toxic by ingestion, inhalation, and skin contact. Inhalation of the vapour over a prolonged period may cause digestive disturbances, nervous disorders, skin eruptions, and damage to kidneys and liver. Dermatitis may result from prolonged contact with weak solutions. Phenol is carcinogenic, mutagenic, and teratogenic. Redistilling phenol is extremely hazardous and should be avoided.

1.3 Assaying yields and quality

The yield and quality of purified nucleic acid is best determined by electrophoresis. Check the size and yield of DNA by running a sample on an agarose gel in TBE buffer (see ref. 1 for details). The DNA will migrate slowly during electrophoresis while the RNA migrates near the dye front. The precise length of genomic DNA is determined by PFGE. Measure the absorbance of DNA solution at 260 nm to determine the concentration. Make a dilution such that the absorbance readings are between 0.1 and 1.0 to be accurate. For dsDNA an A_{260} of 1.0 is equivalent to a concentration of $\sim$ 50 µg/ml. For RNA an A_{260} of 1.0 is equivalent to $\sim$ 40 µg/ml. A test of the purity is given by scanning the absorbance from 230–320 nm or by measuring the $A_{260/280}$ ratio. Pure DNA has an $A_{260/280}$ ratio of 1.8, while pure RNA has a ratio of 2.0.

• one diploid cell of *Drosophila* contains 0.3 pg of DNA

• the haploid genome size is 1.4×10^8 bp

• this is equivalent to 1.1×10^{-5} pmol/µg of DNA

• or 6.6×10^5 copies of single copy genes/µg of DNA

2. DNA extraction procedures

The standard method of extraction of DNA uses phenol. A tissue homogenate is mixed with aqueous phenol which deproteinizes the sample and can be separated from the aqueous layer by centrifugation. The DNA is usually in the upper aqueous layer from which it can be precipitated with alcohol. The purity of the phenol is important. For DNA extraction it is equilibrated with TE buffer with or without 0.1% hydroxyquinoline. A detailed description of the parameters involved in the use of phenol in extraction of nucleic acids is given by Wallace (2).

There are several alternatives to using phenol in DNA extractions. These include a 'salting out' method first described by Miller (3). The commercial extraction genomic DNA kits from Promega (Wizard Genomic DNA kit), Stratech (Kristal Genomic DNA kit), Life Technologies, and Stratagene exploit modifications of this method. Anion exchange resins have been developed by Qiagen and Pharmacia which bind DNA under low salt conditions and allow elution in high salt. These require a final alcohol precipitation step to retrieve the DNA. Methods based on silica gel membranes or glass particles have proved useful for cleaning up DNA from gel fragments and enzyme reactions (available from several companies including Qiagen (Qiaquick and Qiaprep), Promega (Wizard preps), Bio 101 (GeneClean), Life Technologies (GlassMax)), but many of the products available are unsuitable for genomic DNA extractions. However, there are two products which have been specifically developed for genomic DNA extractions: QIAamp kits from Qiagen and QUICK-Geno™ Genomic DNA kit (Clontech). These are quick, easy to use, and provide high quality DNA of 20–50 kb in size.

DNA can be extracted from flies which were quick frozen in liquid nitrogen and stored at –70 °C.

2.1 Preparation of high molecular weight genomic DNA

A standard phenol:chloroform extraction procedure for preparation of high molecular weight DNA is described in *Protocol 1*. DNA is most conveniently extracted from adult flies or dechorionated embryos. The genomic DNA obtained by this method from adult flies will be contaminated with mitochondrial DNA and possibly yeast DNA. If this is likely to be a problem then it is best to perform the procedure on purified embryo nuclei (see the protocols in ref. 4). Particular care should be taken not to shear the DNA if the final product is to be greater than 100 kb. Therefore remove and discard the lower organic layer during the extractions rather than transfer the upper aqueous layer with a pipette.

DNA for PFGE may be prepared by purifying nuclei from embryos or adult flies, embedding them in agarose before releasing the DNA with proteinase K in the presence of SDS or Sarkosyl (see *Protocol 2*). Another method for preparing small amounts of very high molecular weight DNA utilizes the Nuclitip® system from Amersham. This method employs a membrane for capturing nuclei held in a pipette tip. Cells are lysed to release nuclei which are then collected on the membrane. Cell debris are washed away. The Nuclitip® is broken and the lower half containing the nuclei treated with SDS and proteinase K to release the DNA which is recovered from the Nuclitip® by centrifugation. A neutralizing agent is added which sequesters the SDS thereby preventing its inhibitory effects and the need to precipitate the DNA. About 10–20 μg of DNA larger than 800 kb may be obtained in about one hour. The DNA is suitable for making YACs and PFGE.

Protocol 1. Purification of high molecular weight DNA by phenol extraction

Equipment and reagents

- Homogenization buffer: 0.1 M NaCl, 30 mM Tris–HCl pH 8.0, 10 mM EDTA, 0.5% Triton X-100—add 2-mercaptoethanol to 10 mM just prior to use
- Extraction buffer: 0.1 M Tris–HCl, 0.1 M NaCl, 20 mM EDTA
- For other equipment and reagents see *Table 1*

Method

1. Grind 1 g of flies or dechorionated embryos in liquid nitrogen in a pre-cooled mortar and pestle.[a] Embryos which have been stored at –70°C in 1.5 ml microcentrifuge tubes may be removed as a frozen lump by warming the tube momentarily in an ungloved hand and then piercing the bottom of the tube with a 21 gauge syringe needle.

2. Transfer the powder to a 30 ml glass homogenizer containing 20 ml homogenization buffer on ice and homogenize gently with a PTFE (polytetrafluoroethylene) pestle.

3. Transfer the homogenate to two 15 ml centrifuge tubes and centrifuge at 500 *g* for 1 min only at 4°C. This pellets the debris leaving the nuclei in suspension.

4. Use a pipette to transfer the supernatant to two 15 ml Corex tubes and centrifuge in a swing-out rotor at 2000 *g* for 5 min at 4°C.

5. Pour off the supernatant and resuspend the pellets in extraction buffer. This is most easily done by adding 1 ml of homogenization buffer, mixing with a PTFE pestle, and then bringing the volume to 13.5 ml in each tube with more extraction buffer.[a]

6. Transfer the suspension to two 50 ml screw-capped polypropylene centrifuge tubes. Add 0.15 ml of proteinase K solution (10 mg/ml) and 1.5 ml of 10% SDS solution to each tube. Mix gently by swirling and rocking. The suspension should become viscous as the nuclei lyse releasing the DNA. Do not vortex. Incubate at 37°C for 2–4 h.

Protocol 1. *Continued*

7. Add 15 ml of equilibrated phenol pH 8.0 containing 0.1% hydroxy-quinoline. Mix gently by inversion for 5 min. Centrifuge at 5000 *g* for 5–10 min at room temperature. Remove the lower organic layer with a pipette retaining the upper aqueous layer containing the DNA.

8. Repeat the phenol extraction with 15 ml of phenol. Discard the lower layer.

9. Extract once with 7.5 ml of chloroform and 7.5 ml of phenol. Discard the lower layer.

10. Extract once with 15 ml of chloroform. The interface should be clean and sharp.

11. Transfer the upper aqueous layer to three 50 ml centrifuge tubes (approx. 10 ml in each). Add 0.1 vol. of 2.5 M sodium acetate. Mix gently. Add 2 vol. of absolute ethanol. Mix by gentle swirling. The DNA appears at the interface as a white, stringy precipitate. When the phases are completely mixed remove the DNA as a clump with a glass Pasteur pipette sealed and shaped into a hook in a Bunsen flame.

12. Rinse the DNA in 70% ethanol in a round-bottomed tube. Centrifuge gently and remove the supernatant with a pipette.

13. Air dry the pellet, but do not overdry.

14. Resuspend the DNA in 0.5–1 ml TE by leaving overnight at room temperature, swirling occasionally. Store at 4°C.

[a] For preparation of dechorionated embryos see Chapter 6, *Protocol 1*.
[b] It is important to resuspend the pellet completely before adding the SDS.

Protocol 2. Preparation of DNA for PFGE[a]

Equipment and reagents

- Homogenization buffer: 60 mM NaCl, 10 mM Tris–HCl pH 7.5, 10 mM EDTA, 0.5% NP-40, 0.15 mM spermine, 0.15 mM spermidine
- 0.5 M EDTA, 1% *N*-lauroyl sarcosine (Sarkosyl)
- 1% low melting point (LMP) agarose in TE buffer, melt and cool to 45°C, add proteinase K at 10 mg/ml immediately before use
- For other equipment and reagents see *Table 1*

Method

1. Homogenize 100 flies in 1 ml of homogenization buffer in a glass homogenizer with PTFE pestle at 4°C. Transfer the homogenate to a 2.0 ml centrifuge tube, rinsing the homogenizer and pestle with a further 0.5 ml of homogenization buffer.[b]

2. Centrifuge the pooled suspensions at 500 *g* for 30 sec at 4°C to pellet the coarser debris. Transfer the supernatant to a 1.5 ml tube and centrifuge at 500 *g* for 1 min at 4°C. Transfer the supernatant to a fresh tube.

3. Pellet the nuclei by centrifuging at 10 000 *g* for 7 min at 4°C.

4. Resuspend the crude nuclei in 0.7 ml of homogenization buffer pre-warmed to 37°C, and then mix with an equal volume of 1% LMP agarose in TE at 45°C containing 10 mg/ml proteinase K. Immediately pour the mix into 100 μl block moulds cooled on ice.

5. Once set, transfer the inserts to a 50 ml tube containing 2 ml of 0.5 M EDTA/1% Sarkosyl, and incubate at 55°C for 48 h.

6. Further treatments, insert storage, and restriction enzyme digestion follow standard procedures.

---

[a] From ref. 5.

[b] For embryos, disrupt 10–100 μl of dechorionated, 6–18 hour-old embryos with a PTFE pestle in 150 μl of homogenization solution, then mix with an equal volume of 1% LMP agarose in TE, and pour into moulds. Digest each block in 2–4 ml of 0.1 M EDTA, 1% SDS, 100 μg/ml proteinase K at 45–50°C for 48–72 h (6).

2.2 Medium scale preparation of genomic DNA

Protocol 3 is a scaled down version of DNA extraction by phenol:chloroform. The DNA is suitable for Southern blot analysis and PCR and will be mostly < 50 kb in size. *Protocol 4* employs the QIAamp kit. This is designed for yields of up to 100 μg of DNA. It is quick and easy to use since it does not involve precipitation of the DNA.

Protocol 3. Medium scale phenol extraction of genomic DNA

Equipment and reagents

- Lysis buffer: 100 mM Tris–HCl pH 8.0, 50 mM NaCl, 50 mM EDTA, 1% SDS—add spermine to 0.15 mM and spermidine to 0.5 mM immediately before use

- For other equipment and reagents see *Table 1*

Method

1. Homogenize 50–200 flies (may be quick frozen in liquid nitrogen) with a glass homogenizer with a PTFE pestle in 2 ml of lysis buffer. Transfer to a 12 ml screw-capped tube.

2. Add 20 μl proteinase K solution (10 mg/ml). Incubate at 37°C for 1–2 h with occasional swirling and inversion.

3. Extract once with an equal volume of equilibrated phenol pH 8.0 by gentle inversion. Centrifuge at 4000 *g* for 5–10 min at room

Protocol 3. *Continued*

temperature. Transfer the aqueous upper layer containing the nucleic acids into a new tube with a polyethylene Pasteur pipette.

4. Extract twice with 0.5 vol. phenol and 0.5 vol. chloroform. Recentrifuge after each extraction and retain the upper aqueous layer.

5. Extract the aqueous layer with an equal volume of chloroform. Decant the aqueous layer to a new tube and add a 0.1 vol. of 2.5 M sodium acetate.

6. Mix and add 2 vol. of absolute ethanol. The DNA should immediately precipitate.

7. Centrifuge in at 8000 *g* for 10 min to pellet the DNA. Remove the supernatant by aspirating, being careful not to dislodge or remove the pellet.

8. Dry briefly in a vacuum desiccator and redissolve the pellet in 400 μl of TE. Transfer the solution to a 1.5 ml tube.

9. Add RNase A to 100 μg/ml and leave at 37 °C for 30 min.[a] Extract once with an equal volume of 1:1 phenol:chloroform. Retain the upper, aqueous layer and add 0.1 vol. of 2.5 M sodium acetate, then 2 vol. of cold ethanol. Mix carefully.

10. Centrifuge for at 8000 *g* 5 min at room temperature, pour away the supernatant, and add 1 ml of 80% ethanol. Vortex briefly and recentrifuge for 5 min. Remove the supernatant and dry the pellet. Redissolve in TE.

[a] The RNase treatment may be omitted as long as it is included in the loading buffer when running the DNA on a gel.

Protocol 4. QIAamp® tissue DNA extraction

Equipment and reagents

- QIAamp DNA extraction kit (Qiagen, No. 29304) containing: 50 QIAamp columns, 2 ml collection tubes, ATL buffer, AE buffer, AL buffer, AW buffer, proteinase K
- For other equipment and reagents see *Table 1*

Method

1. Take 50 mg of flies and grind in a 1.5 ml tube in liquid nitrogen.

2. Add 180 μl of ATL buffer and 20 μl proteinase K stock solution. Mix by vortexing. Incubate at 55°C until the tissue is completely lysed (~ 1–2 h).

3. Add 20 μl of RNase A (20 mg/ml), vortex, and incubate 2 min at room temperature.[a]

4. Add 200 μl of AL buffer. Mix thoroughly and incubate at 70°C for 10 min.

5. Add 210 μl of ethanol 96–100% and vortex. A white precipitate may form.

6. Apply all the solution and precipitate to a QIAamp column held in a 2 ml collection tube. Centrifuge at 6000 *g* for 1 min at room temperature.

7. Place the column in a fresh collection tube. Add 500 μl of AW buffer. Centrifuge at 6000 *g* for 1 min at room temperature.

8. Place the column in a fresh collection tube. Add 500 μl of AW buffer. Centrifuge at 6000 *g* for 1 min, and at 12 000 *g* for 2 min at room temperature to dry the column.

9. Place the column in a fresh collection tube. Elute the DNA twice with 200 μl of AE buffer or water pre-heated to 70°C. Incubate a room temperature for 1 min then centrifuge at 6000 *g* for 1 min. A third elution with the 400 μl eluate heated to 70°C increases the yield by 15%. Elution should be in buffer of pH 9.0 for optimum yields.[b] The expected yield is 20–60 μg of DNA from 50 mg of flies.

[a] The RNA must be digested otherwise it will compete with the DNA for binding to the resin, reducing the yield.
[b] With samples of < 1 μg elute with 50 μl.

2.3 Single fly extraction of DNA

This section describes four alternative methods of extracting DNA from single flies. The method in *Protocol 5* is that described in refs 7 and 8. It involves lysis and a heat treatment in the presence of SDS and DEPC prior to precipitating with potassium acetate. The final DNA is contaminated with RNA but this should be ignored until the DNA is loaded on a gel. Adding RNase A to a final concentration of 100 μg/ml with the loading buffer is normally sufficient to remove traces of RNA. Yields with this method can be quite variable. Well fed flies and females with fully mature ovaries give the highest yield (~ 1 μg/fly).

Protocol 6 describes a rapid method for obtaining DNA for use in PCR amplifications. The DNA produced is only suitable for PCR. An alternative method, which does not involve homogenization or enzymatic treatment, is described by Czanck in ref. 9. *Protocol 7* involves a commercial reagent designed for larger scale extractions but scaled down for single fly extractions. DNAzol is a single lysis solution containing guanidine isothiocyanate from which the DNA is precipitated with ethanol.

Protocol 5. Single fly DNA extraction with SDS/DEPC and potassium acetate

Equipment and reagents

• Homogenization buffer: 10 mM Tris–HCl pH 7.5, 60 mM NaCl, 10 mM EDTA, 0.15 mM spermine, 0.15 mM spermidine, 5% sucrose
• For other equipment and reagents see *Table 1*

• Lysis buffer: 300mM Tris–HCl pH 9.0, 100 mM EDTA, 0.625% SDS, 5% sucrose—add DEPC to 1% immediately before use

Method

1. Homogenize a single fly in a 1.5 ml microcentrifuge tube with 50 μl homogenization buffer.

2. Add 50 μl lysis buffer and incubate 15 min at 70°C (or 30–45 min at 37°C).

3. Cool to room temperature and add 15 μl of 8 M potassium acetate. Incubate 30 min on ice.

4. Centrifuge at 15 000 *g* for 5 min at 4°C. Remove the supernatant to a fresh tube, taking care not to disturb the surface lipid or pellet.

5. Add 55 μl of TE equilibrated phenol and 55 μl of chloroform. Vortex and centrifuge at 15 000 *g* for 5 min at room temperature. Transfer the supernatant to a fresh tube. Repeat the extraction.

6. Precipitate the DNA by adding 2 vol. of absolute ethanol and incubating at room temperature for 15 min. Centrifuge at room temperature for 5 min at 15 000 *g*. Discard the supernatant.

7. Wash the pellet with 400 μl of 70% ethanol and dry under vacuum. Dissolve the dried pellet in 20 μl TE.

8. For a restriction enzyme digestion add 1 μl of 1 mg/ml DNase-free RNase during the enzyme incubation.

Protocol 6. Rapid DNA extraction for PCR amplification

Reagents

• Extraction buffer 1: 10 mM Tris–HCl pH 8.2, 2 mM EDTA, 0.2% Triton X-100—add pro-   teinase K to 100 μg/ml just prior to use

Method

1. Place a fly or larva in a 0.75 ml tube and add 500 μl of extraction buffer. Break the tissue with a pipette tip.

2. Incubate at 50–56°C for 30 min in a PCR machine. Vortex occasionally. Heat to 95°C for 10 min to inactivate the protease. Cool to 4°C.

3. Centrifuge at 15 000 *g* for 5 min at 4°C and remove the supernatant to a fresh tube.

4. Use 1–5 μl in a 50 μl PCR reaction.

Protocol 7. DNA from single flies with DNAzol

Equipment and reagents

- DNAzol solution (Life Technologies, No. 10503 027): DNAzol is a non-organic reagent (it does not contain phenol), composed of guanidine isothiocyanate, some proprietary detergents, and reagents—it has a high pH which causes degradation of cellular RNA

- For other equipment and reagents see *Table 1*

Method

1. Homogenize one fly in liquid nitrogen in a 1.5 ml tube with a PTFE pestle. Add 25 μl of DNAzol. Apply four to ten strokes to break the tissue.

2. Centrifuge the homogenate at 10 000 *g* for 10 min at room temperature. Transfer the supernatant to a fresh tube.

3. Add 12.5 μl of 100% ethanol. Mix by inversion three to four times. The DNA is precipitated.

4. Centrifuge at 1000 *g* for 2 min to pellet the DNA (higher *g* force gives a harder packed pellet which is difficult to dissolve).

5. Wash the DNA twice with 100 μl of 95% ethanol by inverting the tube two to four times. If the pellet has become loose, recentrifuge, and then carefully decant the ethanol. Centrifuge again for 5 sec to draw any remaining liquid to the bottom of the tube and remove by aspiration.

6. Air dry for 1–5 min and add 10 μl of freshly made, 8 mM NaOH to the pellet. Gently pipette three to four times, and leave at room temperature for 1 h to completely dissolve.[a]

[a] DNAzol isolated DNA does not resuspend well in water or Tris buffer. The pH of 8 mM NaOH is ~ 9 and is easily adjusted with TE or Hepes once the DNA is in solution. See *Table 2* page 358.

---

3. RNA extraction procedures

The key to successful RNA preparation is to inactivate the endogenous nucleases as quickly as possible and to avoid subsequent contamination with RNases. Where possible all solutions should be autoclaved and treated with

Table 2. Adjusting pH of NaOH solutions (for 1 ml of 8 mM NaOH)

Final pH	0.1 M Hepes	Final pH	1 M Hepes
8.4	66 μl	7.2	30 μl
8.2	90 μl	7.0	42 μl
8.0	115 μl		
7.8	135 μl		
7.5	180 μl		

0.1% DEPC. Glassware can be treated by baking for 4 h at 250 °C. Sterile disposable plasticware is normally RNase-free. It is important to wear gloves to protect the sample from RNase contamination and also to protect oneself from the toxic solutions. Remember that automatic pipettes and other equipment have been touched with ungloved hands and are potential sources of nucleases. Detailed accounts of RNA extraction and precautions to be taken are found in ref. 10.

The basis of most methods is to lyse the tissue in the presence of strong denaturants which deproteinize the sample while protecting the RNA from nuclease digestion. The relative efficiencies of the most commonly used denaturants are guanidine isothiocyanate > guanidine hydrochloride > urea (11, 12). The methods described here involve guanidine thiocyanate. A urea-based method is described in ref. 7. Denaturation is enhanced by including a reductant such as 2-mercaptoethanol or dithiothreitol. The initial disruption of the tissue is best performed by grinding in liquid nitrogen followed by homogenization in the lysis solution. There are a variety of tissue homogenizers available for this purpose. In all cases though their use must be kept to the minimum required to produce a homogeneous suspension.

The methods described are for extraction from all stages of development and from all tissues. The method in *Protocol 8* is to prepare dissected tissues from one or more animals prior to extraction of RNA. It is usually convenient to quick freeze material in liquid nitrogen and store at –70 °C before embarking on a series of RNA preparations. Frozen material should not be allowed to thaw before it is mixed with the denaturing solution.

Protocol 8. Freezing dissected tissues ready for RNA extraction

Equipment and reagents
- Siliconized coverslips
- Watchmaker's forceps
- Liquid nitrogen or dry ice pellets

Method
1. Collect all dissected tissues together in a single drop on a siliconized glass coverslip.

---

2. Reduce the size of the drop to a minimum by withdrawing the liquid with pipette.

3. Pick up the coverslip with forceps and slowly lower into liquid nitrogen or freeze by placing on dry ice.

4. Once frozen, invert the coverslip over a pre-cooled 1.5 ml tube. Touch the surface of the coverslip with a finger to slightly warm the glass, and at the same time slide the coverslip across the neck of the tube allowing the frozen drop to fall into the tube. Close the tube. It is useful to keep sample tubes in an expanded polystyrene rack floating on liquid nitrogen. Avoid dropping the tubes directly into the liquid nitrogen as they can partially fill with the liquid making them liable to explode when they are transferred to −70°C for storage.

3.1 Extraction of total RNA

Protocol 9 involves guanidine isothiocyanate denaturation and extraction with acidic phenol:chloroform prior to precipitation with isopropanol, and is a scaled down version of the method described by Chomczynski and Sacchi (12). *Protocol 10* combines guanidine isothiocyanate denaturation with selective precipitation with LiCl and isopycnic centrifugation using caesium tetrafluoroacetate (CsTFA). The method is quick and produces a pellet of total RNA, while proteins migrate to the top of the tube, and DNA remains in solution. The third method in *Protocol 11* uses a silica gel-based membrane supplied by Qiagen and allows extraction of RNA in 30 min. The tissue is lysed under highly denaturing conditions and the RNA is bound to the membrane under high salt conditions. The membrane is washed to remove unwanted contaminants and the RNA is eluted in water. RNAs longer than 200 nucleotides are purified. Small RNAs such as 5.8S RNA, 5S RNA, and tRNAs will not bind quantitatively under the conditions used.

Protocol 9.   RNA extraction with acid guanidine thiocyanate and phenol[a]

Reagents

• Denaturing solution: 4 M guanidine thiocyanate, 25 mM sodium citrate pH 7.0, 0.5% Sarkosyl, 0.1 M 2-mercaptoethanol. For 52.9 ml mix together 25 g of guanidinium isothiocyanate, 31.1 ml water, 1.76 ml of 0.75 M sodium citrate pH 7.0, 0.88 ml of 30% sodium lauroyl sarcosinate. Heat to 65°C to dissolve the crystals. Add 72 μl/ml of 14.3 M 2-mercaptoethanol to each 10 ml of solution just prior to use. Use in a fume-cupboard.

Method

1. Add 100 μl denaturing solution to 10 mg of tissue in 1.5 ml microcentrifuge tube. Homogenize thoroughly.

2. Add 10 μl of 2 M sodium acetate pH 4.0, and mix thoroughly.

Protocol 9. *Continued*

3. Add 100 μl unbuffered aqueous phenol and mix by inversion.

4. Add 20 μl chloroform to create two phases and mix by inversion.

5. Chill on ice 15 min, then centrifuge at 15000 *g* for 10 min at 4°C.

6. Remove the upper phase into a new tube. If dealing with < 1 mg of tissue, add 1 μl RNase-free glycogen at 20 mg/ml.

7. Add 120 μl isopropanol. Incubate at –70°C for 15 min.

8. Thaw and centrifuge in a microcentrifuge at 15000 *g* for 10 min at 4°C.

9. Wash pellet with 70% ethanol, drain, and dry.

10. Resuspend in 10–50 μl DEPC treated water. Yield 1–5 μg/mg of tissue.

[a] Modified from ref. 12 for use on ≤ 10 mg of tissue. Scale up and use an homogenizer if necessary (e.g. 100 mg tissue/ml of denaturing solution).

Protocol 10. Purification of total RNA using the QuickPrep extraction kit

Equipment and reagents

• QuickPrep purification kit (Pharmacia, No. 27 9271 01) containing: extraction buffer (a buffered aqueous solution containing guanidinium thiocyanate and *N*-lauroyl sarcosine), lithium chloride solution in DEPC treated water, CsTFA

Method

1. To 50 flies or larvae in a 1.5 ml tube, add 0.4 μl of extraction buffer and homogenize the tissue thoroughly with a PTFE pestle.

2. Add 350 μl of LiCl solution and continue the homogenization. The sample can be kept on ice while others are being processed.

3. Add 500 μl CsTFA to each sample. Cap the tubes and mix thoroughly by vortexing. Keep on ice for 10 min. Centrifuge at 15000 *g* for 15 min in a microcentrifuge at room temperature.

4. Without disturbing the RNA pellet remove the supernatant by aspiration with a continuous vacuum. Removal by pipetting can lead to contamination of the final RNA sample. Place the tubes on ice.

5. Add 75 μl extraction buffer, 175 μl and 250 μl CsTFA. Vortex. This removes contaminants from the sides of the tube and the pellet.

6. Centrifuge the samples in a microcentrifuge at 15000 *g* for 5 min at room temperature.

7. Aspirate the supernatant taking care not to disturb the pellet. Place on ice.

8. Add 1 ml of 70% ethanol, and vortex. The samples may be stored at −20°C or −70°C.

9. Centrifuge at 15 000 *g* for 5 min. Aspirate the ethanol and air dry the pellet for 10–15 min at room temperature.

10. Dissolve the RNA in 50–100 μl of DEPC treated water. Rehydrate on ice for 30 min, vortex, and finally heat to 65°C for 10 min. Transfer to an RNase-free tube and store at −70°C.

Protocol 11. Purification of total RNA the Qiagen RNeasy kit

Equipment and reagents

- RNeasy Total RNA kit (Qiagen, No. 74103) containing: 20 spin columns (each column will bind ~ 100 μg of RNA), collection tubes, RLT lysis buffer (contains guanidinium isothiocyanate, add 10 μl/ml of 2- mercaptoethanol before use) RW1 wash buffer (contains guanidinium isothiocyanate), RPE wash buffer (supplied as a concentrate, 4 vol. of 96–100% ethanol must be added), and DEPC treated water

Method

1. Homogenize 30 mg of tissue under liquid nitrogen. Add 350 μl of RLT lysis buffer. Ensure complete homogenization to shear the DNA and to avoid subsequent clogging of the silica gel membrane.

2. Centrifuge lysate for 3 min at full speed in a microcentrifuge. Transfer the supernatant to a fresh tube.

3. Add 1 vol. (350 μl) of 70% ethanol to the homogenized lysate. Mix by pipetting. Do not centrifuge this lysate.

4. Apply the sample to the RNeasy column. Centrifuge at 8000 *g* for 15 sec in the microcentrifuge at room temperature. Discard the flow-through and reuse the collection tube.

5. Wash with 700 μl of RW1 wash buffer. Centrifuge at 8000 *g* for 15 sec at room temperature and discard the flow-through. Replace the collection tube.

6. Wash with 500 μl RPE wash buffer. Centrifuge at 8000 *g* for 15 sec at room temperature. Discard the flow-through and reuse the collection tube.

7. Wash with 500 μl RPE wash buffer and centrifuge for 2 min at full speed to dry the membrane.

8. Transfer the column to a fresh tube and elute RNA with 15–30 μl of DEPC treated water. Centrifuge at 8000 *g* for 60 sec. If the expected yield is greater than 30 μg a second elution step is worthwhile.

3.2 Simultaneous isolation of DNA and RNA

This method uses a the TRIzol Reagent which is a monophasic solution of phenol and guanidine isothiocyanate and is a modification of the method developed by Chomczynski and Sacchi (12). The TRIzol solution maintains integrity of RNA while disrupting cells and cell components. Addition of chloroform followed by centrifugation separates the solution into an aqueous and organic phase. RNA is precipitated from the aqueous phase with iso-propanol. DNA and proteins may be recovered from the interphase and organic phase by ethanol and isopropanol precipitation respectively.

Protocol 12. Isolation of RNA and DNA with TRIzol® Reagent

Equipment and reagents
- TRIzol Reagent (Gibco BRL Life Technologies, No. 15596–026)
- 0.1 M sodium citrate in 10% ethanol
- 8 mM sodium hydroxide pH ~ 9.0

A. *RNA extraction*

1. To 50–100 mg of tissue add 1 ml TRIzol™ Reagent and homogenize thoroughly with a PTFE pestle. The sample volume should not exceed 10% of the volume of TRIzol Reagent.

2. Incubate for 5 min at room temperature and add 0.2 ml chloroform.

3. Shake vigorously for 15 sec and incubate at room temperature for 2–3 min.

4. Centrifuge at not more than 12 000 g for 15 min at 4 °C. This produces a lower red phenol:chloroform layer, an interphase, and a colourless upper aqueous phase. The RNA is in the aqueous layer which is about 60% of the original TRIzol volume used.

5. Transfer the aqueous layer to a fresh tube and keep the remainder for DNA and protein isolation.

6. Add 0.5 ml of isopropanol to the aqueous layer. Mix and incubate 15 min at room temperature for 10 min. Centrifuge at 12 000 g for 10 min at 4 °C. The RNA precipitate forms a gel-like pellet on the side and bottom of the tube.

7. Remove the supernatant and wash the RNA pellet once with 1 ml of 75% ethanol. Mix by vortexing and centrifuge at 7500 g for 5 min at 4 °C. Remove the supernatant.

8. Dry the pellet. Do not overdry as this will greatly reduce its solubility. Dissolve the RNA in DEPC treated water or 0.5% SDS solution by heating to 60 °C for 10 min.

B. *DNA extraction*

1. Precipitate the DNA from the interphase and organic layer by adding 0.3 ml of absolute ethanol per 1 ml of TRIzol Reagent used for the homogenization.

2. Mix the sample by inversion and incubate at room temperature for 2–3 min.

3. Centrifuge at 2000 *g* for 5 min at 4°C to precipitate the DNA. Remove the supernatant.[a]

4. Wash the pellet twice in 1 ml of 0.1 M sodium citrate in 10% ethanol. At each wash incubate the DNA pellet at room temperature for 30 min with periodic mixing. Centrifuge at 2000 *g* for 5 min at 4°C.[b]

5. Resuspend the pellet in 1.5–2 ml of 75% ethanol. Incubate at room temperature for 10–20 min. Centrifuge at 2000 *g* for 5 min at 4°C.

6. Briefly dry the pellet for 5–10 min under vacuum and dissolve in 8 mM NaOH. Add enough solution to give a DNA concentration of 0.2–0.3 µg/ml. Typically add 0.3–0.6 ml to DNA isolated from 50–70 mg of tissue. The use of mild alkali assures full solubilization of the DNA pellet.

7. At this stage the sample still contains insoluble material gel-like material (e.g. fragments of membranes). Remove this material by centrifugation at 12 000 *g* for 10 min.

8. Before the DNA is used for applications involving enzyme reactions the pH must be adjusted with Hepes (see *Table 2* page 358).

[a] Proteins can be extracted from this supernatant by precipitation with isopropanol. A protocol is supplied with the TRIzol Reagent.
[b] An additional wash in 0.1 M sodium citrate, 10% ethanol solution may be required with DNA pellets containing > 200 µg. Do not centrifuge at greater *g* force or the pellet will be difficult to dissolve.

3.3 Direct methods of purification of mRNA

The standard method of mRNA purification is to use oligo(dT) cellulose to bind selectively the poly(A)$^+$ RNA. This can be performed on previously prepared total RNA or directly from the initial tissue homogenate. The method described in *Protocol 13* combines RNA extraction using guanidine isothiocyanate, a dilution step to reduce the viscosity of the sample and cause precipitation of some proteins, and then an incubation with oligo(dT) cellulose to bind the mRNA. The resin is washed by several rounds of centrifugation and resuspension, and finally collected in a spin column before eluting the bound mRNA in a warm, low salt buffer.

Protocol 14 employs Oligotex™ as an alternative to oligo(dT) cellulose. Oligotex™ consists of polystyrene latex particles with $(dT)_{30}$ oligonucleotides covalently linked to their surface. The principle of the method is the same as that in *Protocol 13*. 1 mg of the Oligotex particles (10 μl) can bind 10 pmol of poly(A)$^+$ containing RNA. 1 pmol of an average sized (1930 bases) mRNA corresponds to approximately 0.6 μg. Thus 1 mg of Oligotex will bind a maximum of 6 μg of average sized mRNA. The Oligotex particles can be regenerated with an alkali treatment to remove residual RNA, followed by equilibration in TE at pH 7.0. The particles must finally be resuspended at 10% (w/v) in 10 mM Tris–HCl pH 7.5, 0.5 M NaCl, 1 mM EDTA, 0.1% SDS, 0.1% NaN$_3$.

The third method described in *Protocol 15* uses magnetic bead technology. This method uses a LiDS/LiCl-based lysis buffer, which is a milder denaturant than guanidine thiocyanate. The RNA is stable for more than 30 minutes at room temperature in this lysis buffer without being degraded by RNases. The mRNA is purified from the lysate by binding to Dynabeads Oligo (dT)$_{25}$ which are uniform, superparamagnetic, polystyrene spheres with 25 nucleotide long chains of deoxythymidylate covalently attached to their surface. The beads are attracted to the side of the tube by placing it in a rack with an integral magnet. This allows the solution to be poured away while retaining the beads. The method is particularly simple allowing mRNA purification in as little as 15 minutes. The Dynabeads have the advantage that they can be regenerated up to four times, if they are regenerated with a NaOH treatment after each use. The beads cannot be regenerated if a guanidine thiocyanate-based lysis buffer is used (13).

Protocol 13.   Preparation of mRNA using the QuickPrep Micro mRNA kit

Equipment and reagents

- QuickPrep Micro mRNA purification kit containing: oligo(dT) cellulose (25 mg/ml in storage buffer containing 0.15% Kathon® CG), extraction buffer (a buffered aqueous solution containing guanidinium thiocyanate and *N*-lauroyl sarcosine), high salt buffer (10 mM Tris–HCl pH 7.5, 1 mM EDTA, 0.5 M NaCl), low salt buffer (10 mM Tris–HCl pH 7.5, 1 mM EDTA, 0.1 M NaCl), elution buffer (10 mM Tris–HCl pH 7.5, 1 mM EDTA), glycogen solution (5–10 mg/ml in DEPC treated water), 2.5 M potassium acetate solution, MicroSpin columns

Method

1. Place 50–100 mg of larvae or flies in a 1.5 ml tube. Add 0.4 ml extraction buffer and homogenize with a suitable pestle.

2. Dilute the sample by adding 0.8 ml of elution buffer. Mix thoroughly. This dilutes the guanidine concentration and causes precipitation of some proteins.

3. Prepare a cleared homogenate by centrifuging the sample for 1 min at 16 000 *g* in a microcentrifuge at room temperature.

4. Transfer 1 ml of oligo(dT) cellulose suspension to a 1.5 ml tube. Allow to settle and aspirate the supernatant.

5. Add the cleared homogenate to the resin and gently resuspend by inverting. Leave for 3 min on a rocking table or similar device.

6. Centrifuge in a microcentrifuge for 10 sec at 16 000 *g*. Remove the supernatant by aspiration.

7. Add 1 ml of high salt buffer. Resuspend and centrifuge at 16 000 *g* for 10 sec, aspirate the supernatant. Repeat this high salt wash four times.

8. Add 1 ml of low salt buffer. Resuspend and centrifuge at 16 000 *g* for 10 sec, aspirate the supernatant. Repeat this wash once more.

9. Add 0.3 ml of low salt buffer and resuspend the resin. Transfer the slurry into a MicroSpin column placed in a 1.5 ml tube. Centrifuge for 5 sec and discard the flow-through from the tube.

10. Add a further 0.5 ml low salt buffer to the column and centrifuge again. Repeat this wash two more times removing the flow-through each time. Move the column to a new tube with a screw cap.

11. Add 0.2 ml of elution buffer at 65 °C. Centrifuge for 5 sec. The eluate contains 80–90% of the recoverable mRNA. Add a further 0.2 ml of heated elution buffer. Recentrifuge for 5 sec and collect the eluate in either the same tube or a fresh one. Store the mRNA at −70 °C. The maximum yield is 6 μg of RNA. Purity varies depending on the sample with at least 50% and typically > 90% being polyadenylated RNA.

Protocol 14. Direct purification of mRNA using Qiagen Oligotex™

Equipment and reagents

- Oligotex direct mRNA kit containing: spin columns, microcentrifuge tubes, OL1 lysis buffer (contains guanidinium isothiocyanate, add 15 μl of 2-mercaptoethanol per 0.4 ml before use), dilution buffer, Oligotex suspension (10% (w/v) in 10 mM Tris–HCl pH 7.5, 0.5 M NaCl, 1 mM EDTA, 0.1% SDS, 0.1% NaN₃), OW1 wash buffer (may form a precipitate on storage, if so warm to redissolve before use), OW2 wash buffer (10 mM Tris–HCl pH 7.5, 150 mM NaCl, 1 mM EDTA), elution buffer (5 mM Tris–HCl pH 7.5)

Method

1. Grind 50 mg of tissue under liquid nitrogen. Add 600 μl of OL1 lysis buffer. Homogenize thoroughly. Transfer to a 2 ml tube.

2. Add 2 vol. (1200 μl) of dilution buffer. Centrifuge at 14 000–18 000 *g* for 3 min to pellet cell debris and protein which has precipitated on dilution. Transfer supernatant to a fresh tube.

3. Add 70 μl of Oligotex suspension, mix thoroughly, and incubate for

Protocol 14. *Continued*

    10 min at room temperature. Centrifuge for 5 min at 16 000 *g* and aspirate the supernatant. Leave about 50 μl of the supernatant in the tube to avoid any losses.

4. Resuspend the pellet in 350 μl of OW1 wash buffer by pipetting. Transfer to a spin column, sitting in a 1.5 ml tube. Centrifuge at 16 000 *g* for 30 sec. Transfer the column to a fresh 1.5 ml tube.

5. Pipette 350 μl of OW2 wash buffer onto the column. Centrifuge at full speed for 30 sec. Discard the flow-through and repeat this wash once. Transfer the column to a fresh 1.5 ml tube.

6. Add 20 μl to 100 μl of pre-heated (70°C) elution buffer to the column. Pipette up and down to resuspend the resin before centrifuging at 16 000 *g* for 30 sec. Repeat the elution for maximum yield. The first eluate may be used for the second elution if the final volume is to be kept low.

Protocol 15.   Purification of mRNA using Dynabeads® Oligo(dT)$_{25}$

Equipment and reagents

- Dynabeads mRNA Direct™ Kit containing: 1 ml (5 mg) of Dynabeads Oligo(dT)$_{25}$, lysis/binding buffer (100 mM Tris–HCl pH 8.0, 0.5 M LiCl, 10 mM EDTA pH 8.0, 1% LiDS, 5 mM DTT), LiDS washing buffer (10 mM Tris–HCl pH 8.0, 0.15 M LiCl, 1 mM EDTA, 0.1% SDS), washing buffer (10 mM Tris–HCl pH 8.0, 0.15 M LiCl, 1 mM EDTA), elution buffer (2 mM EDTA pH 8.0), reconditioning solution (0.1 M NaOH), storage buffer (250 mM Tris–HCl pH 8.0, 20 mM EDTA, 0.1% Tween 20, 0.02% sodium azide)
- Dynal MPC magnetic rack

Method

1. Resuspend the Dynabeads Oligo(dT)$_{25}$ thoroughly before use. Transfer 0.25 ml to a fresh 1.5 ml tube. Place in a Dynal MPC rack. After 30 sec remove the clear supernatant. Remove the vial from the rack and pre-wash the beads on lysis/binding buffer. Replace the tube in the Dynal MPC rack.

2. Grind 20–50 mg of tissue in liquid nitrogen. Add 1.0 ml lysis/binding buffer and homogenize thoroughly.

3. Centrifuge the lysate for 30–60 sec in a microcentrifuge to sediment the debris.

4. Transfer the supernatant to a fresh tube. The supernatant may be viscous because of the released DNA. This viscosity must be reduced by passing the liquid through a 21 gauge syringe needle to improve the yield of RNA.

5. Remove the pre-wash lysis/binding buffer from the Dynabeads

Oligo(dT)$_{25}$ while in the Dynal MPC rack. Remove the vial from the rack and add the lysate.

6. Mix the beads with the lysate on a roller for 3–5 min at room temperature.

7. Place the vial back in the Dynal MPC rack for 2 min and remove the supernatant.

8. Remove the vial from the rack and wash twice with 0.5–1.0 ml of LiDS washing buffer.

9. Remove the vial from the rack and wash twice with 0.5–1.0 ml of washing buffer without SDS.

10. If the mRNA is to be eluted then add 10–20 μl of elution buffer and incubate at 65°C for 2 min. Place the tube in the Dynal MPC rack for 1 min and carefully transfer the clear supernatant containing the mRNA to a fresh tube.[a]

[a] Some reactions may be performed with the mRNA still bound to the beads, e.g. solid phase cDNA synthesis. See the Dynal Technical Handbook for possible applications.

4. Purification of mitochondrial DNA

Mitochondrial DNA (mtDNA) is of particular interest in evolutionary and population studies of *Drosophila* and other organisms. Ultracentrifugation involving caesium chloride gradients has been commonly used for isolation of mtDNA (14). This approach does not immediately lend itself to small amounts of starting material but small scale ultracentrifugation methods have been described for purification of mtDNA (15–17). Methods not involving CsCl density centrifugation have been described, but these require prior purification of the mitochondria, produce poor yields, or produce mtDNA which is heavily contaminated with nuclear DNA (18–21). More successful protocols have been based on the procedures for isolation of circular plasmid DNA from bacteria (22–24).

Protocol 16 describes the alkaline lysis and SDS precipitation methods for mtDNA preparation. From 50 μg of flies enough mtDNA is obtained for one restriction enzyme digest in about one hour. The final yield is about 1 μg of mtDNA/g of live flies. The purity of the mitochondrial suspension appears not to be very important. Further purification by centrifugation or by sucrose gradient centrifugation is not needed. Contaminating nuclear DNA is precipitated with the potassium dodecyl sulfate complexes. The mtDNA must be circular so fresh flies should be used. Flies must be well fed and mature.

Protocol 17 is a heat–freeze method for mtDNA preparation. The mitochondria are first crudely purified the away from the nuclei. The mitochondria are then heated and frozen causing them to release the mtDNA. No alkali is used in this method as the authors claim that this can cause degradation of the mtDNA. The procedure yields enough mtDNA from 12–15 flies to be seen on an agarose gel. End-labelling the DNA fragments after restriction enzyme digestion allows each sample to be used for more analyses.

Protocol 16. Alkaline lysis method for preparation of mtDNA

Equipment and reagents

- Homogenization buffer A: 0.25 M sucrose, 10 mM EDTA, 30 mM Tris–HCl pH 7.5
- 0.18 M NaOH, 1% SDS freshly prepared
- Homogenization buffer B: 5% sucrose (w/v), 60 mM NaCl, 10 mM EDTA, 10 mM Tris–HCl pH 8.0

- Tris–EDTA buffer: 10 mM Tris–HCl pH 8.0, 0.15 M NaCl, 10 mM EDTA
- SETS buffer: 5% sucrose, 10 mM EDTA, 300 mM Tris–HCl pH 8.0, 1.25% SDS—add DEPC to 0.8% just prior to use
- Other reagents are as in *Table 1*

A. *Preparation of mtDNA from a crude mitochondrial pellet[a]*

1. Homogenize 50 mg of anaesthetized adult flies ($\sim$ 50 flies) in 1 ml of chilled homogenization buffer A with a PTFE pestle.

2. Transfer the homogenate to a chilled 1.5 ml microcentrifuge tube. Centrifuge at 1000 *g* for 1 min at 4°C in order to pellet the nuclei and cellular debris.

3. Remove the supernatant and centrifuge at 12 000 *g* for 10 min at 4°C to pellet the mitochondria.

4. Discard the supernatant and resuspend the mitochondria in 50 μl of Tris–EDTA buffer.

5. Add 100 μl of 0.18 M NaOH, 1% SDS. Vortex the mixture briefly. Keep the tube on ice for 5 min.

6. Add 75 μl of ice-cold 5 M potassium acetate solution. Vortex briefly and keep the tube on ice for 5 min.

7. Centrifuge at 12 000 *g* for 5 min at 4°C. Recover the supernatant and add 0.5 vol. of TE equilibrated phenol, and 0.5 vol. of chloroform. Vortex briefly.

8. Centrifuge at 12 000 *g* for 2 min at room temperature. Transfer the upper aqueous phase to a fresh tube and add 2 vol. of ethanol. Mix by vortexing briefly and let it stand at room temperature for 15 min.

9. Centrifuge at 12 000 *g* for 5 min at room temperature. Wash the mtDNA pellet with 1 ml 70% ethanol, recentrifuge, and briefly dry the pellet.

10. Dissolve the mtDNA in TE. Add DNase-free RNase to remove RNA.

B. *Direct purification of mtDNA[b]*

1. Grind 150 flies or larvae in 1.75 ml of pre-cooled homogenization solution B.

2. Add 1.75 ml of SETS buffer and mix thoroughly by inversion. Incubate at 65°C for 30 min.

3. Cool on ice and add 300 μl of 2 M NaOH. Keep on ice for 6 min.

4. Add 600 μl of 5 M potassium acetate. Mix well and chill at −20°C for 8 min.

5. Centrifuge at 13000 g for 10 min at 4°C. Transfer the supernatant to a 15 ml tube.

6. Add 4 ml of phenol equilibrated with Tris–HCl pH 7.5. Mix by shaking. Keep on ice for 5 min. Centrifuge to separate the phases and transfer the upper layer to a fresh tube.

7. Extract with an equal volume of chloroform. Precipitate the DNA with 2 vol. of ethanol.

[a] Derived from ref. 23.
[b] Derived from refs 22 and 24.

Protocol 17. Heat–freeze method for preparation of mtDNA[a]

Equipment and reagents

- Homogenization buffer: 0.3 M sucrose, 10 mM EDTA, 30 mM Tris–HCl pH 7.5
- SET buffer: 5% sucrose, 10 mM EDTA, 300 mM Tris–HCl pH 8.0
- Other reagents as in *Table 1*

Method

1. Freeze 100 mg of adult *Drosophila* at −20°C for 20 min.

2. Homogenize with a glass homogenizer with a PTFE pestle in 1 ml of homogenization buffer on ice.

3. Centrifuge the homogenate two or three times at 1000 g for 5 min at 4°C until all the particulate matter, nuclei, cellular debris, etc., have been pelleted out.

4. Transfer the supernatant to a fresh tube and centrifuge at 12000 g for 20 min at 4°C, to pellet the mitochondria.

5. Discard the supernatant and resuspend the mitochondrial pellet in 100 μl of homogenization buffer.

6. Add 300 μl SET buffer, mix well, and then add 20 μl of 20% SDS. Mix gently. Incubate for 15 min at 65°C. Cool on ice.

7. Add 100 μl of 5 M potassium acetate. Mix gently. Incubate for 15 min at −20°C.

8. Centrifuge at 13000 g for 5 min at 4°C and transfer the supernatant to a fresh tube.

Protocol 17. *Continued*

9. Extract the DNA with 0.5 vol. of phenol and 0.5 vol. of chloroform. Recentrifuge for 5 min and transfer the aqueous upper layer to a new tube.

10. Repeat the extraction with an equal volume of chloroform. Transfer the supernatant to a fresh tube and precipitate the DNA with 2 vol. of absolute ethanol. Precipitate the DNA by centrifugation.

11. Wash the pellet with 70% ethanol and dry. Redissolve the pellet in 20 μl of TE.

[a] This method is that described in ref. 25.

References

1. Sambrook, J., Fritsch, E. F., and Maniatis, T. (ed.) (1989). *Molecular cloning: a laboratory manual*, 2nd edn. Cold Spring Harbor Laboratory Press, NY.
2. Wallace, D. M. (1989). In *Methods in enzymology* (ed. S. L. Berger and A. R. Kimmel), Vol. 152, p. 33. Academic Press, London.
3. Miller, S. A., Dykes, D. D., and Polesky, H. F. (1988). *Nucleic Acids Res.*, **16**, 1215.
4. Ashburner, M. (1989). In *Drosophila: a laboratory manual*, p. 242. Cold Spring Harbor Laboratory Press, NY.
5. Laurenti, P., Graba, Y., Rosset, R., and Pradel, J. (1995). *Gene*, **154**, 177.
6. Locke, J. and McDermid, H. E. (1993). *Chromosoma*, **102**, 718.
7. Jowett, T. (1986). In *Drosophila: a practical approach* (ed. D. B. Roberts), 1st edn, p. 275. IRL Press, Oxford.
8. Ashburner, M. (1989). In *Drosophila: a laboratory manual*, p. 108. Cold Spring Harbor Laboratory Press, NY.
9. Czank, A. (1996). *Trends Genet.*, **12**, 457.
10. Berger, S. L. (1989). In *Methods in enzymology* (ed. S. L. Berger and A. R. Kimmel), Vol. 152 p.215. Academic Press, London.
11. Chirgwin, J. M., Przybyla, A. E., MacDonald, R. J., and Rutter, W. J. (1979). *Biochemistry*, **18**, 5294.
12. Chomczynski, P. and Sacchi, N. (1987). *Anal. Biochem.*, **162**, 156.
13. Jakobsen, K. S., Haugen, M., Sæbøe-Larssen, S., Hollung, K., Espelund, M., and Hornes, E. (1994). In *Advances in biomagnetic separation* (ed. M. Uhlén, E. Hornes, and Ø. Olsvik), p. 61. Eaton Publishing.
14. Lansman, R. A., Shade, R. O., Shapira, J. F., and Avise, J. C. (1981). *J. Mol. Evol.*, **17**, 214.
15. Fauron, C. M.-R. and Wolstenholme, D. R. (1976). *Proc. Natl. Acad. Sci. USA*, **73**, 3623.
16. Ashburner, M. (1989). In *Drosophila: a laboratory manual*, p. 101. Cold Spring Harbor Laboratory Press, NY.
17. Carr, S. M. and Griffith, O. M. (1987). *Biochem. Genet.*, **25**, 385.
18. Solignac, M., Monnerot, M., and Mounolou, J. C. (1983). *Proc. Natl. Acad. Sci. USA*, **80**, 6942.

19. Powell, J. R. and Zúñiga. M. C. (1983). *Biochem. Genet.*, **21**, 1051.
20. DeSalle, R., Giddings, L. V., and Templeton, A. R. (1986). *Heredity*, **56**, 75.
21. LaTorre, A., Moya, A., and Ayala, F. J. (1986). *Proc. Natl. Acad. Sci. USA*, **83**, 8649.
22. Afonso, J. M., Pestano, J., and Hernández, M. (1988). *Biochem. Genet.*, **26**, 381.
23. Tamura, K. and Aotsuka, T. (1988). *Biochem. Genet.*, **26**, 815.
24. Rozas, J., Hernández, M., Cabrera, V. M., and Prevosti, A. (1990). *Mol. Biol. Evol.*, **7**, 103.
25. Pissios, P. and Scouras, Z. G. (1992). *Experientia*, **48**, 671.

List of suppliers

This core list of suppliers appears in all books in the Practical Approach series. If there are any relevant suppliers that you would like to add to this list for the book you are working on, please send them with your chapter.

Agar Scientific Ltd., 66a Cambridge Road, Stansted, Essex CM24 8DA, UK.
Amersham
Amersham International plc., Lincoln Place, Green End, Aylesbury, Buckinghamshire HP20 2TP, UK.
Amersham Corporation, 2636 South Clearbrook Drive, Arlington Heights, IL 60005, USA.
Amicon Inc., 72 Cherry Hill Drive, Beverly, MA 01915, USA.
Anderman
Anderman and Co. Ltd., 145 London Road, Kingston-Upon-Thames, Surrey KT17 7NH, UK.
BabCO, Berkeley Antibody Company, 4131 Lakeside Drive, Suite B, Richmond, CA 94806, USA.
Beckman Instruments
Beckman Instruments UK Ltd., Progress Road, Sands Industrial Estate, High Wycombe, Buckinghamshire HP12 4JL, UK.
Beckman Instruments Inc., PO Box 3100, 2500 Harbor Boulevard, Fullerton, CA 92634, USA.
Becton Dickinson
Becton Dickinson and Co., Between Towns Road, Cowley, Oxford OX4 3LY, UK.
Becton Dickinson and Co., 2 Bridgewater Lane, Lincoln Park, NJ 07035, USA.
Bio
Bio 101 Inc., c/o Statech Scientific Ltd., 61–63 Dudley Street, Luton, Bedfordshire LU2 0HP, UK.
Bio 101 Inc., PO Box 2284, La Jolla, CA 92038–2284, USA.
Bio-Rad Laboratories
Bio-Rad Laboratories Ltd., Bio-Rad House, Maylands Avenue, Hemel Hempstead HP2 7TD, UK.
Bio-Rad Laboratories, Division Headquarters, 3300 Regatta Boulevard, Richmond, CA 94804, USA.
Biosolve, BioGene Ltd., Bio/Gene House, Greenbury Farm, Bolnhurst, Bedfordshire MK44 2ET, UK.
Boehringer Mannheim
Boehringer Mannheim UK (Diagnostics and Biochemicals) Ltd., Bell Lane, Lewes, East Sussex BN17 1LG, UK.

Boehringer Mannheim Corporation, Biochemical Products, 9115 Hague Road, PO Box 504 Indianapolis, IN 46250–0414, USA.

Boehringer Mannheim Biochemica, GmbH, Sandhofer Str. 116, Postfach 310120 D-6800 Ma 31, Germany.

British Drug Houses (BDH) Ltd., Poole, Dorset, UK.

Cappel, Organon Teknika Corporation, 1230 Wilson Drive, West Chester, PA 19380, USA.

Difco Laboratories

Difco Laboratories Ltd., PO Box 14B, Central Avenue, West Molesey, Surrey KT8 2SE, UK.

Difco Laboratories, PO Box 331058, Detroit, MI 48232–7058, USA.

Dow Corning

Dow Corning Corp., Dow Corning Center, Midland, MI 48686–0994, USA.

Dow Corning GmbH, 65201 Wiesbaden, Germany.

T. P. Drewitt, London E4 9EN, UK.

Du Pont

Dupont (UK) Ltd., Industrial Products Division, Wedgwood Way, Stevenage, Hertfordshire SG1 4Q, UK.

Du Pont Co. (Biotechnology Systems Division), PO Box 80024, Wilmington, DE 19880–002, USA.

European Collection of Animal Cell Culture, Division of Biologics, PHLS Centre for Applied Microbiology and Research, Porton Down, Salisbury, Wiltshire SP4 0JG, UK.

Falcon (Falcon is a registered trademark of Becton Dickinson and Co.)

Fine Science Tools

Fine Science Tools Inc., 373-G, Vintage Park Drive, Foster City, CA 94404–1139, USA.

Fine Science Tools GmbH, Quinckestrasse 4, D-69120 Heidelberg, Germany.

Interfocus, Ltd.(UK distributors) 14/15 Spring Rise, Falconer Road, Haverhill, Suffolk CB9 7XU, UK.

Fisher Scientific Co., 711 Forbest Avenue, Pittsburgh, PA 15219–4785, USA.

Flow Laboratories, Woodcock Hill, Harefield Road, Rickmansworth, Hertfordshire WD3 1PQ, UK.

Fluka

Fluka-Chemie AG, CH-9470, Buchs, Switzerland.

Fluka Chemicals Ltd., The Old Brickyard, New Road, Gillingham, Dorset SP8 4JL, UK.

General Valve Corp., 19 Gloria Lane, PO Box 1333, Fairfield, NJ 07006, USA.

Gentra Systems, Inc., 15200 25th Ave N., Suite 104, Minneapolis, MN 55447, USA.

Gibco BRL

Gibco BRL (Life Technologies Ltd.), Trident House, Renfrew Road, Paisley PA3 4EF, UK.

Gibco BRL (Life Technologies Inc.), 3175 Staler Road, Grand Island, NY 14072–0068, USA.

Glenhurst Ltd., 51 & 52 Long Lane, West Smithfield, London EC1A 9EL, UK.

Grass Instruments Company, 101 Old Colony Avenue, Quincy, MA 02269–0516, USA.

Arnold R. Horwell, 73 Maygrove Road, West Hampstead, London NW6 2BP, UK.

Hybaid

Hybaid Ltd., 111–113 Waldegrave Road, Teddington, Middlesex TW11 8LL, UK.

Hybaid, National Labnet Corporation, PO Box 841, Woodbridge, NJ 07095, USA.

HyClone Laboratories, 1725 South HyClone Road, Logan, UT 84321, USA.

International Biotechnologies Inc., 25 Science Park, New Haven, Connecticut 06535, USA.

Invitrogen Corporation

Invitrogen Corporation, 3985 B Sorrenton Valley Building, San Diego, CA 92121, USA.

Invitrogen Corporation, c/o British Biotechnology Products Ltd., 4–10 The Quadrant, Barton Lane, Abingdon, Oxon OX14 3YS, UK.

Jackson ImmunoResearch Laboratories Inc., 872 W. Baltimore Pike, PO Box 9, West Grove, PA 19390, USA.

Kodak: Eastman Fine Chemicals, 343 State Street, Rochester, NY, USA.

Life Technologies Inc., 8451 Helgerman Court, Gaithersburg, MN 20877, USA.

Merck

Merck Industries Inc., 5 Skyline Drive, Nawthorne, NY 10532, USA.

Merck, Frankfurter Strasse, 250, Postfach 4119, D-64293, Germany.

Millipore

Millipore (UK) Ltd., The Boulevard, Blackmoor Lane, Watford, Hertford-shire WD1 8YW, UK.

Millipore Corp./Biosearch, PO Box 255, 80 Ashby Road, Bedford, MA 01730, USA.

Molecular Probes, Inc., PO Box 22010 Eugene, OR 97402–0414, USA.

National Diagnostics, 305 Patton Dr., Atlanta, GA 30336, USA.

New England Biolabs (NBL)

New England Biolabs (NBL), 32 Tozer Road, Beverley, MA 01915–5510, USA.

New England Biolabs (NBL), c/o CP Labs Ltd., PO Box 22, Bishops Stort-ford, Hertfordshire CM23 3DH, UK.

Nikon Corporation, Fuji Building, 2–3 Marunouchi 3-chome, Chiyoda-ku, Tokyo, Japan.

Nursery Supplies, Exeter Street, Bourne, Lincolnshire PE10 9NJ, UK.

Perkin-Elmer

Perkin-Elmer Ltd., Maxwell Road, Beaconsfield, Buckinghamshire HP9 1QA, UK.

Perkin Elmer Ltd., Post Office Lane, Beaconsfield, Buckinghamshire HP9 1QA, UK.

Perkin Elmer-Cetus (The Perkin-Elmer Corporation), 761 Main Avenue, Norwalk, CT 0689, USA.

Pharmacia Biotech Europe, Procordia EuroCentre, Rue de la Fuse-e 62, B-1130 Brussels, Belgium.

Pharmacia Biosystems

Pharmacia Biosystems Ltd. (Biotechnology Division), Davy Avenue, Knowlhill, Milton Keynes MK5 8PH, UK.

Pharmacia LKB Biotechnology AB, Björngatan 30, S-75182 Uppsala, Sweden.

Pierce, PO Box 117, Rockford, IL 61105, USA.

Polysciences Inc.,Warrington, PA 18976, USA.

Promega

Promega Ltd., Delta House, Enterprise Road, Chilworth Research Centre, Southampton, UK.

Promega Corporation, 2800 Woods Hollow Road, Madison, WI 53711–5399, USA.

Qiagen

Qiagen Inc., c/o Hybaid, 111–113 Waldegrave Road, Teddington, Middlesex TW11 8LL, UK.

Qiagen Inc., 9259 Eton Avenue, Chatsworth, CA 91311, USA.

Wm. Ritchie & Son, 29 Bidder Street, London E16 4ST, UK.

SAS Institute

SAS Institute, Wittington House, Henley Road, Medmenham, Marlow, Buckinghamshire SL7 2EB, UK.

SAS Institute, JMP Sales Dept., SAS Campus Drive, Cary, NC 27513, USA.

Schleicher and Schuell

Schleicher and Schuell Inc., Keene, NH 03431A, USA.

Schleicher and Schuell Inc., D-3354 Dassel, Germany.

Schleicher and Schuell Inc., c/o Andermann and Company Ltd.

Scientific Instrument Services Inc., Rt. 179 RD2, PO Box 198, Ringoes, NJ 0855, USA.

Shandon Scientific Ltd., Chadwick Road, Astmoor, Runcorn, Cheshire WA7 1PR, UK.

Sigma Chemical Company

Sigma Chemical Company (UK), Fancy Road, Poole, Dorset BH17 7NH, UK.

Sigma Chemical Company, 3050 Spruce Street, PO Box 14508, St. Louis, MO 63178–9916, USA.

Sorvall DuPont Company, Biotechnology Division, PO Box 80022, Wilmington, DE 19880–0022, USA.

Stag Instruments, Oxford, UK.
Stratagene
Stratagene Ltd., Unit 140, Cambridge Innovation Centre, Milton Road, Cambridge CB4 4FG, UK.
Stratagene Inc., 11011 North Torrey Pines Road, La Jolla, CA 92037, USA.
Teledyne Electromechanisms, Hudson, NH, USA.
United States Biochemical, PO Box 22400, Cleveland, OH 44122, USA.
Vector Laboratories Inc., 30 Ingold Lane, Burlingame, CA 94010, USA.
Watkins and Doncaster, PO Box 5, Cranbrook, Kent TN18 5EZ, UK.
Wellcome Reagents, Langley Court, Beckenham, Kent BR3 3BS, UK.

Index of *Drosophila* genes, mutations, chromsomal abnormalities and balancers.

A gene is indexed as its symbol cross referenced to its full name, together with its genetic and cytogenetic localization. These localizations are taken from FlyBase and are cited as the best localization without any qualifying comments.

Genes

Ace, see *Acetyl cholinesterase*
Acetyl cholinesterase (3-52.2; 87E3) (*Ace*) 310
Adh, see *Alcohol dehydrogenase*
al, see *aristaless*
al² 8, 68, 122
Alcohol dehydrogenase (2-50.1; 35B2) (*Adh*) 102, 261
amn, see *amnesiac*
amnesiac (1-63; 19A1-A2) (*amn*) 280, 287, 301–2, 310, 313, 315
antennapedia (3-47.5; 84B2) (*antp^{Hu}*)
aristaless (2-0.4; 21C2) (*al*) 16, 38-9, 115–6, 275

b, see *black*
b^{81a2} 126
B, see *Bar*
B^s 112, 124
Bar (1-57.0; 16A1) (*B*) 8, 66, 73
barr, see *barren*
barren (38B2) (*barr*) 136
Bc, see *Black cells*
Bd^s, see *Beaded-Serrate*
Beaded–Serrate (3-91.9; 97F) (*Bd^s*) 13
bithorax (3-58.8; 89E1) (*bx*)
Bl (see *Bristle*)
black (2-48.5; 34D4-D6) (*b*) 16, 38–9, 68–70, 74–5, 78, 81–2, 115–16, 119–23, 128
Black cells (2-80.6; 55A1-A4) (*Bc*) 140
BRC, see *Broad-Complex*
Bristle (2-54.8; 38E3-E9) (*Bl*) 78, 81, 122
Broad-Complex (1-0.28; 2B5) (*BRC*) 175
brown (2-104.5; 59E1-E2) (*bw*) 80, 119–21, 123–4, 128
bw, see *brown*
bw^D 13
bw^v 122
bw^{VI} 68
bw^{v32g} 122, 128
bx^{34e} 9

c, see *curved*
ca, see *claret*
cab, see *cabbage*
cabbage (1-36.6; 11A2-A3) (*cab*) 287, 301
cac, see *cacophony*
cacophony (1-36.6; 11A2) (*cac*) (now known as *nba*) 312
cinnabar (2-57.5; 43E7-E8) (*cn*) 8, 68–70, 74–5, 78, 81–2, 115–16, 119–22, 124, 128
claret (3-100.7; 99B5-B9) (*ca*) 9
cn, see *cinnabar*
cn² 68
couch potato (3-62; 90D4) (*cpo*) 136, 273, 312
cpo, see *couch potato*
cro, see *croaker*
croaker (45E1-E4) (*cro*) 312
ct, see *cut*
Curly (2-6.1;23A3-23B2) (*Cy*) 8, 68, 123, 125–6
curved (2-75.5; 52D3-D9) (*c*) 38–9
cut (1-20; 7B3) (*ct*) 153, 275
Cy, see *Curly*
CyO 12, 16, 36, 114, 125, 138–40, 169, 172–3, 249–50

D, see *Dichaete*
D³ 9
decapentaplegic (2-4.0; 22F2) (*dpp*)
dg, see *don giovani*
Dichaete (3-40.7; 70D2) (*D*)
diminutive (1-4.0 3D5) (*dm*) 67
disco, see *disconnected*
disconnected (1-53.1; 14B3-14B4) (*disco*) 310
diss, see *dissonance*
dissonance (14C1) (*diss*) (now known as *nonA*) 312
dm, see *diminutive*
dnc, see *dunce*
don giovani (1- ; 1B1-5C2) (*dg*) 312

Index

Index

Index

General Index

Genes, mutations, chromosomal abnormalities, and balancers are indexed separately.

adaptation 266
anaesthetics
  carbon dioxide 31–2
  ether 31
antibody
  binding 217–18
  detection 218–19
  properties 216
  specificity confirmation of 218
  storage 216–17
  type 215
associative learning 266
autosynaptic chromosomes 117

bacteria
  control of 29, 270
balancer chromosomes 7–9
  uses 10–12
brain
  X-gal staining 146–7, 149–50
blastoderm
  cellular 188
  syncitial 188

cages
  large 41–3
  population 244
  small 40–1
cell(s)
  death, induced 154
  lines
    blood 321
    cloning 329–30
    culture conditions 322–8
    embryonic 319–21
    freezing and thawing 328
    imaginal disc 321
    neural 321
    table of 339–46
    transfection 331–6
cephalic furrow 190–2
chemical mutagen(s)
  ethyl methanesulfonate (EMS) 56–60
    administration 58–60
    chromosomal rearrangements 56
    denaturing solution 59
    frequency of induced mutations 57
    missense mutations 57

    mode of action 56
    mosaic mutants 57
  formaldehyde 62–3
    administration 63–4
    frequency of induced deletions 62–3
  lists of 55
  N-ethyl-N-nitrosourea (ENU) 60–1
    administration 61
    frequency of induced mutations 60
    mode of action 60
  triethylenemelamine (TEM) 61
    administration 61–2
    frequency of induced deletions 61
chromosome(s)
  aberrations
    origin 6
  autosynaptic 117
    construction of marked 120–1
    generation of 118–19, 122–4
    selective screens for 127–8
  compound 18
    inheritance pattern 19
  dextrosynaptic 118
  extraction 10–11, 244–5
  isogenic 12, 138
  laevosynaptic 117
  substitution 12–13
courtship 312–15
  conditioning 312–14
    measurement of 314
  mutations affecting 312

deficiency, *see* deletion(s)
deletion(s)
  description 13
  generation from autosynaptic stocks 118–20
  generation from h;A autosynaptic stocks 121–2
  generation from inversions 115–16
  generation from translocations 110–11, 113
  generation from transpositions 114
  induction 56, 60–2, 64
  isolation from specific regions 77–82
    with defined end-points 99–101
    by pseudo-dominance 78
    by reverting a dominant 79–81
  mapping 14–15
DNA
  preparation